MONOMODE FIBER-OPTIC DESIGN

WITH LOCAL-AREA AND LONG-HAUL NETWORK APPLICATIONS

MONOMODE FIBER-OPTIC DESIGN

WITH LOCAL-AREA AND LONG-HAUL NETWORK APPLICATIONS

Donald G. Baker

VNR VAN NOSTRAND REINHOLD COMPANY
New York

Copyright © 1987 by Van Nostrand Reinhold Company Inc.

Library of Congress Catalog Card Number 86-24636

ISBN 0-442-21107-4

Printed in the United States of America

Van Nostrand Reinhold Company Inc.
115 Fifth Avenue
New York, New York 10003

Van Nostrand Reinhold Company Limited
Molly Millars Lane
Wokingham, Berkshire RG11 2PY, England

Van Nostrand Reinhold
480 La Trobe Street
Melbourne, Victoria 3000, Australia

Macmillan of Canada
Division of Canada Publishing Corporation
164 Commander Boulevard
Agincourt, Ontario MIS 3C7, Canada

16 15 14 13 12 11 10 9 8 7 6 5 4 3 2 1

Library of Congress Cataloging-in-Publication Data

Baker, Donald G., 1935–
  Monomode fiber-optic design.

  Includes index.
  1. Fiber optics. 2. Optical communications. 3. Local
area networks (Computer networks) I. Title.
TA1800.B36 1987   621.39'92   86-24636
ISBN 0-422-21107-4

PREFACE

Fiber optics is a transmission technique that uses electrical signals to modulate a light source and thereby produce an optical signal proportional to the electrical signal. These optical signals contain information that is transmitted via a glass waveguide to a light-sensitive receiver. Fiber optics has a distinct advantage over copper networks for some applications. The objective of this book is to explore monomode, as opposed to multimode, applications of fiber optics to local area networks (LANs), which have become a rather important aspect of this technology because of the ever-increasing growth of LANs.

Monomode fiber optics requires the use of coherent light sources such as laser diodes, YAGs, and HeNe lasers, to name just a few. It has some distinct advantages over multimode that this text will investigate in a cursory manner. (The author's previous book on multimode fiber optics, *Fiber Optic Design and Applications*, published by Reston, would be helpful but not necessary to augment this text.)

Monomode (or single-mode) fiber optics is the present direction of the state-of-the-art because of its superior performance. Since a few problems existed that limited the growth of monomode technology at the time this book was being written, several sections of the text will be devoted to examining the shortcomings as well as the performance advantages of this technology.

Although this book is meant to serve as a reference for engineers with sufficient background in mathematics and electronics, it also contains introductory material in each chapter to acquaint operations personnel with the subject matter. Since the mathematics includes calculus and differential equations, the book can be used as a text for senior-level engineering courses or five-day seminars on an accelerated-course basis. The information presented will give the reader a reliable foundation in the subject that may be built on through pursuing journals and advanced studies.

For the less mathematically inclined, the fiber-optics segment of the text explores monomode waveguides through the use of diagrams, and, to round out the text, some theoretical approaches using various solutions to Maxwell's equations are examined.

Laser physics are presented to acquaint the reader with fundamentals of coherent light sources, with an emphasis on long-wavelength sources. The outlook for long-wavelength sources will be explored because the transition in popularity from 800-nm to 1300-nm sources was rather rapid. Longer wavelength sources are also becoming lower in cost and more commonly available. Their importance lies in the lower loss exhibited by the transmission medium when they are implemented in transmitters.

To give the reader an indication of what is available in the marketplace, several types of monomode sensors will be addressed that are on the edge of the state-of-the-art; they include magnetic, pressure, temperature, gyro, and strain types. Other sensors will also be examined but not in a rigorous manner.

The various components available to implement monomode-cable plant design include star couplers, tee's, switches, terminations, splices, etc. The objective in presenting them is to produce a loss-budget analysis capable of being programmed on a pocket calculator or personal computer by the practicing engineer.

Linear and digital circuit-design techniques for transmitters and receivers are presented to give the reader familiarity with both common and specialized technologies. A table of commercially available transmitters and receivers provides a useful list of manufacturers and their specifications.

Two chapters are devoted to integrated fiber-optic components and their implementation into integrated fiber optics. These chapters examine future and state-of-the-art issues and pay passing attention to integrated-optics fabrication.

An introduction to acquaint the reader with the basics of LAN technology is presented in Chap. 8. Applications of monomode technology to LANs that will be common in the near future (three to five years) are here discussed. This segment of the book also examines the first three layers of the seven-layer International Standards Organization (ISO) approach to design. (Any book on LANs will supply ample information to complete the network design.) Definitions and examples of each layer are presented. Some of the more common LANs—such as DECNET, Ethernet, SNA, etc.—are discussed to make the reader aware of what is available in the marketplace. Such background information will provide the interested reader with sufficient information to delve into the monomode fiber-optics approach to the design of various LANs. In fact, it is the monomode-fiber approach to LAN design that distinguishes this text from most others on the market.

At present, most monomode networks are designed for long-distance transmission, but as commercial users, such as the telephone company, begin to expand their applications, fiber-optic components will become more cost-effective, thus further increasing usage. The 140-km upper limit of long-distance networks is being rapidly increased as a result of research. The examination of long-distance networks undertaken here will give the reader a feel for some of the problems confronting these designers. A few of the commercial long-distance networks are examined to determine their performance and limitations.

The final chapter addresses some of the Angular Division Multipexing (ADM) issues in network design. This is a new technique for designing short-distance multichannel networks in which all channels operate at the same wavelength.

This technique is still in the experimental or pilot stages but is already being accepted as an approach to LAN design. Its cable plant presents more complex problems than either the monomode or multimode types, and these differences are explored.

The conclusion of the book offers an outlook on various aspects of fiber optics and up-to-the-minute applications. It addresses some of the security applications of fiber optics in conjunction with waveguide design. Many new techniques for secure cable plants are explored.

This book will be an asset for any engineer who wishes to project himself into a new and rapidly changing technology. Many exciting phenomena are still being discovered, which makes the field excellent for doctoral dissertations.

DONALD G. BAKER

CONTENTS

MONOMODE
FIBER-OPTIC
DESIGN

WITH LOCAL-AREA
AND LONG-HAUL NETWORK
APPLICATIONS

1 | INTRODUCTION

Multimode fiber optics have been implemented in local area networks (LANs) rather successfully since the 1970s. One may ask, why use single mode technology? The next fundamental question is what are the basic differences between the two technologies? Answering the second question will also answer the first.

When examining a fiber from the outside, the single and multimode waveguides look identical. The difference lies in the core. The light-carrying core of these waveguides is typically five or six times the wavelength of transmission. For an example, operating at a wavelength of 1.0 μm, the core diameter would be 5.0 μm or 6.0 μm, whereas for multimode transmission the core diameter would be 50 μm or 100 μm, depending on the application. In the next chapter, the waveguides will be examined in more detail. The major reason for using single-mode fiber optics is that the repeater spacing and bandwidth involved are both larger than they are for the multimode technology. These improvements are the driving forces behind the new thrust in research efforts.

The first low-loss single-mode fiber was fabricated in 1970. Some of the problems facing these early technologists were finding suitable methods for injecting light into the waveguides, on the one hand, and devising connectors, on the other.

Present-day LANs operate with repeater spacings of 10 to 20 kilometers (km), and the transmission bandwidth limit at these distances is 100 to 50 Mbits, respectively. With single-mode technology, the bandwidth limit is approximately 60 times greater, i.e., 6 to 3 gigabits for the respective repeater spacing of 10 to 20 km. The electronics would be the limiting factor, but a great deal of research is being conducted with gallium-arsenide (GaAs) integrated circuits (ICs) to extend semiconductor bandwidths. One may observe that, as semiconductors are improved, researchers will strive to improve the fiber optics, and vice versa. Of course, a limit will eventually be reached when the cost for only minor improvements in performance will become prohibitive (the point of diminishing return).

Waveguides are only a part of the technology. Sources such as light emitting diodes (LEDs), semiconductors, lasers, and other types must also be considered when discussing monomode technology. Both spectral width and wavelength are important parameters that should be investigated. The first generation of commercial fiber-optic systems operated between wavelengths of 800 and 900 nm. These wavelengths were produced by gallium-aluminum-arsenide diode lasers. Fiber-optic waveguides exhibit low loss at such wavelenths, but it was discovered that as wavelength is increased, scattering losses in silicon fibers decrease. It then follows that dispersion losses will approach zero if the wavelength is made long enough (1300 nm). The major problem with lasers that produce a long wavelength was their fabrication. Present-day technology, however, produces two long-wavelength lasers, both of which are becoming more common. These are 1300-nm and 1550-nm devices. Losses at these wavelengths have been recorded at 0.5 dB/km and 0.18 dB/km, respectively. Such low attenuations allow for long transmission distances.

An example of how satellite transmission, as compared to fiber-optic links, performs, the delay of satellite transmission is approximately 250 msec because of RF (radio frequency) propagation times between the satellite and the earth. This same delay in a fiber-optic link would be equivalent to 50,000 kilometers of waveguide. At present, Bell Laboratory has been working on a 5,000-km transatlantic communication link. Fiber optics may eventually replace satellite communications for some applications because such systems offer a higher measure of security and are lower in cost to implement.

A topic to be addressed in regard to sources is spectral width. As we shall observe later, narrow spectral widths are necessary to single-mode operation ("monomode operation"). These single-mode lasers are characterized by a single spectral line having a width of 1 to 10 Å (Angstrom units). Since very few lasers meet this criteria, we will have to deal with less-than-perfect sources. Nevertheless, as one may observe, performance of these sources is quite outstanding in spite of their less-than-perfect nature.

One of the other advantages of single-mode fiber optics is the ability it gives design engineers to take advantage of the phase characteristics of single-mode waveforms. The phase of the waveform may be shifted in electric and magnetic fields both directly and indirectly, thus allowing certain advantages over multimode technology. This topic will be addressed in Chap. 6, where fiber-optics components are investigated in more detail.

Single-mode technology is used in integrated fiber optics not only because the size of optical paths is smaller but because such phase shifts in the waveforms can be used to enhance performance.

After digressing about the attributes of fiber optics, attention must be focused on the networks themselves. The local area networks (LANs) to be discussed will be those that lend themselves well to single-mode fiber-optics technology.

Some samples of common local-area network topologies are depicted in Fig. 1-1. In Fig. 1-1(a)—a block diagram of a ring LAN—each terminal and host computer is considered a node. When node one (terminal 1) communicates with the host computer or another terminal, it will transmit from port T; any intermediate node must then act as a repeater and pass on the data. If the data is passed completely around the ring, the transmitting node will receive its data back at receive port R. The node may therefore check the received data for

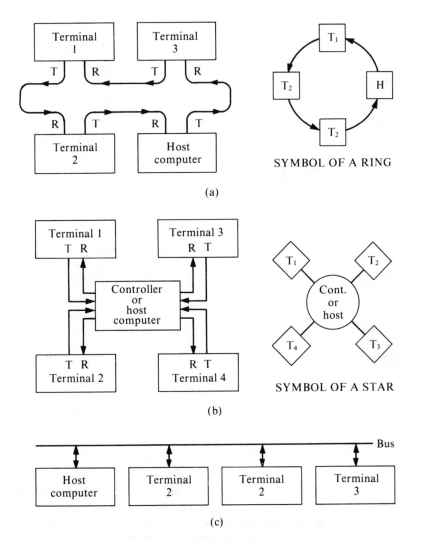

Fig. 1-1. Common local area network topologies: (a) Ring with symbol, (b) star with symbol, (c) bus, and (d) star and ring composite network and symbol.

errors. This is only one method for error checking and transmission on the ring. Other, more efficient techniques will be investigated later.

The next topology to be considered is the star shown in Fig. 1-1(b). This configuration, which was used on the early computer-network designs, is still used today. Note that the star network controller or host computer may communicate with all terminals simultaneously. Star networks are usually opera-

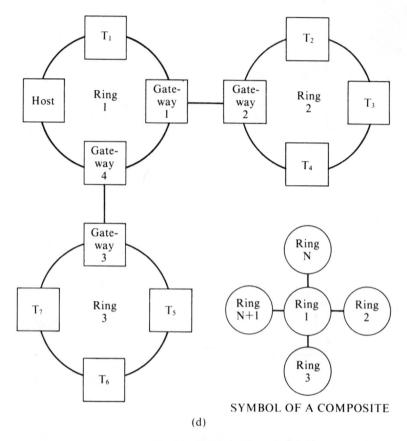

SYMBOL OF A COMPOSITE

(d)

Fig. 1-1 (*Continued*)

tionally faster than rings if both operate at the same transmission rate. Ring topologies are usually lower in cost, however, because of their simplicity. The star controller may be a simple microprocessor, a minicomputer, or a mainframe computer, depending on the size and complexity of the network.

Bus topology is another common LAN configuration. Taking advantage of techniques commonly used in fiber optics or electronics, the terminals and host computer may all communicate virtually simultaneously [see Fig. 1-1(c)].

Networks often fail to follow any of the topologies exactly but are composed of combinations of rings, stars, and buses instead. Nevertheless, some of these network elements may be separated from the main network and analyzed. Let us examine an example of a composite network as shown in Fig. 1-1(d). The network begins to resemble a star if rings are considered nodes.

Why use LANs at all to design networks? LANs use virtual connections when two devices are communicating on the network, i.e., devices communicate by sharing the channel. The channel sharing is performed through the use of time slots, or time division multiplexing (TDM). As an alternative, frequency allocation (frequency division multiplexing) and bus systems may be implemented. These topics will be covered in greater detail in Chap. 8.

Let us consider an example of how virtual connection will benefit a network design. Figure 1-2(a) represents a nonvirtual connected network. It has a total of 12 connections, whereas the virtual network in Fig. 1-2(b) has only 8. The equation describing the number of connections in Fig. 1-2(a) is Eq. 1-1, whereas Eq. 1-2 represents the mathematical relationship in the network of Fig. 1-2(b).

$$N_c = N(N - 1) \qquad (1\text{-}1)$$

$$N_c = 2N \qquad (1\text{-}2)$$

where

N_c = number of connections
N = number of nodes

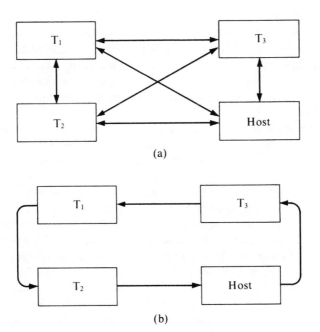

(a)

(b)

Fig. 1-2. Virtual and nonvirtual networks: (a) Point-to-point nonvirtual network, and (b) point-to-point virtual network.

For small-scale networks (networks with only 5 or 10 nodes), no great advantage is gained from virtual connections. Suppose, however, that a LAN with 200 nodes is to be designed. The number of nonvirtual and virtual connections is calculated by using Eqs. 1-1 and 1-2 as follows:

$$N_c = N(N - 1)$$

$$= 200\,(200 - 1)$$

$$= 39,800 \text{ nonvirtual connections}$$

$$N_c = 2N$$

$$= 2 \times 200$$

$$= 400 \text{ virtual connections}$$

As can be observed, the virtual network saves a great deal of wiring. Also, one may note, the bundle of wires connected to each machine consists of 199 wires for the nonvirtual case and only two for the virtual. Cost could become astronomical because of the copper or fiber-optic cables required and the labor to install them.

If telephone networks operated in a way like that in Fig. 1-2(a), bundles of copper wire would be required to connect each home. Subscribers to telephone networks use the equipment for only a very small percentage of the time. Thus, the more equipment and resources that can be shared, the more efficient will be their use, and the lower the cost to each subscriber. The main thrust of LAN technology, therefore, is to provide users of resources an equitable method of sharing them.

To provide network services to each subscriber on a network requires some sort of plant facilities, i.e., a cable to connect subscribers. Next, some rules are needed to allow the subscriber (or user) to connect in an efficient, orderly manner. Such rules are normally called "protocol." The third item in the network is some method of maintaining a routing scheme and a clear definition of the required topology. When a node in the network has multiple sources of data or receives data to be distributed to multiple subscribers, then a method of multiplexing and demultiplexing it is necessary. Some software will also be required to allow smooth transition between the hardware and applications program.

As one may observe, there are a host of complexities involved in configuring LANs. All is not lost, however, because several networks are in the process of becoming standardized, such as Ethernet, DECNET, SNA, IEEE802, etc. Some are manufacturers' de facto standards; others are standards recognized both nationally and internationally.

Why, then, the need for standards? The main reason is compatability, e.g., printers that can be purchased and are able to operate with any microcomputer. When devices meet a certain set of standards and the microcomputer manufacturer designs equipment that also complies with it, the printer will function correctly. Prior to LANs, one manufacturer had approximately 600 different interfaces for various types of equipment. Each new network required new interfaces. Ethernet, on the other hand, requires the same interface for each implement added to the network instead of a new interface for each new network implemented.

As one will note from our description of what is required in a LAN, fiber optics consideration is given to the physical cable plant. The other components of the LAN will function in such a way that the physical connections will be transparent to the network hierarchy. All of these network requirements will be investigated in more detail. So that the reader will observe the subtleties in this presentation of the marriage of two technologies, fiber optics (single mode) and local area networking are represented in some of the designs.

Sensors are another important segment of single-mode technology because these devices improve accuracy in some applications and reduce size and cost in others. Examples of sensors might include a device for measuring rainfall that is mounted on aircraft. Such an indicator would be useful for reconnaissance type vehicles gathering weather data. Another device capable of monitoring pressure variation would be useful for barometric-pressure sensors, strain gauges, etc. Temperature measurements can also be made using fiber-optic sensors. One of the major fiber-optic sensors is the gyro. Gyros constructed of fiber optics are superior in accuracy and smaller in size than present mechanical units. The differences in volume between the mechanical and fiber-optic units is approximately an order of magnitude.

Several of the fiber-optic sensors are in the laboratory stages at present, but by the 1990s many of these devices will be commonplace. Fiber-optic gyros in particular will be useful because they have no moving parts and can withstand gravitational loads of 100 g's. Spacecraft, high-speed aircraft, missiles, etc., all have g loads of this magnitude.

Due to the many strides made by industry in single-mode, fiber-optic technology, most signal transmission will be through fiber-optic mediums. More vendors are entering the single-mode market, a trend that translates into acceptance of single-mode technology.

Copper-cable and multimode fiber-optic technology will disappear slowly because present installations may be in place for 20 years. Also, industry is slow to replace old, well-established technology with new technology. The next chapter will further clarify what single-mode technology is and explore some of the differences between it and multimode fiber optics.

2 | WAVEGUIDE ANALYSIS

The structure of waveguides that support single-mode propagation of optical signals will be examined in this chapter. These structures have some unique properties that make their performance superior to that of multimode types. Eventually, this technology will emerge as a low-cost medium that will no doubt replace copper and multimode fiber-optic waveguides in signal transmission applications.

Single-mode waveguides will support single-wavelength transmission only. Laser sources used in this technology may produce other spectral wavelengths that are unsupported, but these will soon die out. It is very difficult to find a truly single-mode laser, and these devices are fairly costly.

SINGLE-MODE WAVEGUIDE ANALYSIS

A place to begin the analysis is with a simple waveguide diagram to acquaint the reader with the single-mode structure. Figure 2-1 is a diagram of a single-mode structure. The difference between this waveguide and those shown in multimode textbooks is its core and clad size. Single-mode cores are 6 to 8 μm in diameter as opposed to multimode cores, which are generally 35 μm and larger; common values are 50 μm and 100 μm, with clad outside diameters of 125 μm and 140 μm, respectively. During the analysis, we will examine this disparity in greater detail.

In Fig. 2-1, the wavelength enters the waveguide at an angle θ that is less than the critical angle, θ_c. The propagation of light shown in the figure usually fills the core and then spills over into the cladding. Only the supported modes will not damp out and propagate in the core, however, and leaky modes can propagate in the cladding. Since a ray analysis cannot be applied to single-mode fiber without intolerable errors, solutions of Maxwell's equations for electromagnetic fields are necessary to perform a viable analysis.

The objective is to determine the conditions required for single-mode propagation. First, we will present Maxwell's equations for a homogeneous, nonconducting, space-charge-free, isotropic medium (the variables with overscores indicate vectors):

$$\text{curl } \overline{E} = -\frac{\partial \overline{B}}{\partial t} \qquad \text{curl } \overline{H} = \frac{\partial \overline{D}}{\partial t} \qquad (2\text{-}1)$$

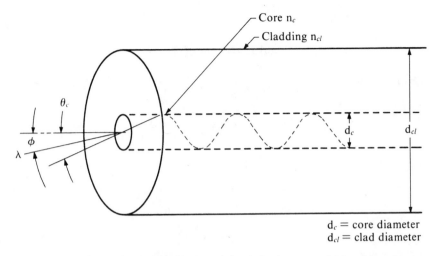

Fig. 2-1. Single-mode waveguide structure with a helical ray propagating down the core.

where

$\overline{E}$ = electric field
$\overline{H}$ = magnetic field

$$\text{div } \overline{D} = 0 \qquad \text{div } \overline{B} = 0 \qquad (2\text{-}2)$$

where

$$\overline{D} = e_r e_o \overline{E} \qquad \overline{B} = \mu_o \mu_r \overline{H} \qquad (2\text{-}3)$$

Then, taking the curl of both sides of the $\overline{E}$ field of Eq. 2-1, we have

$$\text{curl curl } \overline{E} = -\frac{\partial}{\partial t} (\text{curl } B) \qquad (2\text{-}4)$$

Substituting the value of $\overline{B}$ in Eq. 2-3 into Eq. 2-4 yields

$$\text{curl curl } \overline{E} = -\mu_o \mu_r \frac{\partial}{\partial t} (\text{curl } \overline{H})$$

Then by substituting the curl $\overline{H}$ value in Eq. 2-1 it follows that

$$\text{curl curl } \overline{E} = -\mu_r \mu_o \frac{\partial^2 \overline{D}}{\partial t^2} \qquad (2\text{-}5)$$

In the vector identity

$$\text{curl curl } \overline{E} \equiv \overline{\nabla}(\text{div } \overline{E}) - \overline{\nabla}^2\overline{E}$$

it happens that

$$\text{div } \overline{E} = \text{div } \left(\frac{\overline{D}}{e_r e_o}\right) = 0$$

Thus

$$\text{curl curl } \overline{E} = -\overline{\nabla}^2\overline{E} = -\mu_r\mu_o\frac{\partial^2\overline{D}}{\partial t^2}$$

and

$$\overline{\nabla}^2\overline{E} - \mu_o\mu_r\frac{\partial^2\overline{D}}{\partial t^2} = 0$$

Eliminating $\overline{D}$, we have

$$\overline{\nabla}^2\overline{E} - \mu_o\mu_r e_o e_r\frac{\partial^2\overline{E}}{\partial t^2} = 0 \qquad (2\text{-}6)$$

Using a similar analysis we have,

$$\overline{\nabla}^2\overline{H} - \mu_r\mu_o e_r e_o\frac{\partial^2\overline{H}}{\partial t^2} = 0 \qquad (2\text{-}7)$$

Here,

$$\overline{\nabla} = \left(\frac{\partial}{\partial r}\right)\overline{a}_r + \frac{1}{r}\left(\frac{\partial}{\partial \phi}\right)\overline{a}_\phi + \left(\frac{\partial}{\partial z}\right)\overline{a}_2$$

where $\overline{a}_r$, $\overline{a}_\phi$, and $\overline{a}_2$ are unit vectors for radial components. Then,

$$\overline{\nabla}^2 \equiv \left(\frac{\partial^2}{\partial r^2}\right) + \frac{1}{r}\left(\frac{\partial}{\partial r}\right) + \frac{1}{r^2}\left(\frac{\partial^2}{\partial \phi^2}\right) + \left(\frac{\partial^2}{\partial z^2}\right)$$

Since the desired solution to these equations is in cylindrical coordinates $(r\phi z)$

and propagation is along the Z axis (the axis of symmetry), we can let $\psi = f(r, \phi, z)$, where ψ is a dummy variable. Then,

$$\left(\frac{\partial^2 \psi}{\partial r^2}\right) + \frac{1}{r}\left(\frac{\partial \psi}{\partial r}\right) + \frac{1}{r^2}\left(\frac{\partial^2 \psi}{\partial \phi^2}\right) + \left(\frac{\partial^2 \psi}{\partial z^2}\right) - \mu_r \mu_o e_r e_o \left(\frac{\partial^2 \psi}{\partial t^2}\right) = 0 \quad (2\text{-}8)$$

Now, the propagation velocity, v_{pv}, is

$$v_{pv} = c = \frac{1}{\sqrt{e_o e_r \mu_o \mu_r}}$$

where c is the speed of light in a vacuum. When $e_r = \mu_r = 1$,

$$v_{pv} = \frac{1}{\sqrt{e_o \mu_o}}$$

Since the propagation velocity in an isotropic medium where n is the index of refraction is

$$v_{pv} = \frac{c}{n}$$

then

$$n = \sqrt{e_r \mu_r}$$

Equation 2-8 then becomes the scalar wave equation of the form shown in Eq. 2-9, as follows:

$$\left(\frac{\partial^2 \psi}{\partial r^2}\right) + \frac{1}{r}\left(\frac{\partial \psi}{\partial r}\right) + \frac{1}{r^2}\left(\frac{\partial^2 \psi}{\partial \phi^2}\right) + \left(\frac{\partial^2 \psi}{\partial z^2}\right) - \frac{n^2}{c^2}\left(\frac{\partial^2 \psi}{\partial t^2}\right) = 0 \quad (2\text{-}9)$$

Solutions to this equation for both $\overline{E}$ and $\overline{H}$ are similar because the transverse electromagnetic (TEM) waves are mutually perpendicular, and both are perpendicular to the axial component (direction of propagation, or Z axis).

$$\psi = \psi(r, \phi)\, e^{-j(\omega t - \beta z)} \quad (2\text{-}10)$$

where $\omega = 2\pi f$, the angular frequency, and β is the propagation constant. Either $\overline{E}$ or $\overline{H}$ may be substituted in Eq. 2-10 for ψ as a solution to Eq. 2-9.

The previous discussion has dealt with a general solution to Maxwell's equation. The primary concern in single-mode analysis is the analysis of stepped-index waveguides. If the waveguides are assumed to have circular symmetry, then the solution can be expressed in a more convenient form, that is, in Hankel* or Bessel functions of the third kind.[1] These are shown in Eqs. 2-11 and 2-12.

$$F(q) = 2\pi \int_0^\infty f(r) J_o(2\pi qr) \, r \, dr \qquad (2\text{-}11)$$

$$f(r) = 2\pi \int_0^\infty F(q) J_o(2\pi qr) \, r \, dr \qquad (2\text{-}12)$$

Equation 2-11 is referred to as the Hankel transform (of zero order) of $f(r)$, and it is the reciprocal for the case when the Bessel function Kernels are cosine and sine.

Solutions of Eq. 2-9 are subject to the discontinuity at $r = a$. The indexes of refraction of the core and clad are n_1 and n_2, respectively. For this analysis, they are both considered uniform. The solutions are given in Eqs. 2-13 and 2-14, as follows:

$$\psi_z = \psi_{c1} J_K(ur) \cos K\phi \qquad \text{for } r < a \qquad (2\text{-}13)$$

$$\psi_z = \psi_{c2} K_K(wr) \cos K\phi \qquad \text{for } r > a \qquad (2\text{-}14)$$

where ψ_{c1} and ψ_{c2} are constants of the electric or magnetic field and

$$u^2 = \left(\frac{2\pi n_c}{\lambda}\right)^2 - \beta^2 = \beta_c^2 - \beta^2 \qquad (2\text{-}15)$$

$\beta = $ propagation constant of guided wave

$$\beta_c = \frac{2\pi n_c}{\lambda} = \frac{n_c w}{c}$$

$$\lambda = \frac{c}{f}$$

$$w^2 = \beta^2 - \left(\frac{2\pi n_{cL}}{\lambda}\right)^2 = \beta_{cL}^2 - \beta^2 \qquad (2\text{-}16)$$

---

*Named for the German mathematician, Hermann Hankel, 1839–1918.

$$\beta_{cL} = \frac{n_2 w}{c}$$

$$K = \text{integer values}$$

$$K_K(wr) \propto \frac{e^{-wr}}{\sqrt{wr}} \tag{2-17}$$

where $r > a$, $wr \gg 1$, and $a =$ core radius.

The stipulation stated in Eq. 2-17 is considered an evanescent wave, i.e., at large distances from the core the fields die out.

A graph of Bessel function arguments for lower order modes within the waveguide core is plotted in Fig. 2-2. The part of this figure of most interest for single-mode operation is the first root of the Bessel function argument of J_o. Prior to this root, only a single mode, HE_{11}, is present, and, as also may be noted, no transverse electric or magnetic fields (TE_{ek} and TM_{ek}, respectively) are present. The core argument within the Bessel function is shown in Eq. 2-18:

$$0 \leq V < 2.405 \tag{2-18}$$

where

$$\frac{u_{km}^2 r^2}{a^2} = \beta_1^2 - \beta_2^2 = \frac{w^2}{c^2}(n_1^2 - n_2^2)$$

$$V = u_{km} r = \frac{wa}{c}(n_1^2 - n_2^2)^{1/2}$$

Then

$$V = \frac{2\pi a}{\lambda}(n_1^2 - n_2^2)^{1/2} \tag{2-19}$$

Equation 2-19 is the exact value of V in which Eq. 2-20 is an expression of the approximate value for V:

$$V \approx \frac{2\pi a}{\lambda}(2n_2 \Delta n)^{1/2} \tag{2-20}$$

For a typical set of parameters, let us calculate the size of a waveguide core

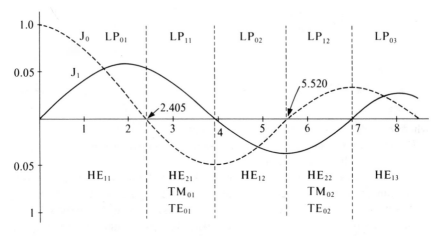

Fig. 2-2. Bessel function solutions indicating various modes of propagation (LP designation refers to linear polarization).

needed to satisfy the constraints using Eq. 2-19. Rearranging Eq. 2-19 and using the inequality in Eq. 2-18,

$$2a < \frac{2.405\lambda}{\pi(n_1^2 - n_2^2)^{1/2}}$$

Let $n_2 = 1.46$, $\Delta n = 0.002$, $\lambda = 0.85$ μm, and $d_c = 2a$, then

$$2a < 8.5 \ \mu m$$

If the core was designed to accommodate a 1-μm wavelength, then $2a < 10$ μm; the 8.5-μm core, however, will accommodate either wavelength, i.e., 0.85 μm or 1.0 μm.

This discussion has primarily centered on the mathematical mechanics of single-mode fiber optics. The physical aspects must now be examined to give the reader a more complete understanding of the mathematical model. The HE_{11} mode (fundamental wavelength) can be decomposed into the E and H transverse fields, which are pulsating at ω frequency. The pulsation frequency is embedded in the propagation, which is a function of u and β. This field description applies to the core material. For fields beyond the core region, an exponential decay occurs. As stated previously, ray diagrams are not accurate and too complex to use to describe monomode behavior; consequently, mathematical solutions are used.

Single-mode waveguides under discussion are considered weakly guided with the restriction, $\Delta n < 1\%$. When this later condition does not hold, the analysis will break down because the linear polarization (LP) designations do not hold because of degenerate mode separation. Throughout this text, this restriction will be adhered to.

At the boundary, a is equal to r, and the following equations must be satisfied (i.e., Eqs. 2-20 and 2-21):

$$u^2 + w^2 = V^2 \tag{2-20}$$

$$u\frac{J_1(u)}{J_o(u)} = w\frac{K_1(w)}{K_2(w)} \tag{2-21}$$

LOSS MECHANISMS

One of the most important properties of optical waveguides is the low attenuation they exhibit. Low-loss waveguides translate into larger repeater spacing (distance in kilometers between transmitter and receiver), and this will reduce the cost of networks. The single-mode waveguide is characterized by losses of 2 to 0.5 dB per kilometer. Attenuation arises from intrinsic material properties, such as Rayleigh scattering and absorption. It is also induced by microbending and irregular geometric properties of waveguides; these latter two losses are the result of waveguide properties only. Discussion of splices and terminations will be deferred to Chap. 5 because of the large amount of material to be covered.

MATERIAL ATTENUATION

Material attenuation effects will be confined to those in doped high-silica glasses (doping is a technique of adding certain impurities to glass to enhance transmission characteristics). Certain dopants, such as fluoride, are becoming of interest because they enhance waveguides at wavelengths of 1.7 μm to 4.0 μm, which thus far are impractical. But, at one period not too many years ago, the same comments were made about 1300-nm and 1550-nm waveguides. Many of the manufacturing processes for waveguides can be found in multimode texts. For the reader who desires some familiarity with (1) inside vaporization depositions (IVD), and (2) outside vaporization depositions (OVD), see Ref. 2.

Pure silica has several infrared (IR) absorption wavelengths; for example, at 9 μm and 13 μm, attenuation is about 10^{10} dB/km. Overtones and combinations of these two fundamental resonances lead to absorption peaks at 3 μm (5×10^4 dB/km) and 3.8 μm (6×10^5 dB/km). The tail of these two absorption peaks has the following typical values of attenuation:

Attenuation	Wavelength
0.02 dB/km	1550 nm
0.1 dB/km	1630 nm
1.0 dB/km	1770 nm
Cutoff	1800 nm

Ultraviolet (UV) displays high attenuation at shorter wavelengths with typical values at the tail of 1 dB/km at 620 nm and of 0.02 dB/km at 1240 nm. The values for the IR and UV absorption are shown in the curves of Fig. 2-3.

Other absorption that affects the material attenuation is due to impurities or dopants added to the silica. Impurities are a particular problem; they may be introduced in water vapor or dust particles. One of the most detrimental to fiber-optic waveguides is the OH^-, and it was, until recently, one of the most diffcult to remove during waveguide fabrication. As little as one part in a million (ppm) hydroxyl ion (due to water vapor) causes resonance at the fundamental wavelength, 2730 nm, and overtones and harmonics. The attenuation at various wavelengths is as follows:

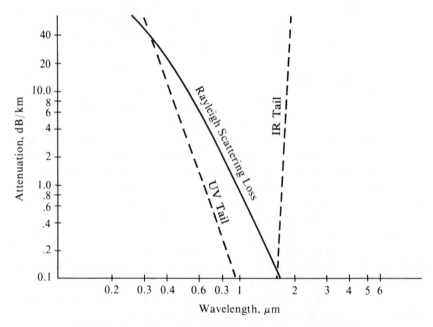

Fig. 2-3. Pure silica material attenuation versus wavelength.

Attenuation	Wavelength
>40 dB/km	2730 nm
40 dB/km	1390 nm
3 dB/km	1240 nm
1 dB/km	950 nm
0.09 dB/km	720 nm

Note that large attenuations occur near 1300 nm and 1550 nm, which are two of the most widely used long wavelengths. OH^- concentration must be kept below 0.1 ppm to reduce these high losses in waveguides; this is easily achieved with today's fabrication processes. For the reader desiring more details of the waveguide manufacturing process, see Ref. 3.

Other impurities such as copper (Cu), iron (Fe), and chromium (Cr) also cause large attenuation peaks. In recent years, however, these impurities have been reduced to such low levels that they are virtually nonexistent. The concentrations are as follows: $Cu < 2.5$ parts per 10^{10}; $Fe < 1$ part per 10^{10}; and $Cr < 1$ part per 10^{10}.

The next material-attenuation source to consider is dopants that enhance either certain properties of glass or the manufacturing process of the waveguide. One important use for dopants is to enhance the radiation properties of glass, a subject that will be discussed later.

Dopants such as GeO_2, P_2O_5, B_2O_3, and F added to glass ease the manufacturing process; they also modify refractive indices. When adding dopants, the objective is not to raise the attenuation of the waveguide over the useful spectral range. For example, P_2O_5 as a dopant causes a large attenuation peak at 3800 nm. Some dopants in the vapor disposition processes are added within certain geometric distances to keep absorption low within the useful spectrum range. Dopants will not be discussed in detail because, as the technology improves, new ones are continually being introduced. For example, the trend to use longer wavelength sources will bring about changes in dopants because the useful spectral range is being changed. At present, the most popular sources fall between 800 and 1800 nm (see Fig. 2-3).

Rayleigh scattering loss is another phenomenon that produces attenuation as a result of microscopic material inhomogeneities. The curve in Fig. 2-3 depicts the relationship between wavelength and Rayleigh scattering loss. The λ^{-4} dependence of the loss makes it highly sensitive to wavelength. This loss becomes insignificant or nonexistent at long wavelengths (approximately 1270 nm for silica). UV and IR absorption also becomes small within this segment of the spectral plot; therefore, sources able to operate in this region are very desirable.

Dopants have an impact on Rayleigh scattering because they increase microscopic inhomogeneities. The refractive index difference, Δn, between core and clad are produced with dopants. In most single-mode fiber, this difference will be very small (less than 1 percent). This technique helps keep Rayleigh scattering low.

Some of the most widely used dopants are GeO_2, B_2O_3, and P_2O_5. Figure 2-4 is provided to illustrate the effects of these dopants. A most noticeable observation from the curves in the figure reveals that Rayleigh scattering is a function of both dopant concentration and material. Furthermore, the figure indicates that as doping concentration decreases (smaller Δn), it has much less impact. This is one of the reasons for keeping concentrations low in single-mode fiber. Thus far, we have discovered that long wavelengths and small Δn reduce Rayleigh scattering, but as one may well surmise, "You don't get something for nothing" in the physical world. Some of the ramifications will be apparent in microbending loss calculations, which will be discussed shortly.

The equations governing Rayleigh scattering will be examined to enhance the analysis further. Equation 2-22 is an expression for Rayleigh scattering that has

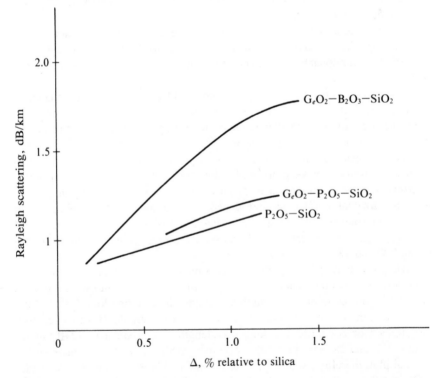

Fig. 2-4. Rayleigh scattering versus Δ for single-mode waveguide (core diameter of 10 μm) with commonly used dopants.

been derived from empirical curves (see Pinnow et al.[4] for further information). The variables for this equation are as follows:

$$\alpha_{RS} = \frac{8\pi^3}{3\lambda^4} \, n^3 P^2 KT\beta_T \tag{2-22}$$

where α_{RS} = scattering attenuation; n = glass index of refraction; P = photoelastic coefficient; β_T = the isothermal compressibility of the material; K = Boltzmann's constant; T = the absolute temperature of the glass, when heated, at which it would come to dynamic equilibrium; and λ = the wavelength of transmission. The equation can be reduced to the form shown in Eq. 2-23, as follows:

$$\alpha_{RS} = K_c \lambda^{-4} \tag{2-23}$$

Here, the value of K_c is a function of all the material constants and attenuation is dependent on the fourth power of the wavelength. An increase in the wavelength of transmission from 800 to 1200 nm will produce a corresponding reduction in Rayleigh scattering attenuation of 10 dB.

A form of this equation developed for GeO_2 single-mode fiber is presented in Eq. 2-24:

$$\alpha_{RSGe} = (0.75 + 66\Delta n_{Ge}) \, \lambda^{-4} \text{ dB/km} \tag{2-24}$$

The other equations related to material attenuation losses that are given in the following equations are categorized as intrinsic scattering functions.

The first, Eq. 2-25, depicts density fluctuations because glass is an imperfect crystal. This equation is a contributor to Rayleigh scattering. The variables are the same as those in Eq. 2-22.

$$\delta_D = \frac{(2\pi)^3}{\lambda^4} \, (n^2 - 1) \, KT\beta_T \tag{2-25}$$

Equation 2-26 depicts the relationship of fluctuations in concentration of doped glass to the scattering-loss components.

$$\delta_c = \frac{2}{3} \frac{(2\pi)^3}{\lambda^4} \, n \left(\frac{dn}{\partial c}\right)^2 (\Delta C)^2_M \Delta V \tag{2-26}$$

The variables in this equation are defined as follows:

n = index of refraction of the glass
$(\Delta C)_M^2$ = mean square of the concentration fluctuation
ΔV = volume over which the concentration varies
δ_c = loss due to concentration.

Stimulated Raman and Brillouin scattering losses are others that appear in single-mode waveguide structures. They are represented in detail in Smith.[5]

WAVEGUIDE ATTENUATION

Losses induced by the waveguide structure itself—such as geometric irregularities, bending loss, microbending loss, and defective joints—are all classified as waveguide attenuation. The latter loss will be discussed in Chap. 5.

Bending of the waveguides induces radiation loss because internal rays are no longer totally reflected. If the bend radius is kept above some critical value, this loss will be minimized. The objective of this investigation is to determine R_c, the critical bend radius, and the parameters that affect it.

When a bend occurs in a waveguide, the wavefront within it is pivoted about the center of curvature at the radius of curvature. The phase velocity of the plane wave in the cladding is greater than that of the core at the bend. As the distance from the center of curvature is increased, the phase velocity will exceed that of the plane wave in the cladding. At this point, the power will radiate away from the guide, thus creating a radiation loss.

A curvature in the waveguide will cause mode coupling between leaky and LP_{01} (guided) modes. The mode coupling occurs in both directions, i.e., leaky modes also couple to LP_{01} modes. Therefore, one would expect oscillatory fluctuations between the modes after the curvature.

Equations 2-27, 2-28, and 2-29 can be used to find attenuation due to bending. These equations will be sufficient for an approximation to the bending loss, but should the reader require a more analytical set of equations, with a discussion of their development, see Jeunhomme.[6]

$$\alpha_b = A_b R_c^{-1/2} e^{-UR_c} \tag{2-27}$$

$$A_b = 30(\Delta n)^{1/4} \lambda^{-1/2} \left(\frac{\lambda_c}{\lambda}\right)^{3/2} \frac{dB}{m^{1/2}} \tag{2-28}$$

$$U \approx 0.7 \frac{(\Delta n)^{3/2}}{\lambda} \left(2.75 - 0.996 \frac{\lambda}{\lambda_c}\right)^3 \frac{1}{\text{meters}} \tag{2-29}$$

where $0.8 \leq \lambda_c/\lambda \leq 2$.

The critical radius of curvature is developed in Jeunhomme for wavelengths in about the 1000-nm range. Equation 2-30 depicts the approximation for R_c, the critical radius. This equation is developed by taking the derivative of Eq. 2-27 and solving for the slope where the maximum change occurs. Note that this is a transcendental equation and that graphical solutions are the easiest method of solving for slope.

$$R_c \approx \frac{14}{U} \qquad (2\text{-}30a)$$

or

$$R_c \approx 20 \frac{\lambda}{(\Delta n)^{3/2}} \left(2.75 - 0.996 \frac{\lambda}{\lambda_c}\right)^{-3} \qquad (2\text{-}30b)$$

These bends are the result of cable installation in which the cable may have been pulled through an electrical conduit. The bends in the conduit itself are of no consequence; it is local bending when fiber-optic cable is laid across other wiring that results in sharp local bends. One problem with conduit installations quite often arises from cable pulled through it. Often, tightly packed conduit not only causes bending, but any local stress on cable that occurs can also cause other problems. Mechanical stress on cable is discussed at the end of this chapter along with fatigue.

Microbending loss is the attenuation caused by the small irregularities that arise during cabling or installing of cabling. The oscillation, similar to bending, between LP_{01} and leaky modes, and vice versa, produces a loss component. Pure bending produces a coupling of leaky modes and LP_{01}, and this results in loss. Bending loss, α_{sm}, is expressed by Eq. 2-31, which contains a variable for the loss of multimode fiber, α_{mm}, as follows:

$$\alpha_{sm} = 2.53 \times 10^4 \alpha_{mm} \left(\frac{W_B}{a}\right)^6 \left(\frac{\lambda_c}{\lambda}\right)^4 \frac{\lambda_c^2}{(\Delta n)^3} \qquad (2\text{-}31)$$

$$\alpha_{mm} = \frac{0.9 a^4 b_n B_h^2}{(\Delta n)^3 R_f^6} \left(\frac{E_c}{E_f}\right)^{3/2} \qquad (2\text{-}32)$$

where:

b_n = number of bumps/km
B_h = height of bumps
R_f = fiber radius
$E_c = 2 \times 10^5$ psi (material modulus of core)
$E_f = 8 \times 10^6$ psi (material modulus of fiber)

Example 2-1: For the following typical values of a fiber:

$$b_n = 200/\text{km}$$

$$B_h = 2 \ \mu\text{m}$$

$$\Delta n = 0.12$$

$$2a = 50 \ \mu\text{m core}$$

$$2R_f = 125 \ \mu\text{m clad}$$

the value of α_{mm} is 3.8 dB/km.

A more useful form of Eq. 2-31 is a normalized form from which the actual value can be calculated, as shown in Eq. 2-32. Equation 2-33 is the normalized ratio of spot and core size.

$$\frac{\alpha_{sm}}{\alpha_{mm}} = 2.53 \times 10^4 \left(\frac{W_o}{a}\right)^6 \left(\frac{\lambda_c}{\lambda}\right)^4 \frac{\lambda_c^2}{(\Delta n)^3} \qquad (2\text{-}32)$$

where:

$$\frac{W_o}{a} = 0.65 + 0.434 \left(\frac{\lambda}{\lambda_c}\right)^{3/2} + 0.0149 \left(\frac{\lambda}{\lambda_c}\right)^6 \qquad (2\text{-}33)$$

A graphic representation of various values of Eq. 2-32 is depicted in Fig. 2-5. Examination of the figure reveals that as the difference in index of refraction, Δn, increases to 1 percent, single-mode wavelengths must be kept below 900 nm or bending losses will begin to become prohibitive. As an example, if $\lambda = \lambda_c$, the bending loss approaches 2.5 times that of single-mode fiber (which can be as large as 1 or 2 dB/km).

Note that the curves in Fig. 2.5 are dimensionless; therefore, for cutoff wavelengths, assume other values for predicting attenuation caused by bending loss at other wavelengths. For example, if cutoff occurs at 1300 nm, then the single-mode loss is equal to the multimode loss at a wavelength of 1430 nm for Δn = 0.5 percent.

Example 2-2: Using the same values for multimode cable as in Example 2-1, except that $\Delta n = 0.01$, then $\alpha_{mm} = 9.43$ dB/km using Eq. 2-32. Monomode cable will have a loss of approximately 3 dB/km when $\lambda/\lambda_c = 0.8$.

As a final topic for the attenuation discussion, nuclear radiation effects should

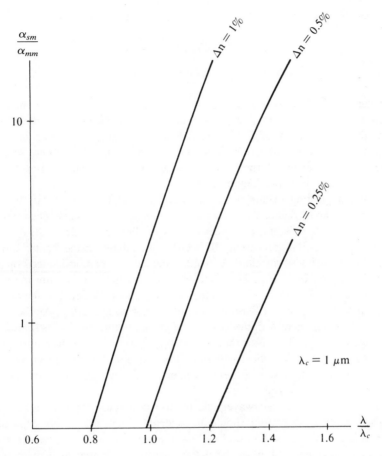

Fig. 2-5. Graphic representation of Eq. 2-32 (α_{mm} = multimode attenuation; α_{sm} = single-mode attenuation).

be considered, primarily because the difference between tactical single-mode networks (military uses) and nuclear reactors (industrial uses) is not so obvious.

Tactical networks are not exposed to radiation most of the time but must withstand large impulses of radiation for short periods. In the industrial situation, radiation may be present over a long span of time at low levels. One example would be the control cables for the refueling grapple (located on top of the reactor and usually exposed to long-lasting doses of radiation).

First, an investigation of the waveguide and how radiation affects its intrinsic properties is in order. Then, military and industrial applications can be considered in more detail.

Before the discussion can begin, some of the terms must be defined:

$$1 \text{ rad} = 1 \text{ joule}/100 \text{ kg of material}$$

$$\text{rads}/\text{hr} = \text{dose or absorbed dose}$$

Nuclear radiation has an ionizing effect on glass that disrupts the chemical bonds within the glass. This radiation will cause absorption of wavelengths in the visible and near infrared spectra. Attenuation varies, depending on glass composition, from 0.1 to 10 or 12 dB/km per rad within the 800-nm spectral range. Longer wavelength transmission systems are somewhat less sensitive to radiation than shorter wavelength systems.

Another interesting feature is that hydroxyl ions (OH^-), which are usually due to water vapor, reduce the radiation sensitivity of silica fibers. The ionizing effects of nuclear radiation on defect centers of the waveguide result in these centers becoming light-absorbing. The OH^- ion inhibits defect-center formations, thus reducing absorption. A diagram depicting radiation-induced optical absorption is shown in Fig. 2-6. Note that the absorption losses are the most favorable in the fiber-optic wavelengths, 900 nm and 1.05 μm. For fibers with these particular dopants, 1300-nm operation is not so appealing. As the OH^- concentrations are increased, the loss will be greater before the fiber is irradiated, but, generally speaking, after the loss becomes lower at selective absorption wavelengths, pure silica fibers will show less damage than the doped variety. The dopants tend to resemble defect centers in the waveguide, thus causing greater permanent damage.

Before selecting a doped waveguide for use near nuclear radiation sources, the wavelength of operation should be selected. Then the dopant-dependent waveguide exhibiting the desired low loss after irradiation should be selected.

Another important feature of the optical waveguide is that it recovers after irradiation but not fully. As an example, a waveguide operating at a 1-μm wavelength may exhibit a loss prior to exposure of 2 dB/km. After being exposed to a large dose of radiation, the loss may increase to 10 dB/km. After some recovery period, depending on the material and the time spent, the loss may be 3 dB/km. The waveguide loss will not return to its original figure of 2 dB/km, but it does sustain a certain amount of damage. The amount of recovery after a pulse of radiation is dependent on the purity of the silica.

Some waveguides, such as synthetic fused silica Suprasil 1 and Suprasil 2, will display increases in loss as the waveguide is further irradiated, but after approximately 10^4 rads, the loss will peak. Any increases in dosage will cause a reduction in loss (bleaching). There are some waveguide materials in which bleaching after irradiation reduces the waveguide loss below its original loss before irradiation.

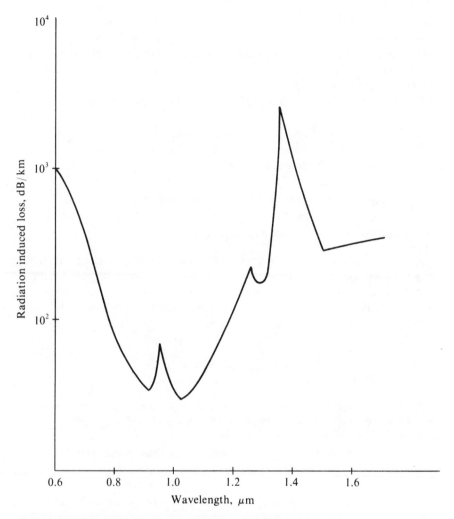

Fig. 2-6. Radiation effects on B-, P-, and Ge-doped waveguide one hour after exposure at 10^5 rads with OH^- present in a 100-ppm concentration.

Several waveguides on the market exhibit radiation hardness at various wavelengths. They are not discussed in this text because fast changing fiber-optic technology will result in the text becoming obsolete before publication.

When designing a fiber-optic communication system that is to withstand radiation, one should be aware that even small doses will affect the induced loss if the pulses occur often. As an example, suppose a nuclear reactor has a refueling control system that has a fiber-optic cable. If the cable is exposed to pulses of radiation during each fueling cycle, the cable will recover, but some residual loss will remain. For a situation where refueling occurs only once a year, the loss of the cable can gradually increase over a span of years. It may finally become so lossy that the control system will fail; in that case, replacing the cable can be quite costly.

In the situation here described, either of two remedies is possible:

1. Shield the cable from the harmful radiation.
2. Select a cable and design a system that will operate with properly increasing loss over the life-cycle of the reactor.

A design of a control system is presented in Chap. 8.

DISPERSION IN SINGLE-MODE WAVEGUIDE

The effect of dispersion is very important in waveguides because it will smear or stretch optical pulses. This condition in turn will limit the bandwidth of operation. In multimode fiber, two types of dispersion exist: intermodal and intramodal. For single-mode waveguides, the restriction is given by Eq. 2-19 and will be repeated here. For $V < 2.405$,

$$V = \frac{2\pi a}{\lambda} \left(n_c^2 - n_{cL}^2 \right)^{1/2}$$

As may be observed from this equation, any of three approaches—reducing the refractive index difference between the core and cladding, decreasing the core radius, or increasing the operative wavelength—can be used to permit single-mode operation. Usually, a variation of all three parameters is used to do so.

One of the reasons for operating a communication system using single-mode technology is that the intermodal dispersion that severely limits bandwidth is not present. The remaining intramodal term is further reduced and this further improves bandwidth.

Some terms that are not addressed in multimode technology must be considered here. *Birefringence* is one of these properties. In the waveguide analysis of LP_{01} modes, the assumption was made that the orthogonal component of electrical and magnetic fields has the same propagation constant because the waveguide is perfectly circular. In actuality, however, some ellipticity of the core exists, as well as some anisotropy in the refractive index distribution because of the stress contributed to differences in propagation constants. This imperfection in the waveguide results in dispersion. When dealing with single-mode fiber-optic sensors, the waveguide is made elliptic to enhance sensor parameters. This subject will be addressed in Chap. 11.

The next term to consider is chromatic dispersion, which is composed of material dispersion, waveguide dispersion, and mixed terms. The equation which depicts chromatic dispersion, Eq. 2-34, looks quite formidable:

$$
\frac{d\tau}{d\lambda} = -\left(\frac{\lambda}{c}\right)\left(\frac{d^2h}{d\lambda^2}\right)[1 + \Delta(Vb)'] - \left(\frac{n_2\Delta}{c\lambda}\right)\left[\left(\frac{N_2}{n_2} - \frac{P}{2}\right)^2 V(Vb)'' \right.
$$

$$
\left. - \left(\frac{N_2}{n_2}\right)P\left(\frac{3(Vb)' - b}{2}\right) + \lambda\left(\frac{dP}{d\lambda}\right)\left(\frac{b + (Vb)'}{2}\right)\right]
$$

(2-34)

where:

$$\tau = \text{group delay}$$

$$P = \left(\frac{\lambda}{\Delta}\right)\left(\frac{d\Delta}{d\lambda}\right) = \text{difference in dispersive properties between core and}$$

cladding

$$\frac{N_2}{n_2} = \text{local variations within the medium}$$

$$b = \text{normalized propagation constant}$$

For the reader who wishes to delve into the development of Eq. 2-34, refer to Jeunhomme[6] (pages 23 to 26).

The following approximations can be made to reduce the complexity of Eq. 2-34: $P \ll 1$ and $(N_2/n_2) \approx 1$. Equation 2-34 then reduces to Eq. 2-35, as follows:

$$
\frac{d\tau}{d\lambda} = -\left(\frac{\lambda}{c}\right)\left(\frac{d^2n}{d\lambda^2}\right)[1 + \Delta(Vb)'] - \left(\frac{n_2\Delta}{c\lambda}\right)\left[\left(1 - \frac{P}{2}\right)^2 V(Vb)'' \right.
$$

$$
\left. - P\left(\frac{3(Vb)' - b}{2}\right) + \lambda\left(\frac{dP}{d\lambda}\right)\left(\frac{b + (Vb)'}{2}\right)\right]
$$

(2-35)

For the condition where $P = 0$, Eq. 2-35 can be further reduced to Eq. 2-36, as follows:

$$
\frac{d\tau}{d\lambda} \approx -\left(\frac{\lambda}{c}\right)\left(\frac{d^2n}{d\lambda^2}\right)[1 + \Delta(Vb)'] - \frac{n_2\Delta V}{c\lambda}(Vb)'' \qquad (2-36)
$$

The chromatic dispersion approximation in Eq. 2-36 is composed of three terms. The first is material dispersion; the second, a mixed term; and the third, waveguide dispersion. Evaluating Eq. 2-36 for a particular set of conditions might be formidable. However, making some approximations like those in Ref. 6 make the task easier. Such approximations are given in Eqs. 2-37 and 2-38.

$$(Vb)' \approx 1.306 - 0.1715 \left(\frac{\lambda}{\lambda_c}\right)^2 \tag{2-37}$$

$$V(Vb)'' \approx 0.08 + 0.549 \left[2.834 - 2.405\left(\frac{\lambda_c}{\lambda}\right)\right]^2 \tag{2-38}$$

For various values of V and λ/λ_c, these equations will display the following percentages of error:

Equation	V	λ/λ_c	Error
2-37	$1.6 \leq V \leq 2.4$	$1 \leq \lambda/\lambda_c \leq 1.5$	1%
	$1 \leq V \leq 3$	$0.8 \leq \lambda/\lambda_c \leq 2$	4%
2-38	$1.3 \leq V \leq 2.6$	$0.9 \leq \lambda/\lambda_c \leq 1.9$	5%

Material dispersion for a germanium-doped waveguide is given in Fig. 2-7 as a function of wavelength for various Δ values. As one may observe from the graph, dispersion can assume positive or negative values with proper selection of wavelength of operation and Δ material. Through a method of tailoring index profiles such as that shown in Fig. 2-8, the W index profile dispersion components can be made to have a cancelling effect.

Making use of the curves in Fig. 2-7 and Eqs. 2-36, 2-37, and 2-38, it is possible to calculate values of dispersion in germanium-doped fibers. A sample calculation is shown below:

When $\lambda = 1300$ nm, $\Delta = 0.01$, and $\lambda_c = 1300$ nm,

$$(Vb)' = 1.1345$$

$$V(Vb)'' = 0.205$$

$$-\frac{\lambda}{c}\frac{d^2 n_2}{d\lambda^2} = -2$$

$$\frac{d\tau}{d\lambda} = 1.2216 \text{ picosec/nm} - \text{km}$$

The assumption made here is that the operation is single-mode, i.e., $V \leq$

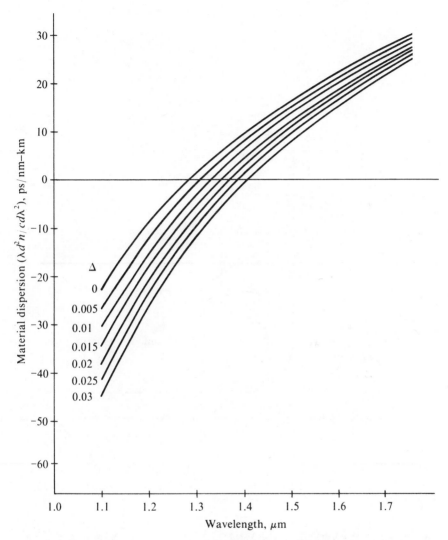

Fig. 2-7. Material dispersion versus wavelength for various Δ values between core and clad for GeO_2/SiO_2 fibers.

2.405. If a value of Δ = 0.005 were used for the waveguide at 1300 nm, zero material dispersion would result. This shift is exploited in some waveguides to produce a near-zero chromatic dispersion. Note, as Δ increases, zero dispersion crossing occurs at longer wavelengths. The curve reflects only a value for the material dispersion and mixed term of Eq. 2-36. The waveguide term is present,

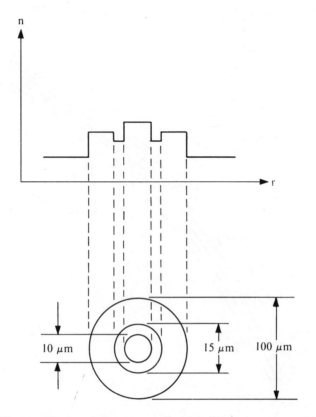

Fig. 2-8. *W* index-of-refraction profile for single-mode waveguide.

but if the sign of the material term can be made positive, then some cancelling will occur. A problem at the end of this chapter will examine this aspect of dispersion.

STANDARD WAVEGUIDE PROFILES

Waveguide performance may be enhanced by using various doping profiles, e.g., in stepped-index waveguides, the single clad and the doping is such that core and clad have a Δ difference. The step index is shown in Fig. 2-9(a). Most single-mode fiber will have this profile because it is more easily manufactured. The next profile is the graded index because the clad has a homogeneous index, but the core gradually changes until it peaks at the center of the core, as depicted in Fig. 2-9(b); a third type is the depressed inner cladding (with thick cladding) shown in Fig. 2-9(c); and the fourth and last profile is the depressed inner

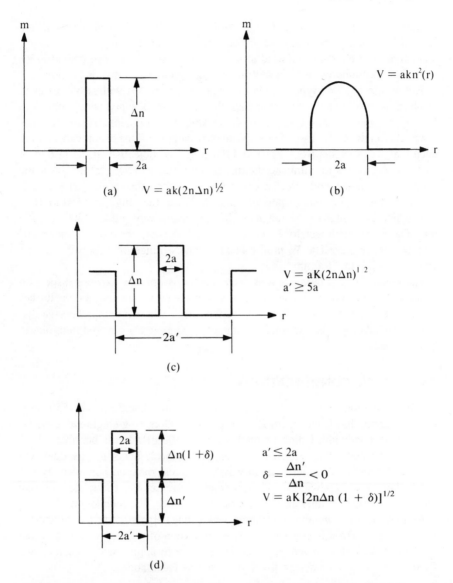

Fig. 2-9. Various index-of-refraction profiles and the describing equations: (a) Stepped-index profile, (b) graded-index profile, (c) depressed inner clad with thick inner clad (T-DIC), and (d) depressed inner clad with narrow inner clad (N-DIC).

cladding (with thin cladding) of Fig. 2-9(d). These last two are also often called *W profiles*. Such profiles are not the only types available, but profiles (a), (c), and (d) are those most often used in single mode. Applications of graded-index profiles are commonly used in multimode waveguide to reduce dispersion.

The profile of "W" type waveguides can be constructed with two dopants to produce the profile. Generally speaking, the inner core is doped with germanium for the core material and fluorine for decreasing the clad index. This profile has lower Rayleigh scatter loss than step-index waveguide. Dispersion cancellation occurs at wavelengths from 1250 to 1700 nm. The objective in the design of such waveguides is to eliminate the dispersion term, $V(Vb)''$, in the chromatic dispersion equation and effect a cancellation on the other terms. The T DIC profile in Fig. 2-9(c) has dispersion cancellation in the range of 1360 to 1650 nm, which is similar to the range of the step-index waveguide, 1300 to 1550 nm. The form has lower loss characteristics, however, because the index differences are higher. The W profile has the widest range of dispersion cancellation, or 1250 to 1700 nm.

The reason for operating at long wavelengths is obvious because dispersion cancellation will allow transmission at a very large bandwidth. For a further discussion of these waveguides, see Jeunhomme[6], where a comprehensive discussion of this topic is presented. It cannot be completely covered here due to lack of space.

BANDWIDTH CONSIDERATIONS

When attenuation and time dispersion are made to coincide (near 1550 nm), time dispersion has been recorded at 10 ps/km.[7] Obtaining single-mode waveguides with bandwidth-length products of 50 to 100 gHz-km is possible.

Most intramode dispersion can be made to cancel, but some residual will always be present. As tolerances in waveguides and material homogeneity improve, however, bandwidths will continually increase. For the reader desiring more information on tolerance effects on waveguides, see Jeunhomme.[8]

As they do for common multimode waveguide, single-mode manufacturers will specify bandwidth length product. A discussion of risetime will be deferred until the total link is examined, that being a more appropriate occasion. See Chap. 8 for a systems design involving risetime calculations.

TEST EQUIPMENT

In recent years, a large number of manufacturers have ventured into the test equipment market. A multitude of simple, hand-held digital voltmeters with optical-to-electrical conversion are available. Most of these devices are rather inexpensive and as simple as most electrical multimeters to use. Converters are also available for use on standard voltmeters, the operator being given a series of calibration curves to convert voltmeter readings to optical power measurements. Most of these meters will read optical power in dBm or μW.

The optical time-domain reflectometer (OTDR) is one of the pieces of test equipment with the most utility. This instrument is tailored after its counterpart, the time-domain reflectometer (TDR), which is a similar device used for making measurements on copper transmission line. The OTDR operates on the principle of pulse echoes in a way similar to radar or sonar. Instead of a gaseous or liquid infinite medium, a confined medium is used, i.e., a glass waveguide.

A simplified OTDR block diagram is shown in Fig. 2-10. This diagram is common to both multimode and single-mode OTDR, but performance of the various blocks is different because of the superiority of single-mode waveguides. The two mechanisms that produce the reflected light are Fresnel reflections and backscattering. Fresnel reflections contribute most of the reflected power at the fiber end, where cleaved orthogonal surfaces exist. Faults in the waveguide usually produce irregular surfaces where Fresnel reflections are small when compared to backscatter.

Let us now discuss some of the equations governing the behavior of the instrument (the block description will be deferred until later). Equation 2-39 depicts the Fresnel reflected power from the end of the waveguide:

$$\text{Reflected power} = P_{RP} = 100 \left[\frac{n_f - n_a}{n_f + n_a} \right]^2 \qquad (2\text{-}39)$$

where:

n_f = refractive index of glass
n_a = refractive index of air

If n_f is assumed to be approximately 1.5, then P_{RP} is 4 percent at the end of the waveguide.

Equation 2-40 depicts backscatter power as a function of time:

$$P_{BS}(t) = \frac{E_o S \alpha_s}{n_f} e^{-2\alpha(ct/2n_f)} \qquad (2\text{-}40)$$

An equation that describes this phenomenon with somewhat more precision is Eq. 2-41:

$$P_{BS}(t) = \left(\frac{\lambda}{2\pi n_c W_o} \right)^2 P_p \alpha_s(x) \tau_p \frac{c}{2N_2} e^{-2(x)ct/N_2} \qquad (2\text{-}41)$$

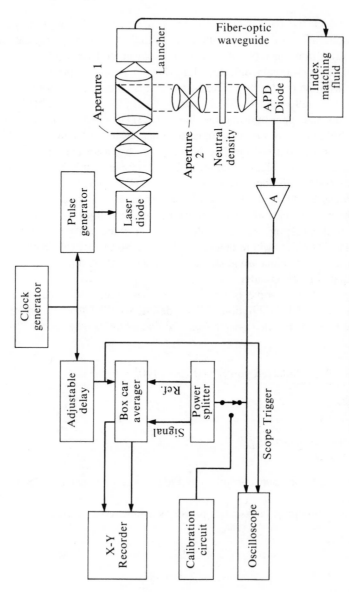

Fig. 2-10. Optical time-domain reflectometer (OTDR).

where:

$P_{BS}(t)$ = backscatter power as a function of time
E_o = transmitted pulse energy
S = scattering portion guided back as backscatter
α_s = fiber scattering loss (Nepers/meter)
P_p = peak optical input power
c = velocity of light in a vacuum
n_c = index of refraction of the core
α = total waveguide attenuation
$x = \dfrac{ct}{2N_2} \approx \dfrac{ct}{2n_f}$ $(\Delta < 1\% \ N_2 \approx n)$
τ_p = launch pulse width

$$10 \log \frac{P_{BS}(t)}{P(t)} \cong -54 + 10 \log \tau_p \qquad (2\text{-}42)$$

where $\tau_p = \tau$ μsec.

A typical OTDR will have the following parameters: a τ_p of 0.2 to 2 μsec and a peak optical power of 1 to 2 mW. Thus, for signal levels for the worst-case, $\tau_p = 0.2$ and $P = $ mW, Eq. 2-41 will yield a power ratio of -61 dB; therefore, for a power input of 1 mW (0 dBm), the receiver sensitivity must be greater than -61 dBm. As one can observe, these signals would be almost impossible to resolve.

The usual method of operation is to combine a number of techniques to improve readout. Averaging circuitry can sample and store signals over a long period of time and resolve signals that are embedded in noise. For relatively short lengths of waveguide, narrow pulse widths may be used, as can be observed from Eqs. 2-40, 2-41, and 2-42. When high resolution is desired, particularly near the launch end, narrow, high-power pulses become necessary because wide pulses may mask any defects close to the launch connector. For gross examination of a long waveguide, large pulse widths are preferred because more energy can be imparted to the waveguide; therefore, more optical energy will be received and less averaging will be necessary, the latter being time-consuming.

Attenuation coefficients can be measured using backscatter techniques. Equation 2-43 depicts the relationship, as follows:

$$\alpha_{av} = \frac{\ln P_2 - \ln P_1}{\left(\dfrac{c}{n}\right)(t_2 - t_1)}$$

Then, since $ct/2n = $ distance $= y$,

$$\alpha_{av} = \frac{\ln P_2 - \ln P_1}{2(y_2 - y_1)} \qquad (2\text{-}43)$$

Now we must return to the block diagram description of the OTDR in Fig. 2-10 and define each of the blocks. Their descriptions follow.

LASER

The laser source must operate at the wavelength at which the waveguide will be used. It must produce high peak radiance and have the capability of being switched at high repetition rates with pulse durations of 200 to 2 nsec. The power output can be anywhere from 1 mW as a minimum to 10 to 15 mW, values that are presently available. Most of our studies will involve wavelengths in the 1200-nm to 1550-nm range.

DETECTORS

Due to the low signal levels produced by backscatter systems, a very high gain detector is necessary. These detectors are usually avalanche photodiode detectors (APD) and must have a rather high degree of linearity to provide the necessary accuracy. By inserting a neutral density filter (calibrated) into the backscatter beam, a comparison between the attenuated and nonattenuated signals will yield a measure of the linearity.

BEAMSPLITTERS AND COUPLERS

Coupling loss is defined as follows:

$$L_c = 10 \log (\phi_1/\phi_2) \qquad (2\text{-}44)$$

where:

$\phi_1 = $ collimated source output power
$L_c = $ coupling loss, in dB
$\phi_2 = $ power directed to the detector after being reflected

A series of beamsplitters and couplers is shown in Fig. 2-11. The backscatter radiation in all of them is assumed to be unpolarized. Figure 2-11(a) is a common garden variety of 50/50 beam splitter. Most beamsplitters have a 50/50 split at some nominal wavelength, and variation in wavelength from this nominal value will result in other than a 50/50 split. Figure 2-11(b) is a beamsplitter

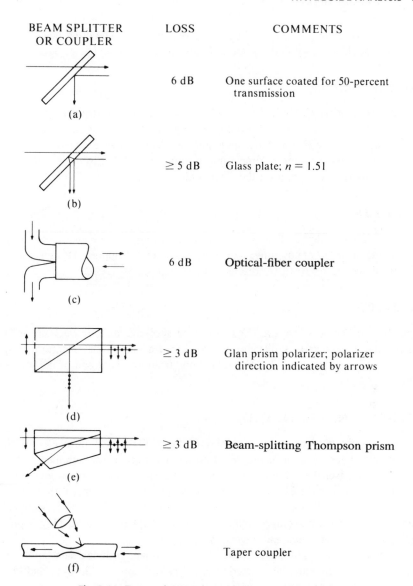

BEAM SPLITTER OR COUPLER	LOSS	COMMENTS
(a)	6 dB	One surface coated for 50-percent transmission
(b)	≥ 5 dB	Glass plate; $n = 1.51$
(c)	6 dB	**Optical-fiber coupler**
(d)	≥ 3 dB	Glan prism polarizer; polarizer direction indicated by arrows
(e)	≥ 3 dB	**Beam-splitting Thompson prism**
(f)		Taper coupler

Fig. 2-11. Types of OTDR beam splitters and couplers.

constructed of plate glass with a high angle of incidence. The other beam splitters and couplers shown here are discussed in more detail in Chap. 6.

BOXCAR AVERAGE

The EG&G Princeton Applied Research Model 162 is an instrument that minimizes drift in the laser source and detector by taking ratios of backscatter and unscanned reference signal. Since an adjustable delay allows the boxcar scan to begin at the desired point in time (time is proportional to distance), the scan may be started at any point on the waveguide.

LAUNCHER

The launcher is a device that is coupled to a waveguide with index matching fluid to maximize light coupling; it also provides mode stripping. Design details of this device are reserved for discussion in Chap. 6.

APERTURES

Apertures are devices that are used to reduce spot size on a waveguide or to restrict the excited modes that are launched into a waveguide. Aperture 2 in Fig. 2-10 is used to reduce Fresnel reflections from the input end of the waveguide.

APPLICATIONS OF THE OTDR

Thus far, only the operating principles of the OTDR have been examined. Among other applications, it may be used for a variety of measurements. The single-mode Laser Precision Corporation OTDR, in particular, has a great deal of measurement utility. This instrument can be operated in either real-time mode or average mode. The latter technique will reduce noise. Signals are sampled digitally, samples are averaged and stored, and an alphanumeric display is provided for operating status and measurement.

The operator can manipulate two CRT screen cursors to index data points on the display. The measurement function circuitry will compute distance and loss information for the waveguide under test.

The source used for the Laser Precision Corp. instrument is a 1300-nm diode with a peak optical power greater than 1 mW. This instrument will measure distances up to 64 km with a resolution of 2 to 16 m depending on pulse duration and power. The attenuation resolution is within ± 0.027 dB, which is fairly impressive.

Chapter 2 has introduced the reader to the transmission medium. Later in the text when waveguides are seen to be an integral part of a system, the OTDR display will provide important measurement parameters. An examination of connector losses, splices breaks, and attenuation characteristics will then be more appropriate.

As improvements occur in OTDR technology, one would expect much higher resolution and accuracy in third-generation equipment. One of the most backward areas in fiber optics is instrumentation, but as the fiber-optic LAN and long-haul, single-mode market grows, single-mode instrumentation will also improve.

A series of problems is presented here to reinforce the reader's comprehension of the subject matter.

REVIEW PROBLEMS

1. Generate a family of curves assuming the following conditions: The waveguides operate in single mode only, i.e., only EH_{11} is present; the index difference is 0.1%, 0.2%, and 0.3%; and the cladding refractive index is 1.50. Plot the core radius versus wavelength.

2. What is the most desirable wavelength to use for transmission to reduce Rayleigh scattering in silica waveguides?

3. Draw a family of curves on semilog graph paper depicting attenuation versus wavelength for germanium-doped fiber when $\Delta n = 0.01$, 0.02, 0.03, 0.05, and 0.1.

4. Derive the critical radius of curvature, R_c, from Eq. 2-27.

5. For SiO_2 doped with Ge, calculate the intramodal dispersion under the following conditions: cutoff wavelength, 1550 nm; operating wavelength, 1300 nm; and index difference of refraction, 0.1.

6. What is the most ideal waveguide profile to use at 1670-nm wavelengths?

7. What is the amount of reflected power from a single-mode waveguide with an index difference of 0.5% and a clad refractive index of 1.50?

8. How does the backscatter power compare to that calculated in question 7 when the following conditions obtain: operating wavelength = 1 m; $n_1 = 1.498$; spot size = 82 nm^2; peak power = 2 mW; pulse width = 200 nsec; $\alpha_s = 1.5$ dB/m (convert Np/m); and total round trip time = 20 nsec.

REFERENCES

1. Bracewell, R. N. *The Fourier Transform and its Applications*. New York: McGraw-Hill, 1978.
2. Gowar, J. *Optical Communications Systems*. New York: Prentice-Hall, 1984, pp. 90–100.
3. Jeunhomme, L. B. *Single-Mode Fiber Optics*. New York: Marcel Dekker, Inc., 1983.
4. Pinnow, D. A.; Rich, T. C.; Ostermayer, Jr., F. W.; and DiDeminico, Jr., M. Fundamental Optical Attenuation Limits in the Liquid and Glossy State with Applications to Fiber Optical Waveguide Materials, Appl. Phys. Letters, Vol. 22, Number 10 (1973), pp. 527–529.
5. Smith, R. G. Optical power handling capacity of low loss optical fibers as determined by stimulated Raman and Brillouin scattering. *Applied Optics* II, 2489 (1972).
6. Jeunhomme, L. B. *Single-Mode Fiber Optics*. New York: Marcel Dekker, Inc., 1983.
7. Kawana, A., et al. *Electronic Letters* 16 (Feb. 28, 1980).
8. Jeunhomme, L. B. *Single-Mode Fiber Optics*. New York: Marcel Dekker, Inc., 1983.

3 | TRANSMITTERS

This chapter is divided into six sections, each building on the others. It will deal primarily with solid-state sources, which are in the forefront of fiber-optic technology at present. Therefore, the first section is dedicated to introducing the reader to the principles of solid-state sources and some of the physics involved. Other sources such as gas, YAGs, and chemical and liquid lasers will also be examined.

Lasers require certain feedback techniques to improve their stability and overall performance, and these will be investigated. After this introduction, the transmitters by means of which sources are implemented in actual circuitry will be studied. These may be either analog or digital. The final section of the chapter is devoted to purely electronic issues, such as interface circuitry for transmitters and encoders and the study of noise sources.

SOLID-STATE LASER PHYSICS

In this section, the general operation of solid-state lasers will be examined without regard to whether they are single-mode or multimode.

A typical laser construction is shown for a gain-guided stripe laser in Fig. 3-1. The first layer is the positive contact, which may also function as a heatsink. In Fig. 3-1(a), the metal contact protrudes through a window etched in the SiO_2 insulation layer to the contact layer. The layers below the contact layer are necessary to confine the optical power and carriers. When power is applied to the laser, current density is greatest in the active, stripe region. Lasers of this type with stripe widths less than 10 μm are commonly referred to as *gain-guided lasers*. The active region shown is the laser cavity. This is the region where light is produced. One of the advantages of stripe lasers is that the lateral confinement of the stripe allows for a lower threshold current because of the high current density in the stripe area. The spectral width of the output is also reduced, thus producing higher energy levels in the wavelength of operation. Also, the emissive area will match the waveguide to which it is being coupled. As a result, coupling for stripe lasers will be much improved over that of other lasers that do not have lateral confinement.

An oxide isolation layer is only one technique to define the stripe. Other confinement techniques are shown in Figs. 3-1(b) and (c). All of these produce similar results, i.e., they confine the active region laterally to improve definition of the cavity. If the cavity did not have stripe confinement laterally, then a

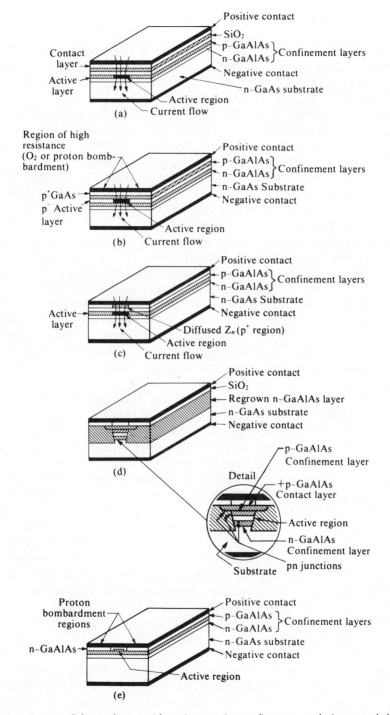

Fig. 3-1. Stripe solid-state lasers with various cavity confinement techniques and their construction.

current density of 20 A/mm^2 produced by current flow through a laser 0.5 mm long and 0.1 mm wide would be 1 A. Stripe confinement of 10 μm will reduce the current required to produce the same current density with 100 mA. Thus, the laser will require less power and produce less heat.

Note that, in Figs. 3-1(a), (b), and (c), confinement is accomplished with doped layers of GaAlAs to make them either P or N material. The PN junctions form the necessary electrical barriers to confine the carriers and optical energy.

The other technique to form a much stronger confinement is shown in Figs. 3-1(d) and (e). These two lasers are referred to as *buried heterostructure* devices (BH). These are index-guided as well, i.e., the regrown areas around the active region have a lower index of refraction and confine the optical power in that region more efficiently. The advantage of this structure is that lower threshold currents are able to produce laser action for an optical output. Buried stripe heterostructures can reduce threshold currents to 10 mA with laser cavity widths as small as 2 μm. The lasers shown in the figures are GaAlAs/GaAs-based structures that lend themselves to short-wavelength operation (800 to 900 nm), but these structures are also used in InGaAsP/InP fabrication of long-wavelength lasers (1300 to 1550 nm). The short-wavelength lasers are now rather inexpensive, but as fabrication techniques and performance improve, the long-wavelength devices will eventually dominate the market. Some of the long-wavelength attributes that warrant this prediction have already been examined in Chap. 2.

The next laser construction to examine is the *phase-locked array*. This type of laser is composed of multiple-stripe lasers all phase-locked and focused to produce optical power output of 100 mW with a threshold current of 250 mA. The number of stripe lasers stacked in this example is ten. Power conversion efficiency is increased in laser arrays to as large as 30 percent. The readers who wishes to delve into the subject further is directed to References 1 through 4.

Laser construction has been briefly discussed without mentioning the theory of operation. First, the radiative recombination process that produces injection luminescence must be examined. A diagram of the process that produces emission or absorption is shown in Fig. 3-2. In Fig. 3-2(a), no luminescence is produced when an electron absorbs energy in the presence of an electromagnetic field and the atom is brought to an upper energy state. If an atom drops from a higher energy state to a lower state, it releases a photon. This uncontrolled, or spontaneous, emission is represented by Fig. 3-2(b), and controlled, or stimulated emission, by Fig. 3-3(c).

If a single quantum of radiation is absorbed, it will react with the atoms in the upper energy level, and a stimulated downward transition will occur, producing emission of photons (luminescence). The mean lifetime of atoms in the excited state is reduced in the presence of radiation. Any radiation produced has the same frequency and phase as the stimulating radiation, and the radiation is

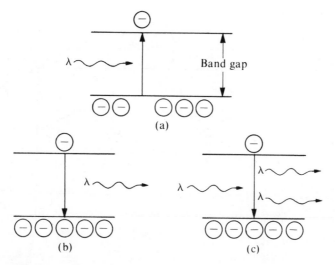

Fig. 3-2. Photon absorption and emission: (a) absorption, (b) spontaneous emission, and (c) stimulated emission.

therefore coherent. Radiation produced in this manner has three important attributes: (1) It has a narrow spectral width, (2) the luminescence is highly directional, and (3) these types of stimulated radiation exhibit the ability to be modulated over large bandwidths.

The radiation wavelength, λ, in meters, is given by the relationship shown in Eq. 3-1 (note that both h and c are constants):

$$\lambda = hc/\Delta\epsilon \tag{3-1}$$

where:

h = Planck's constant = 6.626×10^{-34} Joules-sec
$c = 3 \times 10^8$ meters/sec (speed of light in a vacuum)
$\Delta\epsilon$ = bandgap energy

The wavelength of emission is thus inversely related to bandgap energy. From this, one may surmise it to be a simple matter to make long-wavelength lasers. Relative ease of fabrication and stability of devices, however, have a large impact on laser performance, and these are not reflected in this simple equation.

Some background in the mechanics of laser operation should be examined at this point. Refer to Fig. 3-3 for a graphic view of a laser cavity. As current is passed through the structure, recombinations of injected electrons in the P region (confinement layer) induce radiative carrier recombination in the active region.

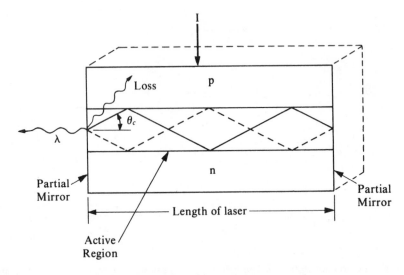

Fig. 3-3. Semiconductor laser cavity construction.

This active region is a resonant cavity to a particular small band of wavelengths. When the current density reaches a certain value, sufficient photons will be produced to reflect back and forth within this cavity. After the photons have attained sufficient energy, they will escape through the semimirrored ends (facets). The radiation that escapes is coherent light (lasing cavity). Figure 3-3 also indicates that some of the photons are lost when their angle with the center axis of the cavity is greater than some critical angle, θ_c. This loss is referred to as *absorption*.

The objective of any laser construction is to confine and control the photons. The five different stripe lasers in Fig. 3-1 perform this task by means of various construction techniques. The stripe laser confines the cavity to a small area of the structure so as to increase the current density of the active area, which is more conductive than the surrounding area. Laser efficiency as expressed by Eq. 3-2 will allow the reader to consider the impact of decreasing laser current:

$$\text{Efficiency} = \frac{P_O}{(I - I_{th})\,(h\lambda/e)} \qquad (3\text{-}2)$$

where:

P_O = optical output power
$h\gamma$ = photon energy

I_{th} = threshold current at which lasing begins

$e = 1.6 \times 10^{-19}$ C of electron charge.

This equation indicates that lasers that can produce high current density with small threshold currents will increase efficiency. As current density is increased, however, the heat generated because of the internal resistance will also increase, and this will reduce device life. One must be aware that the heat generated increases by the square of the current, i.e., increasing the current density by a factor of 2 will increase the power dissipation of the cavity by 4.

A curve showing the optical power versus current relationship is depicted in Fig. 3-4. As input current is increased, spontaneous emission occurs, and the laser output will exhibit noncoherent emission similar to that of an LED. As the current is increased above the threshold, coherent emission will occur. Since stimulated emission has a steep slope above the threshold, small changes in current will produce large changes in power output. This can present a problem, as will be investigated later when thermal effects are discussed.

Let us now examine the spectral width of laser output. Figures 3-5(a) and (b) represent two plots of the longitudinal modes of a multimode and single-mode laser, respectively. The multimode is the most common, with 3-dB spectral widths from 1 to 10 nm. The single-mode laser, on the other hand, represents a single spectral line; its line width is usually less than 1 Å. Some single-mode lasers have line widths as small as 0.1 Å. Readers progressing through the text will become aware of the positive aspects of narrow spectral widths and also the shortcomings.

Let us now examine the spectral plot using an inuitive approach. The cavity resonance should be a function of length (l) between the laser mirrored surfaces

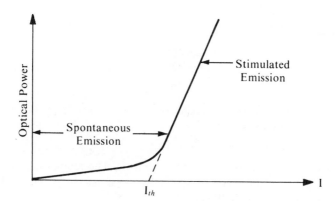

Fig. 3-4. Laser-diode curve of characteristic power versus current.

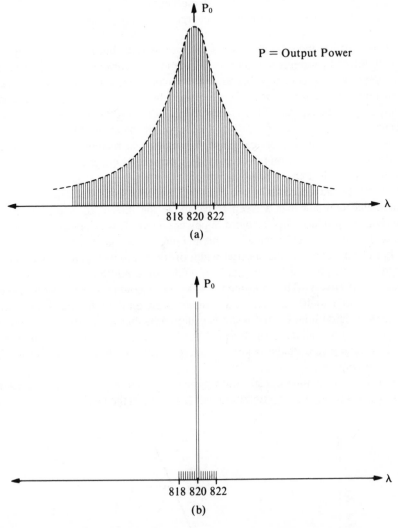

Fig. 3-5. Spectral plots of two lasers commonly used in single-mode transmitters: (a) multimode laser, and (b) single-mode laser.

(facets), wavelength material, and mode number. The relationship in Eq. 3-3 depicts the conditions for cavity resonance provided that cavity gain is sufficient (a topic that will be discussed later):

$$K_m \frac{\lambda_k}{2} = \mu_m l \qquad (3\text{-}3)$$

where:

λ_k = wavelength
K_m = mode number
μ_m = a material constant.

Substituting the relationship for frequency for the Kth mode, i.e., $\lambda_k = c/f_k$, and rearranging the terms of Eq. 3-3, we have

$$f_k = \frac{K_m c}{2 \mu_m l} \tag{3-4}$$

The separation between modes is

$$\Delta f = f_k - f_{k-1} = \frac{K_m c}{2 \mu_m l} - \frac{(K_m - 1)c}{2 \mu_m l}$$

or

$$\Delta f = \frac{c}{2 \mu_m l}$$

When $\Delta f \ll f_k$, then the relationship can be written as follows:

$$\frac{\Delta \lambda}{\Delta f} = \frac{\lambda_k}{f_k}$$

If this equation is rearranged after substituting values for Δf and f_k into it, we have Eq. 3-5, which is the free-space wavelength of separation between modes:

$$\Delta \lambda = \frac{\lambda_k^2}{2 \mu_m l} \tag{3-5}$$

Equations 3-3, 3-4, and 3-5 allow the reader to gain some insight into how geometric and physical parameters affect the spectral width of the laser output.

The next items to be investigated are the thickness and width dimensions of the cavity and how they affect the spectral width and performance of the laser in general.

The analysis must begin with the scalar wave equation and the diagram shown in Fig. 3-6(a). The scalar wave equation for the TE mode configuration is described by Eq. 3-6.

$$\frac{\partial^2 E_y}{\partial x^2} + \frac{\partial^2 E_y}{\partial^2 Z} + \bar{n}^2(x)\, K_O^2 E_y = 0 \tag{3-6}$$

where K_O = free-space propagation constant = $2\pi/\lambda_O$.

Solutions of this equation have been observed experimentally and agree with the theoretical results that use separation of variable techniques for partial differential equations (see Refs. 5 and 6 for more details).

The separation of variables using a solution of the form shown in Eq. 3-7 will yield differential Eqs. 3-8 and 3-9:

$$E_y(x, z) = E_y(x)\, E_z(z) \tag{3-7}$$

$$\frac{d^2 E_z(x)}{dx^2} + \bar{n}^2(x)\, K_O E_z(x) - k^2 = 0 \tag{3-8}$$

$$\frac{d^2 E_y(z)}{dZ^2} - k E_y(z) = 0 \tag{3-9}$$

The solutions for these equations are of the form shown below:

$$
\left.
\begin{aligned}
E_Z &= Q_C e^{iBz} \cos kx \\[1em]
H_z &= -i\left(\frac{k}{\omega\mu_O}\right) e^{iBz} \sin kx
\end{aligned}
\right\} \quad : \ |x| < \frac{d}{2}
$$

$$
\left.
\begin{aligned}
E_y &= e^{iBz} e^{-\gamma(|x|-d/2)} \\[1em]
H_z &= -i\left(\frac{x\gamma}{|x|\,\omega\mu_O}\right) A_e\, d^{iBz} e^{-\gamma(|x|-d/2)} \cos k\frac{d}{2}
\end{aligned}
\right\} \quad : \ |x| > \frac{d}{2}
$$

The equations solving for the field parameters are as follows:

$$2N\pi < \sqrt{\bar{n}_1^2 - \bar{n}_2^2}\, d K_O \qquad \text{even modes } N = 0, 1, 2, \ldots$$

$$N\pi < \sqrt{\bar{n}_1^2 - \bar{n}_2^2}\, d K_O \qquad \text{odd modes } N = 0, 1, 2, \ldots$$

$$\tan k\frac{d}{2} = \frac{\gamma}{k} \qquad \beta = \sqrt{\bar{n}_1^2 K_O^2 - \bar{n}^2} \qquad \gamma = \sqrt{\beta^2 - \bar{n}_2^2 K_O^2}$$

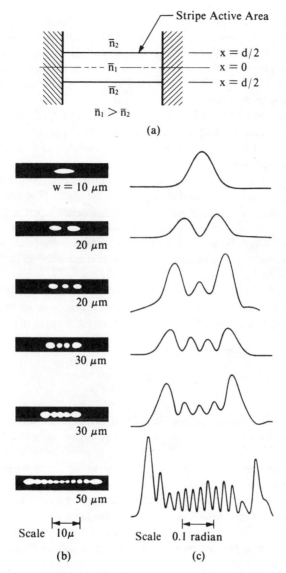

Fig. 3-6. Laser cavity profile: (a) mirrored end, and (b) near-field and (c) far-field patterns as a function of stripe width.

For the situation when $N = 0$ and $\bar{n}_1 - \bar{n}_2$ is a 5-percent step in the cavity of the stripe at the boundary, the waveguide fundamental mode occurs at $N = 0$ and $d \leq 0.7$ μm. The first odd mode will occur at $N = 1$ and $d \leq 0.35$. This is one reason why the active region in double-heterostructure lasers is much less than 1 μm in thickness.

The near-field and far-field radiation effects are shown in Figs. 3-6(b) and 3-6(c). Near-field radiation is the power output density across the facet of the laser, while the far-field distribution is the directional characteristic of the laser along the axis perpendicular to the facet at a distance of several wavelengths from the cavity. For stripe widths of 10 μm or less, a single high-intensity mode occurs. Increasing the drive current above the point where lasing begins will cause a lack of complete confinement within the cavity, and higher modes will begin to appear because sufficient carriers are being injected into confinement areas to produce lasing. As the stripe area gets hotter, self-focusing will occur because of variations in the index of refraction.

Increasing laser stripe widths above 10 μm in the 20- to 30-μm range excites the higher order modes in steadily increasing numbers. The lasing threshold also becomes sharp. For these wider stripes, a decrease in current density occurs at the center of the stripe as compared to its edges as a result of stimulated recombination. The dielectric constant profile, which is inversely related to the carrier density, decreases with distance from the stripe axis, as can be observed in Figs. 3-6(b) and (c).

To calculate far-field radiation angular divergence, the Avry diffraction relation may be used, as stated in Eq. 3-10; this equation assumes that the light is diffracted through a slit of width d, with $\Delta\theta$ being the angular divergence in degrees:

$$\Delta\theta = 1.22 \, K_A \left(\frac{\lambda_O}{d}\right) \qquad (3\text{-}10)$$

Standard diffraction theory cannot be applied to the stripe mirror because the aperture itself, i.e., the facet ends of the laser cavity, is a source of electromagnetic fields. The K_A value can be added to improve accuracy because large errors will occur due to the large refractive index step at the heterojunctions.

The next two rather important items to consider are the conditions necessary for laser action in semiconductors. First, the cavity gain must exceed the transmission and scattering losses to promote self-oscillation, i.e., lasing. Second, the optical cavity must provide sufficient positive feedback to sustain oscillation. The gain equation is derived by Gower[7], and the result is represented by Eq. 3-11:

$$g_{12} = \frac{c^2(n_2 - n_1)}{8\pi f_{21}^2 \tau_{SP}} \, \epsilon_O(f_{21}) \qquad (3\text{-}11)$$

where n_2 and n_1 are the number of atoms/unit volume; $\epsilon_O(f_{21})$ = the normalized spectral width of a beam of radiation at frequency f_{21}; τ_{SP} = mean lifetime of spontaneous emission; f_{21} = frequency of the light beam ($= c/\lambda_{21}$ also).

The derivation of Eq. 3-11 proceeds as follows. First we have Boltzmann's equation:

$$\frac{n_1}{n_2} = e\left(\frac{\Delta\epsilon}{KT}\right) \tag{3-12}$$

where $\Delta\epsilon$ = the bandgap energy = $\epsilon_2 - \epsilon_1$. Then, from Planck's black-body radiation equation, the photon energy is

$$\rho(f) = \rho_{BB}(f) = \frac{8\pi h f^3}{c^3 \left[e^{hf/KT} - 1\right]} \tag{3-13}$$

The units of $\rho(f)$ are electromagnetic energy per unit volume per spectral frequency about f.

The electromagnetic energy density in the beam of the laser energy is

$$\rho(f) = \frac{P\epsilon(f)}{c} \tag{3-14}$$

where $\epsilon(f)$ = normalized power spectral density.

The energy absorbed per second per unit length of cavity is

$$-\frac{d}{dZ}\left(\frac{dE'}{dt}\right) = B\rho(f_{21})(n_1 - n_2)hf_{21} \tag{3-15}$$

If we substitute Eq. 3-14 into Eq. 3-15 and divide both sides by P, the result is Eq. 3-16:

$$-\frac{1}{P}\left(\frac{dP}{dZ}\right) = \frac{B(n_1 - n_2)hf_{21}\,\epsilon(f_{21})}{c} \tag{3-16}$$

This is the equation that relates attenuation per unit length to absorption of carriers. The constant B is one of Einstein's coefficients that is related to properties of atoms and energy states. The attenuation coefficient for the beam is defined as α_{12} and is given by

$$\alpha_{12} = -\frac{1}{P}\left(\frac{dP}{dZ}\right) \tag{3-17}$$

Solving the differential equation for $P(z)$ with $P(0)$ equal to P_{max}, we have

$$P(Z) = P_{max} \, e^{-\alpha 12^z} \qquad (3\text{-}18)$$

The rate of excitation must equal the rate of decay, and this can be expressed using Einstein's coefficients as shown in Eq. 3-19:

$$A_{21}n_2 + B_{21}\rho_{BB}(f_{21})n_2 = B_{21}\rho_{BB}(f_{21})n_1 \qquad (3\text{-}19)$$

Solving Eq. 3-19 for $\rho_{BB}(f_{21})$ and setting it equal to Eq. 3-13 will produce Eq. 3-20.

$$\frac{8\pi h f^3}{c^3[e^{-hf/KT} - 1]} = \frac{[A_{21}/B_{21}]}{[B_{12}/B_{21}][(n_1/n_2) - 1]} \qquad (3\text{-}20)$$

Substituting Eq. 3-12 into Eq. 3-20 results in the expression described by Eq. 3-21:

$$\frac{8\pi h f^3}{c^3[e^{-hf/KT} - 1]} = \frac{A_{21}/B_{21}}{B_{12}/B_{21}[e^{\Delta\epsilon/KT} - 1]} \qquad (3\text{-}21)$$

where:

$$B = B_{12} = B_{21}$$
$$\frac{A_{21}}{B_{21}} = \frac{A}{B} = \frac{8\pi h f^3}{c^3}$$

Also,

$$A = \frac{1}{\tau_{SP}}$$

where τ_{SP} = the lifetime of spontaneous emission.

Solving for the coefficient B and substituting this value into Eq. 3-16 will produce Eq. 3-11.

Equation 3-11 describes that situation when absorption is balanced by spontaneous emission, i.e., $n_1 > n_2$. To obtain lasing action, population inversion, or $n_2 > n_1$, is required; this produces optical amplification, i.e., the rate of stimulated emission exceeds absorption.

If gain were to be inserted into Eq. 3-18, the power would increase without bound, which of course cannot happen. Either burnout of the device will occur, or feedback will limit the gain and produce stable operation.

When stimulation occurs as the emissions are reflected by the facets' mirrored surfaces, some of the energy is lost through absorption. To account for this loss, Eq. 3-18 can be rewritten with g in place of $-\alpha_{12}$ and the loss α_s subtracted from the gain to indicate loss per unit length, as shown by Eq. 3-22:

$$P(Z) = P_{max}e^{(g_{12} - \alpha_s)Z} \tag{3-22}$$

For $Z = 2l$, one round trip within the cavity of length l, Eq. 3-22 can be rewritten as shown in Eq. 3-23:

$$\frac{P(2l)}{P_{max}} = e^{2(g_{12} - \alpha_s)l} \tag{3-23}$$

The conditions for oscillation are given as shown below:

$$R_1 R_2 \frac{P(2l)}{P_{max}} > 1$$

i.e., reflected power must be greater than the initial power:

$$P(Z = 0) = P_M$$

$$R_1 R_2 e^{2(g_{12} - \alpha_s)l} > 1$$

If both mirrors have equal reflectivity, then $R_1 = R_2 = R$ and

$$(g_{12} - \alpha_s) = \frac{1}{l} \ln \frac{1}{R}$$

The gain required to sustain oscillation with optical feedback is then

$$g_{12} > \alpha_s + \left(\frac{1}{l}\right) \ln \left(\frac{1}{R}\right) \quad \text{for } R_1 = R_2$$

or $$\tag{3-24}$$

$$g_{12} > \alpha_s + \left(\frac{1}{2l}\right) \ln \left(\frac{1}{R_1 R_2}\right) \quad \text{for } R_1 \neq R_2$$

Laser frequency response is similar to a second-order low-pass filter with a damping coefficient that is less than critical.

Let us examine an equation for an RLC parallel resonant circuit and examine some of its knowns. The equation is of the form,

$$\frac{P(\omega)}{P(0)} = \frac{\omega_0^2}{S^2 + 2\delta S + \omega_0^2} \qquad (3\text{-}25)$$

Replacing the LaPlace operator by $j\omega$ for nontransient behavior, we have the equation,

$$\frac{P(\omega)}{P(0)} = \frac{\omega_0^2}{(\omega_0^2 - \omega^2) + j2\delta\omega}$$

The values for ω_0 are a function of photon lifetime and spontaneous emission lifetime, as shown below:

$$\omega_0^2 = \frac{1}{\tau_{SP}\tau_{Ph}K}$$

As biasing current increases, bandwidth increases and the threshold current remains fixed, i.e., K' decreases with increasing bias, assuming that $I_b > I_{th}$. Then,

$$I_b = \left(\frac{1}{K}\right)I_{th} + I_{th}$$

or

$$K = \frac{I_{th}}{I_b - I_{th}}$$

When resonance is reached—i.e., $\omega = \omega_0$—then $|\omega_0/2\delta| = 2$ for a 3-dB overshoot at resonance (where δ is the damping ratio). Consequently,

$$\delta = \frac{1}{4}\sqrt{\frac{(I_b - I_{in})}{I_{th}\tau_{SP}\tau_{Ph}}} \qquad (3\text{-}26)$$

where I_{in} is the laser input current. Here, the 3-dB value was substituted based on experiences with lasers. Not reflected in the equations are the material-dependent parameters, but all parameters such as I_{th}, I_b, etc., are affected by the types of materials used for construction of the laser.

A curve showing the typical response of a laser is presented in Fig. 3-7. Overshoot occurs at ω_0, which is usually 2 to 4 dB, and the roll-off is 40 dB/decade. At a 3-dB overshoot, the damping ratio, δ, is 0.4. This is not a desirable damping ratio, however, especially for baseband digital switching. The waveforms will ring at the leading and trailing edges. Ideally, the value of δ should be kept less than 0.7, which is critical damping. Compensation can be introduced that will limit bandwidth, provided that total device bandwidth is not required. These techniques will be discussed in the section on transmitter design and analysis later in this chapter.

Environmental issues for lasers will now be addressed. One of the most problematic parameters for lasers is temperature. As temperature is increased, the operating point will shift, causing a decrease in optical power output and an increase in current. Most lasers have either temperature compensation, optical feedback, or both. Often, an optical detector is positioned near one of the facets to monitor optical output power through feedback. The bias current is altered to produce a stable optical output power. For a thermal-feedback system, a temperature-sensing element (thermistor) is embedded in the laser package. When temperature variations occur, the thermal feedback circuit will compensate the laser temperature with a thermal-electric device to correct the condition. These feedback circuits will be discussed in this chapter in the section on transmitter analysis and design. The investigation here will address the material parameters that affect environmental behavior in the devices.

A series of curves has been provided in Figs. 3-8(a) through 3-8(d) that depicts the impact of temperature on devices of various wavelengths. The first figure,

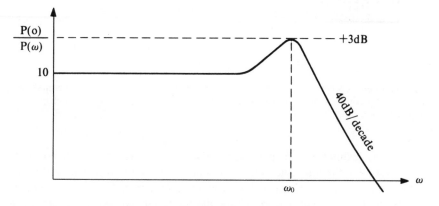

Fig. 3-7. Laser frequency response curve.

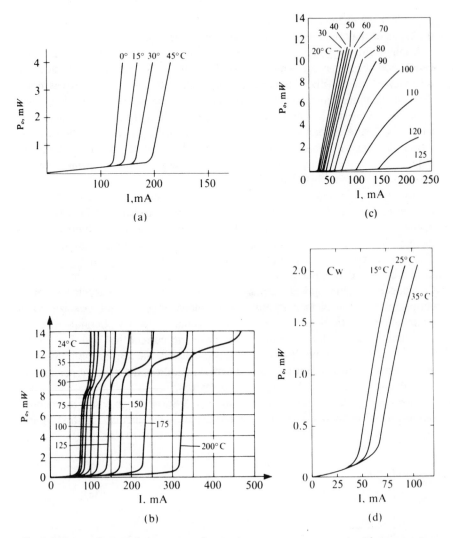

Fig. 3-8. Laser characteristic curves of output power versus current with temperature variation: (a) and (b) 820-nm stripe laser, (c) 1.3-m stripe laser, and (d) 1.5-m stripe laser BH-DFB.

Fig. 3-8(a), exhibits the output characteristic of a typical (820-nm) nonstripe laser. Note that the threshold current of approximately 148 mA will produce an output of 2 mW at 10°C; then, for an increase in temperature of 5°C (i.e., at 15°C), the laser current will fall below the threshold and lasing will stop. For the other possible excursion—i.e., a drop in temperature—the output may rise to 5 or 6 mW, at which burnout can occur. As one can readily observe, there is a need for temperature compensation.

The curve of Fig. 3-8(b) is a short-wavelength (820-nm) stripe-laser characteristic. These devices have lower threshold currents, and the slope of the characteristic curve is steeper. The threshold current will be in the neighborhood of 25 to 50 mA, depending on the stripe width. These lasers also have what is referred to as "kinks" (nonlinearity in the characteristic curves). Reduced output below the "kink" or some form of compensation is needed to make the output linear with current excursion.

For a long-wavelength laser (1.3-μm) such as that depicted in Fig. 3-8(c), not only does the threshold current change, but also the slope of the characteristic curves. This situation requires more complex compensation circuitry. The other long-wavelength laser, that in Fig. 3-8(d), has a bit more stability, but the output power is smaller.

LONG-WAVELENGTH SOURCES

Operation of systems at long wavelengths has several favorable attributes that outweigh their disadvantages. Table 3-1 presents this data in tabular form. A second table, Table 3-2, depicts the wavelength range of various substrate and active-region materials.

As can be observed from Table 3-2, the GaInAsP/InP material is one of the most suitable crystal materials for long-wavelength sources. The materials for sources with wavelengths exceeding 1.55 μm are not useful for high-silica transmission mediums because of the high attenuation of the infrared (IR) band edge. This does not imply they are not useful. There is a great deal of research devoted to the development of fibers of compositions other than silica for use with wavelengths in the 3-μm and higher regions. The objective of longer-wavelength mediums is to increase bandwidth and keep losses low. However, the longer wavelengths will permeate fibers with larger cores because single-mode wavelengths are greater. Alignment of fibers during the termination process will be less critical, and this will be reflected in lower losses across connectors.

TABLE 3-1. Critical Parameters for Laser Sources.

Wavelength, λ, nm	Bandwidth	Attenuation (lowest), dB/Km	Ease of Manufacture	Dispersion
820	400–800 MHz-km	2.5	Very easy	High
1300	1–2 GHz-km	0.27	Becoming more common	Medium
1500	2–10 GHz-km	0.16	Difficult	Low

TABLE 3-2. Materials for Sources.

Active-region material	Substrate Material	Wavelength, μm
GaInAsP	InP	1.1–1.67
	GaAs	0.65–0.90
GaInSbP	InP	0.9–1.3
	GaSb	20–30
GaInAsSb	InAs	1.6–35.0
	GaSb	1.67–40.0
AlGaAsSb	GaSb	1.1–1.67
InAsSbP	InAs	1.6–38
AlGaAs	GaAs	0.7–0.88

Before leaving the subject of laser physics, a number of important topics relating to injection lasers should be qualitatively addressed.

Gain-guided lasers, which are the most common sources used in communication system transmitters, are characterized by soft turn-on, i.e., a rounded knee in the output (see Fig. 3-4). Threshold currents for these devices range from 50 to 100 mA. Gain-guided lasers are multimode devices, and the focus is not the same in the planes perpendicular and parallel to the active region.

Index-guided lasers emit a single spectral line and are characterized by sharp turn-on (the knee is sharper) and low threshold currents (in the 20- to 50-mA range). The single spectral line and fundamental spatial mode will be maintained under aging, temperature, and power output, but the wavelength can change. Under modulation, these devices ordinarily emit multispectral lines. These devices will mode hop from line to line, producing noise in the communication system as a result. When exposed to optical feedback, the system noise saturation is further aggravated. At the time this book was being written, two manufacturers could produce a stable version of the index-guided laser. Several approaches have been undertaken to stabilize the device, such as a third mirror coupled into the cavity or a grating along the active-region plane. The latter device is commonly referred to as a distributed feedback laser (DFB). With DFB, spectral line limits have been produced in the range of ± 5 to $\pm 10\%$.

At present, index-guided lasers using AlGaAs material produce outputs of 40 mW at CW. By the time this book is in print, a doubling in power output is expected. Power outputs for these devices will eventually reach a 1-watt value. Bandwidth is expected to increase to 2 to 10 GHz-km. Due to the higher output of communication lasers, the development of small lasers that can be directly driven is expected, e.g., a 2 μm × 50 μm laser package that can produce sufficient power with 2 V at 1 mA to drive fiber-optic waveguides directly. With sources such as these integrated optics would be one step closer to a reality.

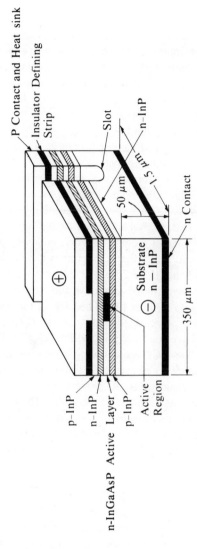

Fig. 3-9. Edge-emitting diode for long-wavelength operation (1300 nm): The active region thickness is 0.05 μm, with surrounding heterojunction layers of 2 to 3 μm.

EDGE-EMITTING DIODES

Edge-emitting diodes are useful for short-range single-mode systems because of their highly directional light outputs. The beam width in a plane parallel with the active layer is approximately 120°, and the width perpendicular to the active layer is approximately 30°; in comparison, lasers have beam widths in the active layer plane of 10° to 20° and perpendicular to the active layer plane of 30° to 50°. The edge emitters project an ellipse whose major axis is orthogonal to the projection of the major axis projected by the laser.

The cavity formed by the stripe of the edge-emitting diode has a thickness of almost a magnitude smaller than the equivalent laser cavity dimension. The slot in the LED shown in Fig. 3-9 contracts the active region to the point of its being too short to support laser action. The active layer tends to absorb the light generated, but this layer is made very thin. The confining layers, on the other hand, do not absorb light energy, but rather guide the light. The materials of which they are made have a larger bandgap energy, and absorption is greater at shorter wavelengths (820 nm) rather than longer ones (1300 nm). This tends to narrow the spectral width from 35 to 25 nm at 820-nm wavelengths and from 100 to 70 nm at 1300-nm wavelengths. Note that the coupling of edge-emitting LEDs is quite good compared to most LEDs because of the narrow beam angle of the projection in the plane perpendicular to the active layer plane. When used with lenses, these LEDs have high coupling efficiencies in 50-μm multimode waveguides. For single-mode fiber, the coupling losses would be much higher, but, since waveguide attenuation is correspondingly much lower, they can be used in short-range LAN applications. At present, the 1300-nm edge-emitting diodes are very expensive and difficult to heatsink. These problems will be rectified, however, as the number of device users increase and fabrication processes improve. The other types of LEDs will not be covered in this text, but the reader is urged to examine multimode-fiber-optics books for further information on these sources.

GAS LASERS

This section will give an overview of gas lasers. These devices are not used in communication systems because they are physically large and power outputs range from milliwatts to megawatts.

HeNe Laser

One of the first gas lasers, a helium–neon (HeNe) unit designed by A. Javan, R. W. Bennet, and D. R. Herroit, was constructed in the fall of 1960 at Bell Laboratories. A schematic is shown in Fig. 3-10. The laser consisted of a discharge tube 100 cm long, with a 1.5-cm inside diameter. This tube was filled

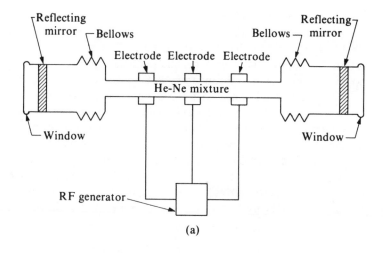

(a)

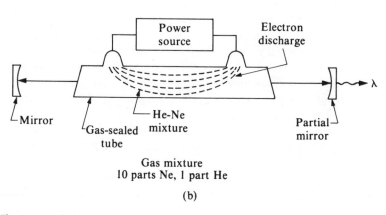

(b)

Fig. 3-10. Helium–neon laser construction: (a) earlier model, and (b) later model.

with a HeNe gas mixture at a partial vacuum of 1 mm of Hg for the He and 0.1 mm of Hg for the Ne. Reflector plates were fitted at each end to form a part of the laser cavity and had to be parallel within parts of an arc second. Bellows were provided to act as a mechanical tuning element for mirror (reflector plate) surfaces. The windows at the ends were slightly reflective and produced some unwanted reflections.

The operation of the HeNe laser is as follows: Molecular or atomic energy exchanges are induced in the gas mixture when He is raised to a higher energy level as a result of its collision with electrons. An electric discharge (from the RF generator) imparts energy to the He atoms—an exchange commonly referred to as "pumping." After the He atoms collide with the Ne atoms, the neon is raised to a higher, more unstable energy state. When the Ne atoms then drop back to their stable state, the laser emits red light. This red light will reflect back and forth between the two end reflectors until enough energy is imparted to the photons to allow them to escape through the mirrored end refractors.

Although HeNe lasers are not used for communication systems because of their physical size, they are nevertheless practical for laboratory experimentation purposes. HeNe devices have a rather pure spectral wavelength of operation and a power output of from $\frac{1}{2}$ mW to megawatts. Also, these lasers are comparatively inexpensive, which makes then rather appealing for laboratory use when studying single-mode lasers. They are often lower in cost than many of the present semiconductor lasers, particularly long-wavelength devices with equivalent power outputs. Semiconductor lasers with outputs of a watt or so, however, will be available at lower cost in the near future (two to six years).

CO_2 Lasers

Carbon dioxide (CO_2) lasers are another type found in industry. These are generally useful because of their large output power in the gigawatt region. They also exhibit high efficiency—approximately 30 percent. Since they are of particular interest to the military for use in new beam-type weapons, the military community has spurred research on them in recent years. Their operation is similar to that of the HeNe laser, with nitrogen replacing helium and CO_2 taking the place of neon.

CO_2 lasers have a large number of uses in the industrial sector. Some of them have been designed with sufficient power output to vaporize any known substance. One immediate use for this capability is in mass spectrometers that vaporize sample material and analyze the resultant spectrum to determine the chemical elements present. Another application is in machining operations. As an example, line tracers can follow a pattern, and a laser can be used to cut out that pattern from metal, glass, or other materials. A diagram of such a system is shown in Fig. 3-11.

The laser milling machine is, of course, simplified here, but machining operations can be rather high-speed if the laser power output is high. The line tracer head will follow the line of the pattern, and the steering motor moves the head through the use of a drive wheel. As the steering motor moves the line tracer head about the platen to follow the drawing outline, x and y information is taken from a digital or analog resolver in the line tracing head. After signal processing, the x and y signals produced will drive the machine table in the correct direction to cut the part. The electronic control can be regulated to cut parts that are much larger than the actual pattern (10:1 or larger) and parts smaller than the pattern as well.

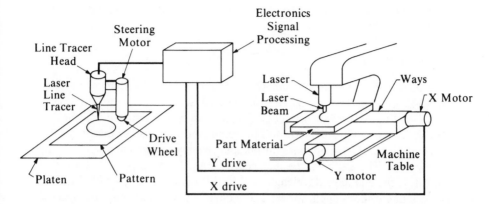

Fig. 3-11. Laser machining operation using line-tracer milling-machine techniques.

One of the problems with this type of milling or machining operation is that the electronics accuracy is much greater than the mechanical accuracy of the bench, the former being as high as one ten-thousandth of an inch, whereas the latter is five thousandths of an inch at best.

The steering head and platen may be replaced with a microcomputer terminal and the part drawn in three dimensions using graphic techniques. Cutting on the machine table can be performd by two lasers that intersect at the point where machining takes place. An x, y, and z series of stepping motors moves the work piece, thereby affording the system a full six degrees of freedom.

The system discussed in Fig. 3-11 is analog and relatively inexpensive to construct. It can be easily retrofitted to milling-machine plasma-cutting equipment or other similar equipment.

Rare Gas Lasers

Other gas lasers that hold promise are the rare gas types. Argon lasers, for example, emit a green to blue light in the visible spectrum and ultraviolet. These particular lasers are useful for medical purposes because the red pigment in blood will absorb green light and result in local heating. Such lasers are commonly used as a scalpel. They have also been used successfully to reattach detached retinas, stop ulcer bleeding, and any number of other medical uses.

Fluidic and Chemical Lasers

The other types of lasers, which will not be addressed, but the reader should be aware of them, are fluidic and chemical lasers. These two types are not used in communications, but they have a large number of industrial applications that cannot be discussed for want of space.

YAG LASERS

These devices are nonsemiconductors constructed with an yttrium–aluminum–garnet (YAG) rod. The rod has mirrors located at each end to form the laser cavity (see Fig. 3-12 for construction details). An external source is located at one end of the cavity, as shown. For communication lasers, this will be a semiconductor LED that provides the pumping energy required to produce lasing in the cavity. The LED can be a standard GaAsAl device. For YAGs doped with Nd^{3+}, the emission wavelength will be 1.06 μm, a useful length for fiber-optic communications applications. These devices have a rather narrow spectral width—0.5 nm for a cavity diameter of 5 to 6 μm. Note that the cavity diameter is approximately the same as a single-mode waveguide core, which makes coupling a bit easier. Part of the waveguide itself can be constructed as the YAG rod.

YAGs have one disadvantage: their fluorescence lifetime, which may be only as long as 100 μsec. This will limit direct modulation frequencies to 4 to 5 kHz. This modulation frequency is totally unacceptable, but all is not lost; indirect techniques can be used to modulate YAGs at frequencies greater than 1 GHz.

Attributes such as high coupling efficiency for single-mode waveguide, narrow spectral width, high modulation frequencies (externally modulated), and operation at near 1-μm wavelength make this device appealing for communications applications. For higher power outputs, multiple LEDs can be used as external pumping sources.

External modulators, in effect, operate as a switch that blocks or allows light

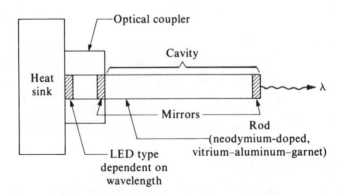

Fig. 3-12. YQG laser construction.

to pass while the laser itself is operating continuously. If direct modulation were implemented, the pumping source itself would be turned on and off. The external modulators are usually either electro-optic or acousto-optic.

Electro-optic modulators alter the birefringence in the material through a change in electric field that affects light polarization. This modulator switches an electric field and requires very small amounts of power. The acousto-optic devices are ordinarily used at low modulation frequencies, 50 MHz being the maximum. Insertion loss is 25 to 30 percent for these devices, whereas acousto-optic devices have a maximum insertion loss of 10 percent and a minimum as low as 5 percent in the 600- to 3500-nm range.

There have also been efforts to develop rods with shorter fluorescence decay time. The materials proposed are in the ultraphosphate gorup. One of the problems is that standard end pumping cannot be used; a more complex and costly side pumping technique has had to be substituted. However, as one may observe, YAGs are here to stay as communication sources.

With the rapid improvement of laser sources, prices are dropping. Therefore, before the reader selects a source, he should make a search of the literature. The technology is changing very rapidly, and often within six months pricing is obsolete. Cost performance tradeoffs will be discussed later in the text to assist the designer in his source selection.

LASER FEEDBACK TECHNIQUES

The feedback techniques to be discussed here will be adequate for most designs; should further precision be desired, refer to Baker.[8] The two primary feedback mechanisms used in lasers are shown in Figs. 3-13(a) and (b). The mechanism in Fig. 3-13(a) will tap a small amount of laser output power through the use of an optical coupler (or the facet opposite the drive can be used), and this optical power will be routed to a detector, as shown (λ_f). The optical feedback signal will be amplified (electrically), depicted in the figure by e_{fb}. The signal is then averaged to produce e_{Afb}. A small amount of the electrical input signal to the laser drive amplifier is averaged and compared with the feedback signal; this signal will control the bias current of the laser. Initially, the optical power output is set, and since it decreases as a result of heating, the bias control signal (e_{bias}) will increase the laser bias current to compensate. If the laser output power increases because of ambient or other conditions, then e_{bias} will be such that the laser bias current, which is the difference between the average signal, e_{as}, and average feedback, e_{fb}, will decrease, thereby reducing the average output power.

The optical power feedback will compensate for temperature, but, as one can observe, when power output decreases as a result of heating and a shift in bias occurs, then increasing biasing current will cause further self-heating in the laser.

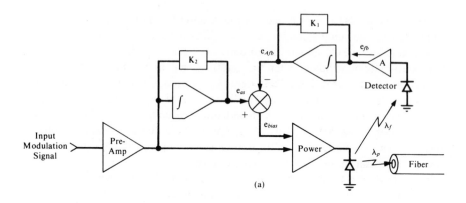

(a)

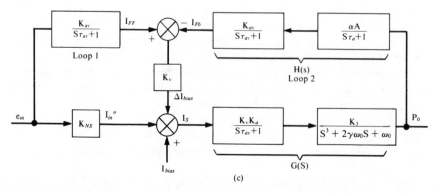

(b)

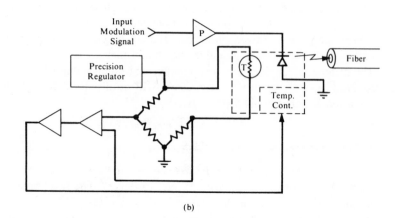

(c)

Fig. 3-13. Servo diagrams of an optical feedback system for bias control and temperature compensation techniques: (a) Optical-feedback bias-control, (b) temperature control with thermal feedback, and (c) optical-feedback with bias-current control.

Eventually, burnout will occur in the laser if it is not externally cooled and self-heating is allowed to continue.

External heating and cooling control is shown in Fig. 3-13(b). The circuit shown in the figure depicts thermal control and the temperature-sensing elements embedded in the laser package; this configuration is common because thermal mass must be kept low if the temperature compensation is to operate with small thermal time constants.

An oven is another type of external temperature control that can be used to maintain the laser operating point at a fixed value. Ovens control the ambient temperature, i.e., the laser package temperature is maintained at a constant temperature. Ovens maintain the package temperature within a few degrees. They also have a long thermal time constant because of the thermal mass package, plus the oven itself. Ovens have been used to stabilize crystal temperatures for precision frequency sources for many years; a large number of suppliers exists therefore.

A discussion of feedback to linearize LEDs will be presented in the last section of this chapter. When LEDs are used as sources, large signal excursions cause nonlinear operation because of the normal nonlinearity of the diode characteristic curves.

The laser, in conjunction with its feedback, can be considered an electronic servo system. Let us first examine the power amplifier frequency response. The laser source can be represented by Eq. 3-27; as the reader will note, this is a version of Eq. 3-25:

$$Z_{TS}(S) = \frac{K_S}{S^2 + 2\delta\omega_0 + \omega_0^2} \tag{3-27}$$

$$Z_{TA}(S) = \frac{K_1}{S\tau_a + 1} \tag{3-28}$$

This is the transfer function for the source, $Z_{TS}(S)$, as a function of the LaPlacian operator, S. Equation 3-28 is the power-amplifier transfer function showing frequency response with the assumption that it has a single pole, i.e., a 3-dB roll-off. The constants shown in both equations account for gain and units where needed.

The two averaging circuits will be identical. This is a valid assumption because the amplifier gain can compensate for any deficiencies. The time constants are identical because the waveforms will have identical shapes. Equation 3-29 therefore depicts the averaging circuit, which is very similar to a low-pass filter for a power-supply output:

$$Z_{AV}(S) = \frac{K_2}{S\tau_{av} + 1} \tag{3-29}$$

A block diagram of the optical feedback system for the laser transmitter is shown in Fig. 3-13(c). This set-up looks quite formidable but, if the voltages of two averaging blocks are converted to current, then the transfer function for each loop can be written as depicted in Eqs. 3-30 and 3-31. The transfer function for the transmitter is described by equation 3-32.

$$\text{Loop 1:} \quad e_{in}\left(\frac{K_{av}K_{v1}}{S\tau_{av} + 1} + K_{vs}\right) = I_{in} = I''_{in} + I_{ff}$$

$$\text{Loop 2:} \quad I_{in} + I_{bias} - I_{fb} = I_s$$

(3-30)

Let

$$I_{in} + I_{bias} = I'_{in}$$

Then

$$I'_{in} - I_{fb} = I_s$$

Using $Z(S) = G(S)/G(S)H(S) + 1$, the feedback equation for loop 2, we have the expression with $G(S)$ and $H(S)$ that is defined in Fig. 3-13(c). The transfer function for the transmitter is then:

$$Z_{TL2}(S) = \frac{K_v K_a K_s (S\tau_{av} + 1)(S\tau_a + 1)}{K_{av}\alpha A K_v K_a K_s + (S\tau_a + 1)(S^2 + 2\delta\omega_0 S + \omega_0^2)(S\tau_{av} + 1)(S\tau_a + 1)}$$

(3-31)

where $Z(S) = P_o/I_{in}$.

The transfer function for the entire circuit is $Z_{TL1}Z_{TL2} = Z_T(S)$, as shown below:

$$Z_T(S) =$$

$$\frac{(K_{av}K_{v1} + K_{v2} + S\tau_{av}K_{v2})(K_v K_a K_3)(S\tau_{av} + 1)(S\tau_a + 1)}{(S\tau_{av} + 1)[K_{av}\alpha A K_v K_a K_3 + (S\tau_a + 1)(S^2 + 2\delta\omega_0 S + \omega_0^2)(S\tau_{av} + 1)(S\tau_a + 1)]}$$

(3-32)

After this mathematical derivation of a transfer function, Eq. 3-32 may be used to examine the circuit for instabilities by finding the roots of its denominator using numerical values. Whenever a negative root appears, the circuitry will become unstable. An example is shown below (for the fifth root):

$$\left(S_1 + \frac{1}{\tau_1}\right)\left(S_2 + \frac{1}{\tau_2}\right)\left(S_3 + \frac{1}{\tau_3}\right)\left(S_4 + \frac{1}{\tau_4}\right)\left(S_5 - \frac{1}{\tau_5}\right)$$

$$= S^5 + aS^4 + bS^3 + cS^2 + dS + K$$

If the time constants for certain blocks are large compared to those of others, these may often be ignored. For example, if source frequency ω_0 is 100 MHz and the bandwidth of the power amplifier is 10 MHz, then the source frequency response can be ignored. When the time constants for any of the blocks are close in value, they interact to produce roots that differ from one another.

The thermal feedback system of Fig. 3-13(b) can be analyzed in the same manner as that in Fig. 3-13(a). Since the thermal time constant is the largest factor, it will dominate the system.

Some examples of both systems will be investigated during the transmitter design sections of this chapter.

TEMPERATURE CONTROL

A temperature control circuit is shown in Fig. 3-14(a), which is a rather simple circuit. The zener diode provides a stable reference voltage across the bridge. The difference voltage is represented by Eq. 3-33:

$$V_2 \frac{R_X}{R + R_X} - V_2 \frac{R_T}{R + R_T} = \Delta V_{diff} \qquad (3-33)$$

It is important to note that ΔV_{diff} is directly proportional to any variations in V_2. With proper biasing, however, the zeners can be made stable with temperature excursions.

A more pressing problem is the nonlinearity of thermistors, the temperature-sensing device. These devices may be partially linearized through networks found in most thermistor application notes. Nickel strain gauges, which exhibit linear resistance change with temperature, can be used as temperature sensors, but they do not have the sensitivity of thermistors.

A large number of other devices can be used as temperature sensors such as silicon or germanium diodes and capacitors. The type of sensing element will depend on the application.

The differential amplifier detail is shown in Fig. 3-14(b). This particular circuit can be designed to use most general-purpose operational amplifiers (op amps).

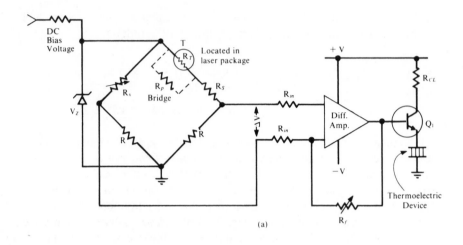

(a)

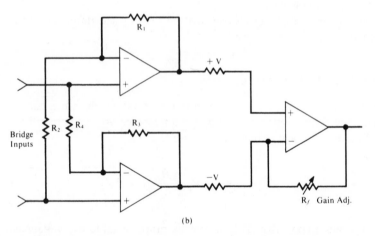

(b)

Fig. 3-14. (a) Thermal compensation circuitry, and (b) differential amplifier detail (instrument amplifier).

For the differential amplifier shown in Fig. 3-14(b), the resistors R_{in} may be eliminated. Other amplifiers such as the RCA CA3028 would require them.

The values of biasing resistors R_2 and R_4 can be calculated from the relationships shown below, including the gain equation:

$$R_2 = R_4 = \frac{V_{in} \times 5\%}{I_{bias}}$$

$$\text{Gain} = \frac{V_{out}}{\Delta V} = \frac{R_1}{R_2} + \frac{R_3}{R_4}$$

The 5-percent value for calculating R_2 and R_4 is a rule of thumb for determining biasing resistance, i.e., 5 percent of the minimum value of V_{in}. Note that the gain is a contribution of one-half from each half of the circuit. This particular circuit has excellent noise immunity and common mode rejection.

As an example of these calculations for a typical circuit, let $I_{bias} = 10 \ \mu A$, $V_{in} = 100 \ mV$, and the output voltage = 20 V. The total gain is 200, with each half producing a gain of 100. Then,

$$\frac{R_1}{R_2} = \frac{R_3}{R_4} = 100$$

The biasing resistance for R_2 and R_4 is

$$R_4 = R_2 = \frac{100 \ mV \times 5\%}{10 \ \mu A} = 500 \ \Omega$$

Thus, $R_1 = R_3 = 100 \times 500 \ \Omega = 50 \ k\Omega$.

Since the power supply for the op amp must be capable of supplying adequate voltage to prevent saturation, it must be capable of supplying $\Delta V_{max} \times$ Gain $<$ $(V_+ - V_-)$, assuming zero offset.

The power amplifier that consists of Q_1 is a relatively straightforward emitter-follower circuit. The current-limiting resistor R_{CL} is not always necessary, but many of these modules require control of both current and voltage for accurate temperature control. These thermoelectric devices serve as heat pumps. One side is hot and the other cold, with the temperature difference being 60°C in most designs.

OPTICAL FEEDBACK CIRCUITRY

The optical feedback circuitry controls the bias current of the laser and differs from thermal feedback. Thermal feedback indirectly controls the laser operating point by shifting the temperature, whereas optical feedback operates directly on the bias current, and this in turn changes the operating point.

A typical transmitter is depicted in Fig. 3-15(a). Transistors Q_1, Q_2, and Q_3 with their associated resistors make up an integrated circuit similar to the CA3028 manufactured by RCA. Transistor Q_3 functions as a constant-current source for the differential pair Q_1 and Q_2. Transistor Q_4 is driven by amplifier A_1. These two devices control the biasing current of the laser that results from the summation of the photodetector feedback, part of the input signal, and a reference voltage. The circuit will operate provided the input signals are dc and slowly varying or some further filtering is added.

Figure 3-15(b) is a block diagram of a more practical laser transmitter. In this

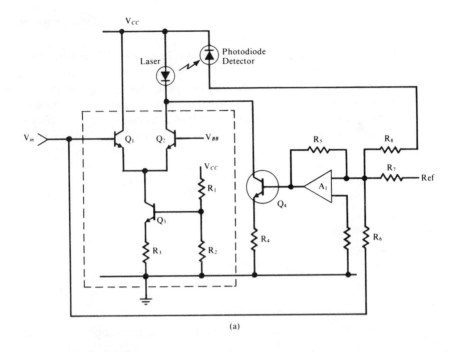

(a)

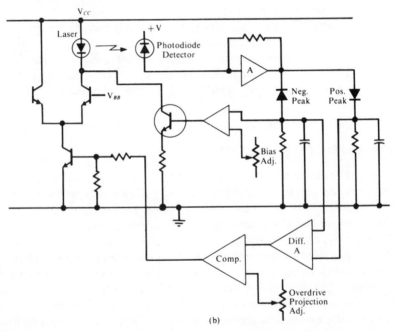

(b)

Fig. 3-15. (a) Optical feedback and laser drive circuitry, (b) improved optical feedback and laser drive circuitry with overdrive protection, and (c) optical feedback with overdrive protection and differential-pair-controlled laser current.

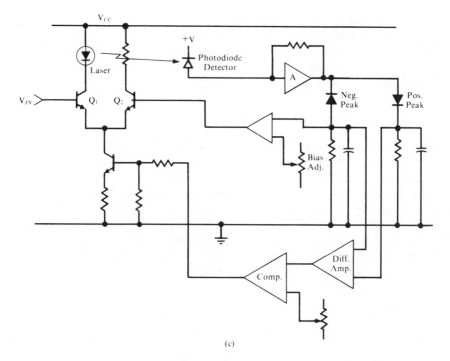

(c)

Fig. 3-15. (*Continued*)

configuration, the negative peak detector provides the feedback to control the bias of a laser diode. This circuit is a dual control for the laser. If the peak-to-peak voltage gets large enough, the constant-current source will be acted on by the second loop. This second loop can be used to turn off the current source should the laser be overdriven to the point of burnout of the device. Overdriving the laser will not always result in catastrophic failure. The power output will usually be reduced below some minimum value that is considered unacceptable. Several manufacturers consider this to be 20 percent below the minimum power output stated in the specification sheet. Throughout this text, this is the criterion that will be used.

This circuit provides two adjustment points. The first, at the input of A, is used to set the biasing level of the laser, whereas the second is to set the comparator threshold. The latter will set the over-voltage value.

Another possible control technique is illustrated in Fig. 3-15(c), which allows Q_1 to drive the laser and Q_2 to function similarly to an automatic gain control (AGC). The current will remain constant, but, as the laser temperature changes and thereby causes a corresponding shift in operating point, Q_2 will either shunt some of the current from the laser or begin to turn off, thus providing more current to it. One of the advantages of using differential-drive amplifiers for the laser with a constant-current source is that the technique reduces current spikes on the voltage supply lines when digital signals are used on the inputs. Note that no mention has been made of receiver performance for optical feedback. A discussion on receivers is reserved for Chapter 4.

DIGITAL TRANSMITTER DESIGN WITH LASERS

This section will present the necessary techniques to design laser transmitters. One of the advantages of digital transmission is that the laser is either off or on; nevertheless, we will also examine the NRZ case in which the laser is biased between off and on at idle. For a 50-percent duty cycle and Manchester encoding, the average output power will be the same at idle as during data transmission.

Digital transmitters will be presented in two parts. The first encompasses the transmitter drive and source circuitry and the second, the encoding techniques. To begin the design, a block diagram should be produced that establishes the necessary requirements for the transmitter and encoding.

The power output must be determined based on system constraints. Later in the text, the technique for system considerations for laser power output will be discussed. The requirements are listed below:

Emission wavelength:	1550 nm
Power coupled into waveguide:	−10 dBm
Threshold current:	< 100 mA
Drive current:	< 150 mA
Mode of operation:	Single mode
Bandwidth of operation:	< 50 MHz

An ideal candidate for this application is the Lasertron QLM 1530M laser diode module. The module consists of a low-threshold laser diode constructed with GaInAsP/InP and buried heterostructure geometry, thermoelectric cooler, thermistor, and GaInAs/InP monitor detector. The output coupling is accomplished using tapered and lensed single-mode Corning fiber with a minimum power output of 0.25 mW (−6 dBm). The specifications for the particular model are given in Table 3-3.

The laser will be driven by a differential amplifier similar to the design in Fig. 3-16(a) except that the ground is replaced with a −5-V supply, Q_1, Q_2, and Q_3. The constant-current bias will be adjustable with a potentiometer because the bias current can have a 5:1 variation between the minimum and maximum. The value of R_3 must be large enough to swamp out variation in $r_{e'b}$,

TABLE 3-3. Specifications for the Lasertron QLM 1530M (Laser Temperature of 25°C).

	Min.	Typ.	Max.
Emission wavelength, nm	1500	1530	1560
Spectral width, nm (FWHM)	—	3.2	6.4
Fiber-coupled power at max drive I			
Classification 01 (mW, peak):	0.5	0.6	—
Classification 02 (mW, peak):	0.25	0.3	—
Fiber-coupled power at threshold, mW	—	0.005	0.020
Extinction ratio (rated P 5/threshold P)	20	—	—
Threshold current, mA	—	30	55
Drive current to rated power, mA	10	40	50
Forward voltage at rated power, V	1.2	1.4	1.8
Monitor photocurrent, μA (at rated P)	50	100	500
Monitor dark current, μA (at -5 V)	—	0.1	0.25
Monitor risetime into 50 ohms, nsec	—	—	1.0
Cooler current at 65°C ambient T, mA	500	625	800
Cooler voltage at 65°C ambient T, V	1.1	1.24	1.50
Cooler response time, sec	—	30	—
Thermistor resistance at 25°C, kohms	9.8	10.0	10.2

the base-to-emitter resistance. The calculation (at room temperature) is shown below:

$$r_{e'b} \approx \tfrac{1}{2} \, \Omega$$

The value of R_3 selected is 100 Ω. When the typical drive current (40 mA) is used, the voltage at the base of Q_3 is approximately -0.4 V.

Figure 3-16(a) depicts the laser drive circuitry that not only performs as a driver but responds to optical feedback to correct biasing currents. Q_1, Q_2, and Q_3 are matched transistors on a single substrate (IC). The differential pair has identical current passing through each of the transistors, Q_1 and Q_2. Although this is rather wasteful from a power point of view, it eliminates large switching spikes when the laser is modulated by digital *signals*. The bias current can be adjusted to an initial value that depends on the laser characteristics, which, in turn, depends on manufacturing tolerances. As temperatures rise because of the increase in ambient, the power output will begin to decrease. The detector preamplifier output will also decrease, and thus the peak detector output voltage will drop. The output voltage of amplifier A will increase and turn on Q_4, resulting in an increase in laser current. The sensitivity to power output changes is dependent on the gain of amplifier A. An analysis of the photodiode detector

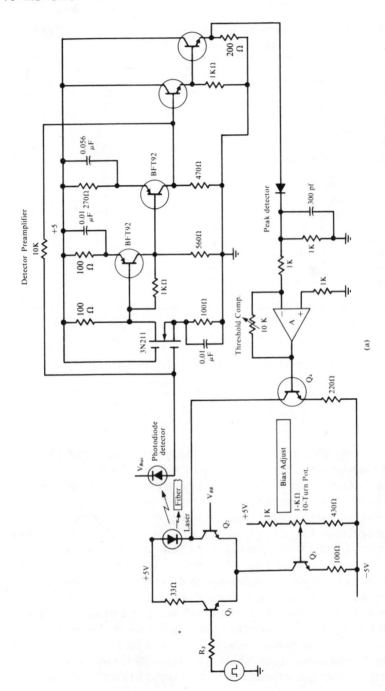

Fig. 3-16. (a) Laser transmitter with feedback, (b) waveforms of data and encoder data, (c) Manchester encoder, (d) differential Manchester encoder, and (e) differential Manchester encoder logic detail.

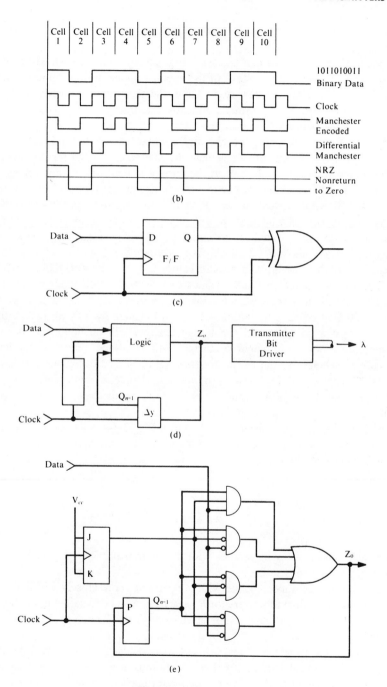

Fig. 3-16. (*Continued*)

preamplifier will be presented in Chap. 4, which considers the design aspects of receivers.

If the peak detector time constant isn't large enough, the dc level will continuously track the waveform instead of the average optical power output. On the other hand, if the time constant is too large, the slow reaction to power output will cause unacceptable performance. Ideally, the time constant should be five or ten times the period of the bit rate.

The temperature compensation circuitry is rather straightforward. Let us examine Figs. 3-14(a) and (b) to calculate the appropriate value for the components. At the wavelength of interest, the current is 625 mA, with a voltage of 1.24 V, and a 65°C ambient will be required to operate the thermoelectric device correctly. This value appears in the QLM 1530M Lasertron specifications. The voltage at the base of transistor Q_1 will be equal to the thermoelectric voltage plus one diode drop from base to emitter (1.84 V). If the drop across the transistor is 1.5 V and a 5-Vdc supply is used, then the drop across the limiting resistor is 1.66 V. The limiting resistor value, R_{CL}, is 1.66 V/0.625 A or 2.65 Ω. Thermistors change resistance values exponentially with temperature. For example, the thermistor resistance will drop from 10 kΩ at 25°C to approximately 800 Ω at 65°C. The bridge voltage will change by a factor of 2 with a temperature change of 40°C. The bridge changes will be nonlinear, but the thermoelectric device temperature changes will be almost linear with voltage and current. Therefore, the thermistor must be linearized over the temperature excursion with a shunting network or resistor. Doing so will desensitize the thermistor, but the amplifier gain can be made to compensate for this desensitization.

The shunt linearizing resistance is shown in Fig. 3-14(a) by dashed lines. A resistor can be added in series with the thermistor and the parallel resistor with the following restrictions:

$$R_{series} \gg R_{TH} \| R_P$$

$$R_P < R_{TH}$$

where R_P is the resistor in parallel with the thermistor, R_{TH}; R_{TH} is at 25°C; and R_{series} is the series resistor.

To match the resistance change with temperature more closely with the thermoelectric cooler, potentiometers for R_P and R_{series} can be installed in the bridge and adjusted for the particular temperature excursion. The potentiometer values should be 1 kΩ and 10 kΩ ten-turn for the R_P and R_{series}, respectively. The power dissipated by the thermistor must be kept low, or the thermistor will develop self-heating. The safe operating region of a thermistor is less than one-sixteenth of a watt. For further information on thermistor biasing techniques, refer to the

Fenwal thermistor catalog and handbook or catalogs of other manufacturers of these devices.

The last, but not the least important, item to consider is the encoder. Encoding data in Manchester code will be the first technique discussed. The data waveform is broken up into cells 1 bit wide, as shown in Fig. 3-16(b). An examination of the Manchester waveform in this same figure reveals the technique rather graphically. When the binary data is a binary 1, as in cells 1, 3, 4, 6, 9, and 10, the first half of the cell in the Manchester waveforms is a 1 with a zero for the second half; for the binary 0 case, the first half of the bit cell is zero and the second half is a 1.

The encoding of the waveform can be accomplished using an exclusive OR gate. Data should be passed through a D flip-flop to synchronize it with the clock signal. The encoder is shown in Fig. 3-16(c). The exclusive OR gate will drive the laser with the circuitry shown in Fig. 3-16(a). These encoders can be purchased from a number of manufacturers such as Harris Semiconductor, National Semiconductor, E^2C, Signetics, etc. Most of the devices, however, operate at 1 M bits/sec except for E^2C, which has devices that operate at 10 M bits/sec or more. For 100 M bits/sec, gate arrays are the best alternative for high volume and fusible link gate arrays for low volume.

The differential Manchester encoder is a bit more complex than the Manchester version. Upon examination of the waveform in Fig. 3-16(b), it can be noted that transitions occur only at the center of the bit cell for binary 1s and at the center and end of the cell for binary 0s. Thus the waveform has no reference point as it does in the Manchester case.

A block diagram of the differential encoder is shown in Fig. 3-16(d) and its circuit in Fig. 3-16(e). The details of the logic design will not be presented here but can be found in Baker.[8] The reader should note from the block diagram that this encoder, as compared to the Manchester type, has feedback. When feedback is present, it implies that the output is affected by some previous output, which happens to be the actual case here.

The NRZ waveform is a standard binary waveform with the binary data centered about the dc value. One of the problems with the NRZ waveform is that if large numbers of zeros or ones are detected at the receiver, dc wander will occur. The result is a distortion of the binary data by the receiver amplifiers because of a shift in the dc level. This condition can be remedied by the use of bit stuffing, i.e., for long strings of ones or zeros, a zero or one will be injected to force a transition in the waveform. These extra bits can then be removed at the receiver with no corruption of the data having occurred. If Manchester or differential Manchester encoding is used, however, a sufficient number of transitions will have occurred to prevent dc wander at the receiver.

One word of caution here: you don't get something for nothing. The worst case conditions for both waveforms will cause transitions that are either a replica of the clock or $\overline{\text{clock}}$. For example, if the Manchester encoder is presented with a long string of ones or zeros, the encoder output will resemble the clock for the first case and an inverted version of the clock for the second. Therefore, the data being transmitted is at twice the frequency of the binary data waveform alternating ones and zeros. Hence, encoding data in Manchester or differential Manchester will allow a trade-off between the bandwidth and performance of the receiver, usually linked with error rate, complexity of the receiver, and delay. Usually when bit stuffing is required, the receiver will require a delay, the length of which depends on the number of consecutive zeros or ones that can be tolerated.

Another possibility exists for handling analog data with digital circuitry, i.e., analog-to-digital (A/D) conversion before transmission and digital-to-analog (D/A) conversion after reception at the receiver. Conversion technique will be addressed in more detail during network applications.

It will suffice to alert the reader that digital transmission limitations are a function of the electronics. During the writing of this text, the upper limit on laser sources was 1000 GHz, which could not be switched with any logic circuitry known at the time.

LINEAR TRANSMITTERS

In this section, the application of linear transmitters to single-mode fiber will be investigated. The laser drive electronics of the previous section are suitable for linear operation, but this section will examine transmitters used in RF applications. For simple transmitter design for multimode operation at low frequency, see Baker.[8] Here we will examine transmitters that operate at 100 MHz and up. Although intended for linear operation, these transmitters can also be used for digital NRZ operation provided that the following restrictions are observed:

$$\frac{5}{\tau} < BW$$

where $\tau = 1/B$ (sec/bit) and $B = $ bits/sec.

A linear amplifier is shown in Fig. 3-17. The lasers, LDS 10-PM and LDS 10-PMF, have 3-GHz and 6-GHz direct-modulation bandwidths, respectively. The maximum bandwidth of the amplifier is only 500 MHz, but either of these particular lasers can be selected because they are both single-mode lasers with very narrow spectral width (less than 0.1 Å units). If the single-mode feature is not desired, then cheaper narrow-bandwidth devices can be used, such as the long-wavelength Lasertron devices. The Lasertron device has a bandwidth that matches the amplifier more closely. Several manufacturers produce linear-am-

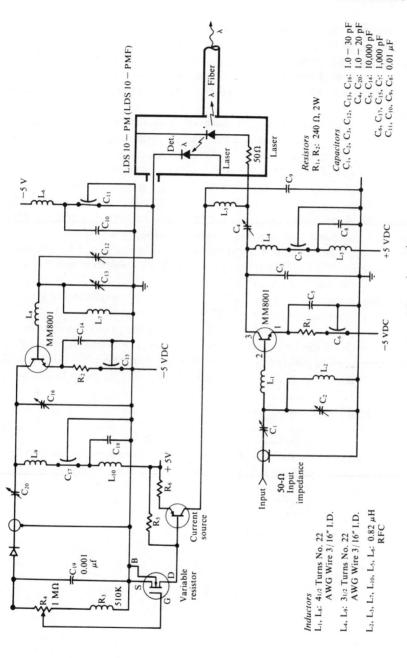

Fig. 3-17. Linear amplifier schematic.

Resistors
R_1, R_2: 240 Ω, 2W

Capacitors
C_1, C_2, C_3, C_{12}, C_{13}, C_{16}: 1.0 – 30 pF
C_4, C_{20}: 1.0 – 20 pF
C_5, C_{14}: 10,000 pF
C_6, C_{17}, C_{15}, C_7: 1,000 pF
C_{11}, C_{10}, C_9, C_8: 0.01 μF

Inductors
L_1, L_8: 4½ Turns No. 22
        AWG Wire 3/16" I.D.

L_4, L_6: 3½ Turns No. 22
        AWG Wire 3/16" I.D.

L_2, L_3, L_7, L_{10}, L_5, L_6: 0.82 μH
        RFC

plifier modules that are rather easier to implement because all of the matching is embedded in the module. Most of the modules have usable gains below 1 GHz. For high-power microwave devices, the device manufacturers' handbooks may be consulted. A complete design procedure with test circuits may be found. The Motorola RF data manual is a good source; it also contains many circuit layouts.

A current source supplies the laser diode with bias current and optical feedback will supply the signal that controls the MOSFET variable resistor of the current source. The noise floor of the laser is -90 dBm at a bandwidth of 1 MHz, with an output power of $+10$ dBm. Therefore, the signal-to-noise ratio is 100 dB; at a normal video bandwidth, it is 93.3 dB. This signal-to-noise ratio is adequate for CATV type applications but a later study of receivers will show this figure to be smaller because of other noise contributions. As the single mode laser is modulated, it exhibits multimode operation depending on how hard it is driven.

Some of the other aspects of linear transmitters, such as AM, FM, and FSK, will be examined in the chapters on networks. The presentation here is meant only to acquaint the reader with the drive electronics.

The circuitry shown in Fig. 3-17 does not include thermoelectric coolers; since it is rather straightforward, it will be left to the reader.

Another topic that will be covered is LED linearizing techniques. The power output of some of the LEDs is comparable to that of a laser except that they lack coherent optical power output. The drivers are simpler because no threshold bias current is needed. A discussion of LED drivers can be found in Baker.[8]

LINEARIZING LED TRANSMITTERS

When LEDs are to be used as transmitter sources, they must be linearized. This section deals with their use in systems consisting of short links, a situation that arises when linear operation is required rather than digital. A block diagram of the linearizing circuit is shown in Fig. 3-18. The LED optical power versus current is given by Eq. 3-34.

$$p(t) = a_1 i(t) + a_2 i^2(t) + a_3 i^3(t) + \cdots a_n i^n(t)$$

$$p(t) \approx a_1 i(t) + a_2 i^2(t) \tag{3-34}$$

$$p(t) \approx a_1 i(t) \left[1 + \left(\frac{a_2}{a_1^2} \right) a_1 i(t) \right] \tag{3-35}$$

where:

$$a_1 \gg a_2$$

$$\left(\frac{a_2}{a_1^2} \right) a_1 i(t) \ll 1$$

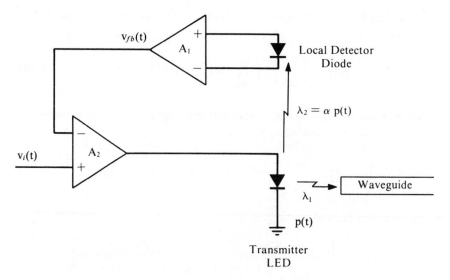

Fig. 3-18. LED linearization circuit.

If A_2 is in amperes/volt and the light captured by the local detector is given by $\alpha p(t)$, then

$$i(t) = A_2(v_i - v_{fb})$$

where:

$$v_{fb} = A_1 \alpha p(t)$$

Thus,

$$i(t) = A_2[v_i - A_1 \alpha p(t)] \tag{3-36}$$

Solving the quadratic Eq. 3-35 for $i(t)$ will yield Eq. 3-37:

$$i(t) = \frac{-a_1 + [a_1^2 + 4a_2 p(t)]^{1/2}}{2a_2} \tag{3-37}$$

which becomes

$$i(t) = \left(\frac{a_1}{2a_2}\right)\left\{-1 + \left[1 + \left(\frac{4a_2}{a_1^2}\right)p(t)\right]^{1/2}\right\} \tag{3-38}$$

Expanding Eq. 3-38 into a series and using only the first two terms, we obtain Eq. 3-39:

$$i(t) = \left(\frac{a_1}{2a_2}\right)\left\{\left(\frac{2a_2}{a_1^2}\right)p(t) - 2\left[\frac{a_2 k(t)}{a_1^2}\right]^2\right\} \tag{3-39}$$

Substituting Eq. 3-36 into Eq. 3-39 and rearranging the terms, we have the following quadratic equation in $p(t)$:

$$0 = p^2(t)\left(\frac{a_2}{a_1^3}\right) - p(t)\left[\left(\frac{1}{a_1}\right) + A_1 A_2 \alpha\right] + A_2 v_i(t)$$

Solving for $p(t)$ and rearranging terms, we obtain Eq. 3-40 as follows:

$$p(t) = \left[\frac{A_2 v_i(t)}{A_2 A_1 \alpha + \left(\frac{1}{a_1}\right)}\right]\left[\frac{A_2 a_2 v_i(t)}{a_1^3[A_2 A_1 \alpha + (1/a_1)]^2} + 1\right]$$

If we let $K = A_2/[A_2 A_1 \alpha + (1/a_1)]$, then

$$p(t) = K v_i(t)\left(\frac{a_2}{a_1^2}\right)\left[\frac{K v_i(t)}{1 + A_1 A_2 a_1 \alpha} + 1\right] \tag{3-40}$$

In Eq. 3-40, the final result, $a_1 A_1 A_2 \alpha$ is considered to be the loop gain. If the loop gain is made large (10 to 20), then the nonlinearities will be negligible. Increases in A_1, A_2, a or feedback coupling, α, will result in loop-gain increases. An underlying assumption in this analysis is that the optical feedback is in the same spatial mode as the optical power coupled to the transmission waveguide.

A technique not discussed here is a method of predistortion of the signal. This is not as reliable as feedback techniques because undesirable harmonics can be generated in the process.

This chapter has introduced the reader to transmitter design. In the following chapter, these designs will be used to further the reader's knowledge of systems aspects of single-mode fiber optics as applied to networks.

REFERENCES

1. Scifres, D. R.; Lindstrom, C.; Burnham, R. D.; Streeper, W.; and Paoli, T. L. Phase-locked GaAlAs laser diode emitting 2.6 W CW from a single mirror. *Elect. Lett.* 19:169 (1983).
2. Ackley, D. E. Single longitudinal mode operation of high power multiple stripe injection lasers. *Appl. Phys. Lett.* 42:152 (1983).

3. Scifres, D. R.; Burnham, R. D.; and Steifer, W. High power coupled multiple stripe quantum well injection lasers. *Appl. Phys. Lett.* 41:118 (1982).

4. Goldberg, L.; Taylor, H. F.; Wetter, J. F.; and Scifres, D. R. Optical injection locking of a GaAlAs phase coupled laser array. 13th International Quantum Electronics Conference, PD-B4, 1984.

5. Amazigo, J. C., and Rubenfeld, L. A. *Advanced Calculus and Its Applications to Engineering and Physical Science.* New York: Wiley, 1980, pp. 314–353.

6. CSELT. *Optical Fibre Communications.* New York: McGraw-Hill, 1981, pp. 429–434.

7. Gower, J. *Optical Communication Systems.* New York: Prentice-Hall, 1984, pp. 295–325.

8. Baker, D. G. *Fiber Optic Design and Applications.* Reston, VA: Reston, 1985, pp. 103–104.

4 | RECEIVERS

The discussion of receivers will be divided into two sections. The first is primarily dedicated to the physics of detection devices; the second, to the receiver circuitry. Although there are several types of detectors, the discussion will be limited to single-mode types, which are solid state. Most single-mode detectors are long-wavelength devices, but this section will cover short-wavelength devices instead because they are readily available and fairly cost effective. Some manufacturers make single-mode or near-single-mode sources at the shorter wavelengths (820 nm).

The first detector to be discussed is the PiN diode (P intrinsic N). To understand the PiN detector, the PN detector behavior will first be described and extended to include the PiN.

PiN DIODE DETECTORS

To excite an electron enough to cross the bandgap within a PN structure, the photon energy must be at least equal to the bandgap energy, ϵ_g. Equation 4-1 describes the relationship, and a diagram depicting bandgap energy is given in Fig. 4-1. Equation 4-2 is important when comparing detectors.

$$\epsilon_g \approx h\gamma \qquad (4\text{-}1)$$

where:

$$\gamma = \frac{c}{\lambda_{TH}}$$

c = speed of light in a vacuum
λ_{TH} = threshold wavelength (which is usually slightly longer than the value in the calculation)

Then,

$$\lambda_{TH} \approx \frac{hc}{\epsilon_g} \qquad (4\text{-}2)$$

Let us now examine Fig. 4-1 in more detail. When an incoming photon is absorbed in the p material (point A), a hole and free electron are created. If this

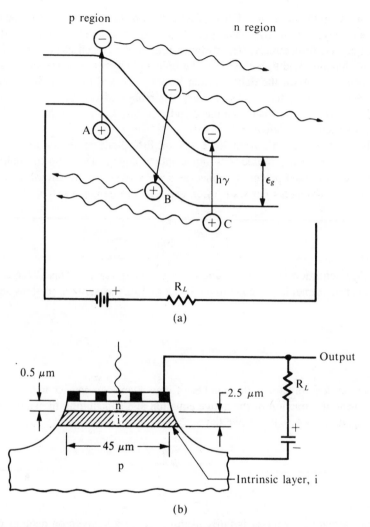

Fig. 4-1. PIN photodiode: (a) Three types of photon absorption that contribute to current flow in the photodiode external circuitry, and (b) PIN photodiode construction.

event occurs within a diffusion length of the depletion layer (the average distance a minority carrier traverses before recombining with an opposite type of carrier), it is the electron, with high probability, that will drift across the layer boundary, thus contributing a charge flow to the external circuit. A similar phenomenon will occur when a photon is absorbed in the n material (point C) except that holes instead of electrons will drift across the layer boundary and create a charge flow in the external circuit. For the last situation, the photon is absorbed in the depletion layer (point B), which produces a hole and an electron that drift to p and n layers, respectively. The distance traversed by each charge, is less than the full junction width, and each of the hole and electron absorptions gives rise to a charge e. Also, the delay current response caused by finite diffusion time is avoided. This latter process is very desirable and can be induced with an intrinsic layer inserted between the p and n material to form a pin structure. A typical structure is shown in Fig. 4-1(b).

The photogenerated current in a PN or PiN diode increases linearly over several orders of magnitude with input optical power. The ratio of photodiode current over optical power is defined as *responsivity*. The relationship is given in Eq. 4-3 in amperes per watt (often expressed in μamps/μwatt):

$$R = \frac{I_{Ph}}{\Phi_{OP}} \tag{4-3}$$

Another definition is in order, that of *quantum efficiency*. This is the ratio of the electrons generated/second to the incident flux, measured in photons/second, or

$$\eta = \frac{I_{PH}}{e\Phi_{OP}/\epsilon_{Ph}}$$

where the denominator is the number of equivalent photons/second.

Clearing the fraction of the above equation and substituting the value, hc/λ, for ϵ_{Ph} and I_{PH}/Φ_{OP} for R, we have

$$\eta = R\frac{hc}{e\lambda} \tag{4-4}$$

where $\eta < 1$.

An alternative form that is fairly useful because it is given on manufacturer's specification sheets is the following:

$$R = \frac{\eta e\lambda}{hc} \tag{4-5}$$

The depletion layer between the edge near the surface and the edge near the substrate is the region where useful absorption of the incident optical power takes place. Some of the optical energy is reflected, and the energy passing through the surface is attenuated exponentially as it propagates through the material. An equation depicting this is as follows:

$$P = (1 - R_f) P_O \left[e^{-\alpha x_1} - e^{-\alpha x_2} \right] \tag{4-6}$$

where R_f = reflected optical power, α = attenuation, x = depth of penetration below the surface, and P_O = optical power at the surface.

The values, x_1 and x_2, represent the two edges of the depletion layer. The optical power P is the actual power converted to photogenerated current. Equation 4-6 may now be rewritten in terms of quantum efficiency, as shown in Eq. 4-7, with the exponential term expanded in a series.

$$\eta = \frac{P}{P_O} = (1 - R_f)\, \alpha \left\{ (x_2 - x_1) - \alpha \frac{(x_2^2 - x_1^2)}{2!} + \alpha^2 \frac{(x_2^3 - x_1^3)}{3!} \right.$$
$$\left. - \alpha^3 \frac{(x_2^4 - x_1^4)}{4!} \cdots \right\} \tag{4-7}$$

Examination of Eq. 4-7 shows that improvement can be easily accomplished by making x_2 as large and x_1 as thin as possible, but this is not very practical.

Materials for PiN diode detectors used in some of the more advanced units are GaAlAs/GaAs in the 700- to 900-nm wavelength region and GaInAs for wavelengths exceeding 1 μm to 1.55 μm. Germanium diodes, which operate well up to 1.8 μm, have been available for several years. The direct bandgap, ternary, and quaternary semiconductors are being exploited for larger wavelength operation. All these devices exhibit narrow bandgaps; often the junctions are Schottky barrier diodes. They also exhibit larger dark currents, i.e., current that flows in the absence of incident radiation. Since dark current, I_d, increases rapidly with temperature, they are often cooled to maintain high sensitivity.

Diode-detector capacitance can be calculated using the standard equation exhibited in most books on field theory, here given by Eq. 4-8:

$$C_d = \frac{\epsilon_0 \epsilon_r A}{w} \tag{4-8}$$

where ϵ_0 = free-space permittivity, ϵ_r = relative permittivity, A = junction area, and w is the width of the lightly doped intrinsic region (i).

A typical value for C_d is 0.1 pf, where $\epsilon_r = 12$, $A = 10^{-7}$ m^2, and $w = 100$ μm. Dark current for silicon detectors will be less than a nanoampere at room temperature. For germanium, it will usually be higher. A typical value for GaAlAs/GaAs devices is 0.2 nA and for GaInAs devices, approximately a magnitude higher, or 2 nA. A typical PiN has already been shown in Fig. 4-1(b).

Schotty-barrier (heterojunction) diodes are used for materials far apart from the threshold wavelength and attenuation coefficient. The Schottky-barrier is a reversed-bias metal semiconductor contact that functions as a back-biased rectifier. Its use eliminates the heavily doped (highly conducting) surface layer that reduces the quantum efficiency of heterojunction types of detectors. A cross section of a Schottky-barrier diode detector is shown in Fig. 4-2(a). Short wavelengths allow the use of thinner layers in PiN detectors because the attenuation is greater. At long wavelengths, PiN detectors must be constructed of different materials or heterojunction techniques must be employed.

Various detector materials exhibit a quantum efficiency that peaks at the wavelengths of interest. Some examples are given in Table 4-1. Silicon detectors do not peak at the wavelengths of interest; germanium detectors, on the other hand, cover the entire range of interest for communications purposes. Therefore, germanium would seem to be the logical choice for high-sensitivity detectors, but efficiency is not the only parameter to consider. SNR, bandwidth, thermal characteristics, etc., must also be taken into account. For example, germanium diodes have larger dark currents than silicon devices do, and they are more sensitive to temperature excursions.

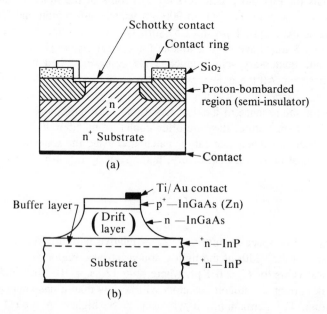

Fig. 4-2. Heterojunction PIN detector construction: (a) Schottky-Barrier diode, and (b) construction details of a PIN diode.

Table 4-1. Quantum Efficiencies of Detector Materials at Various Wavelengths.

Material	Wavelength, nm	Efficiency, %	Comments
Silicon	600	32	Peak
Silicon	400	16	
Silicon	800	16	
Germanium	1300	50	Peak
Germanium	600	35	
Germanium	1600	35	

A great deal of effort to produce a variety of efficient heterojunction detectors has proven fruitful. These ternary structures—for example, $In_{0.55}Ga_{0.47}As$ grown on InP substrates—have produced bandgap energies of 0.75 eV, with long-wavelength cutoff at 1.65 μm. A cross section of the diode is shown in Fig. 4-2(b). The detector may be constructed for either front or back illumination, involving only a slight modification of the contrast configuration.

For short-wavelength PiN detectors, a quantum efficiency of 65 percent has been produced by one manufacturer using a GaAlAs/GaAs heterojunction structure. The operational wavelength is 700 to 900 nm. These devices are rather expensive, but eventually other breakthroughs will reduce the cost of production. State-of-the-art long-wavelength devices include GaInAs detectors with InP substrates that operate from 1.0 to 1.65 μm. These devices will be investigated further when receivers are considered later in the chapter. The technology is changing so rapidly that what is state-of-the-art today will be typical in the near future. The objective here is to encourage the reader to be aware of the new technology and do a literature search before attempting a receiver design that meets or approaches performance limitations.

The frequency response of detectors is one of the most important parameters requiring examination. One of the advantages of using fiber optics instead of copper coaxial cable involves bandwidth considerations. Three mechanisms limit frequency response of photodiodes, as follows:

1. The finite diffusion time of p and n carriers produced in their respective regions
2. The shunt junction capacity that places an upper limit on the intensity modulation frequency, i.e., $\omega_m \approx 1/R_e C_d$, where R_e is the equivalent shunting resistance and C_d is the function capacitance.
3. The finite transit time of the carriers drifting across the depletion layer.

The first two items have been discussed previously. The transit time is apparent when the modulation frequency is high, i.e., the period of the modulation frequency is less than the transit time; therefore, photodiode current i_{Tr} cannot follow the modulations of the optical power.

Transit is affected by drift velocity. If the electric field is increased, drift velocity will also increase until saturation velocity is reached. After saturation velocity is reached, further electric-field increases will not affect drift velocity. Saturation electric fields occur between 10^5 and 10^6 V/M for most detector materials. The drift velocity is usually about 10^5 m/s. For example, if the drift region is 25 μm, then the bias voltage must be at least 25 V when the saturation electric field is 10^6 V/M. The transit time is calculated using Eq. 4-9:

$$t_{Tr} = \frac{d}{v_d} \tag{4-9}$$

where d = drift-width region and v_d = saturation drift velocity. Transit-time calculations for the 25-μm drift region give a t_{Tr} of 0.25 ns. Since Eq. 4-9 was derived on the assumption that a charge flows across the depletion layer, it may be assumed that a current flows. The current is a charge flow per unit time as given by the following:

$$i(t) = \frac{\Delta q}{\Delta t} = \frac{Ne}{t_{Tr}} \tag{4-10}$$

where N is the number of charges that flow.

The carrier generation rate is depicted by the term, $\dot{N}_O (1 + m \sin \omega t)$. These electrons are induced to drift across the depletion layer under the influence of a saturating electric field. The number of electrons generated in the time interval, x/v_{sc}, is $dt = dx/v_{se}$. The total number, N, generated in transit at any given time, t, can be expressed by Eq. 4-11, where l is the total depletion-width region:

$$N = \int_0^l \left(\frac{\dot{N}_O}{v_{se}} \right) \left\{ 1 + m \sin \omega \left[t - (x/v_{se}) \right] \right\} \frac{dx}{v_{se}} \tag{4-11}$$

After the integration, we have

$$N = \frac{\dot{N}_O l_O}{v_{se}} \left\{ 1 + \frac{m v_{se}}{Wl} \left[\cos \omega \left(t - \frac{l}{v_{se}} \right) - \cos \omega t \right] \right\}$$

Performing a half-angle reduction on the result, we obtain Eq. 4-12:

$$N = \frac{\dot{N}_O l}{v_{se}} \left\{ 1 + \left(\frac{2m v_{sc}}{\omega l} \right) \sin \left(\frac{\omega l}{2 v_{se}} \right) \sin \omega \left(t - \frac{l}{2 v_{se}} \right) \right\} \tag{4-12}$$

Current can be expressed in terms of the variables of Eq. 4-12 by substituting Eq. 4-9 into Eq. 4-10 and then Eq. 4-12 into the result, as shown in Eq. 4-13. First,

$$i(t) = \frac{Ne}{t_{Tr}} = \frac{N_e v_d}{d}$$

If we let $v_d = v_{se}$ and $d = l$, then,

$$i(t) = \dot{N}_0 e \left\{ 1 + \left(\frac{2mv_{se}}{\omega l} \right) \sin \left(\frac{\omega l}{2v_{se}} \right) \sin \omega \left(t - \frac{l}{2v_{se}} \right) \right\} \quad (4\text{-}13)$$

Equation 4-13 can be further simplified using the relationship of Eq. 4-9 with a change of variables, i.e., $t_{Tr} = l/v_{se}$

$$i(t) = \dot{N}_0 e \left\{ 1 + m \left[\frac{\sin (\omega t_{Tr}/2)}{(\omega t_{Tr}/2)} \right] \sin \omega (t - t_{Tr}) \right\} \quad (4\text{-}14)$$

The frequency response at high frequency is determined by the amplitude term of Eq. 4-14, which is the equivalent of $\sin x/x$. Using a table or plotting the amplitude of the $\sin \omega (t - t_{Tr})$ term in Eq. 4-14 will result in the following expression at the -3 dB point, i.e., when $\sin x/x = 0.707$:

$$ft_{Tr} = 0.44 \quad (4\text{-}15)$$

where $f =$ the optical modulation frequency. For the previous case, where t_{Tr} was calculated at 0.25 nsec, the maximum modulation bandwidth is 1.76 GHz.

The noise contribution of the PiN detector to the receiver is usually smaller than of most other components, but occasionally optical noise will become significant. Therefore, to complete this discussion of PiN detectors, noise components must be addressed. Most texts present a rather superficial coverage of the subject because of the small contribution of optical noise but, when we delve into the subject of couplers, optical noise will become more significant. The SNR for the detector is given by Eq. 4-16:

$$\text{SNR} = 10 \log \left(\frac{i_S^2}{i_q^2 + i_d^2} \right) \quad (4\text{-}16)$$

where $i_S =$ signal current, $i_q =$ optical noise (shot noise), and $i_d =$ noise due to dark current. The SNR in this equation applies only to the detector without any regard to the receiver.

The optical signal current, i_S, for a PiN detector is found from the following:

$$i_S^2 = \gamma^2 G^2 m^2 P_O^2 \tag{4-17}$$

where $G = 1$, γ = responsivity, m = modulation index, and P_O is the optical power impinging on the detector.

The quantum noise, i_q, is given by the following:

$$i_q^2 = 2e\gamma P_O B_n G^2 = 2e\gamma P_O B_n \tag{4-18}$$

The dark current noise contribution, i_d, is found from the following:

$$i_d^2 = 2eI_d G^2 B_n = 2eI_d B_n \tag{4-19}$$

where B_n = noise bandwidth and I_d = the dark current.

Rewriting Eq. 4-16 with substitutions for the values, I_q^2, i_q^2, and i_d^2, we have the result shown below:

$$SNR = 10 \log \left(\frac{\gamma^2 m^2 P_O^2}{2e\gamma P_O B_n + 2eI_d B_n} \right) \tag{4-20}$$

where $I_d \gg I_L$ (leakage current).

When a complete analysis of a receiver is undertaken, the general case with I_L included will be examined. For most applications, the leakage current can be neglected. Note that the best that can ever be achieved—when I_d can be neglected—is shown by Eq. 4-21:

$$SNR = 10 \log \left(\frac{1}{2} \right) \left(\frac{\gamma m^2 P_O}{eB_n} \right) \tag{4-21}$$

If a receiver can approach this value within 4 or 5 dB, then it can be considered fairly high quality, although at the time this book was being written a laboratory-grade receiver achieved a value within $1\frac{1}{2}$ dB of this limiting SNR. If a specification has an SNR greater than the theoretical value given by Eq. 4-21, moreover, it cannot be built. Equation 4-21 can be calculated from values on a detector specification shut, but it can also be written in an alternative form that depicts more of the optical variables:

$$SNR = 10 \log \left(\frac{\eta e m^2 P_O}{h\gamma e B_n} \right) = 10 \log \left(\frac{\eta \gamma m^2 P_O}{hcB_n} \right) \tag{4-22}$$

AVALANCHE PHOTODIODE DETECTORS (APD)

In the avalanche photodiode detector, gain is associated with the detection mechanism. When shot noise is dominated by electrical noise, an APD will improve the SNR when implemented. Quantum and dark current noise are greater in these devices, but for many applications, electrical noise is much greater than either of these optical components. When receivers are discussed, such devices will be examined in more detail.

The avalanche process multiplies the number of carriers and also produces an increase in RMS V_c noise. There is an optimum multiplication factor that will produce the best signal-to-noise ratio. The avalanche diode is similar in many respects to the photomultiplier tube. When a single carrier pair—i.e., hole and electron—generates eight other carrier pairs, then the multiplication involved, M, is 9.

Shot noise increases as M^n, where $2 < n < 2^{17}$, and experimental behavior that nears the ideal is $M^{2.1}$. The equations deriving M from avalanche theory can be found in Gower.[1] The input is similar to that of a PiN diode with the multiplication factor, $M = G$. With this variable change, Eq. 4-17 can be rewritten as

$$i_S^2 = \tfrac{1}{2} \gamma^2 M^2 m^2 P_O^2 \tag{4-17}$$

The SNR due to the avalanche diode only is again determined by Eq. 4-16 (with the same change of variable, $G = M$):

$$SNR = 10 \log \left(\frac{\gamma^2 M^2 m^2 P_O^2}{2ei_d B_n M^n + 2e\gamma (P_O m + P_B) B_n M^n} \right) \tag{4-23}$$

When the receiver is discussed, so will the terms due to the load resistances. The first term in the denominator results from the dark current, and the second term is quantum noise. The optical power, P_B, is due to the background optical power, which can be neglected in most situations. Note in this equation that M^n is the gain term and usually larger than 2, as was stated earlier. Often this is expressed by

$$M^n = M^2 F$$

where F is the noise figure of the APD. The author, however, prefers the form in Eq. 4-23. One may also observe that as M increases, the denominator increases faster than the numerator, thus reducing the SNR for large values of M.

The minimum detectable value of P_O, P_{min}, occurs when $SNR = 1$, that is, when no load resistance is present. Thus, when

$$SNR = 1 = \frac{\gamma^2 M^2 m^2 P_{min}^2}{2ei_d B_n M^n + 2e\gamma P_{min} m B_n M^n}$$

then

$$P_{min} = \frac{eB_n M^{n-2} + \sqrt{e^2 B_n^2 M^{2n-2} + 2ei_d B_n M^n}}{\gamma} \tag{4-24}$$

This equation reveals many variables that affect the minimum power, including bandwidth, B_n; multiplication factor, M (gain); responsivity, γ; and dark current, i_d. For the condition when dark current is small enough to be neglected, Eq. 4-24 reduces to Eq. 4-25—the quantum limit of detection:

$$P_{min} = \frac{2eB_n M^n (m^{-1} + M^{-2})}{\gamma}$$

or

$$P_{min} \approx \frac{2eB_n M^{n-1}}{\gamma} \tag{4-25}$$

where $M \gg 1$. This value of P_{min} can be approached but never reached because of the other noise contributions.

Bandwidth in avalanche diodes has a few additional terms that are a part of the transit time across the drift region. These additional terms include the time required for the avalanche to develop, τ_A, and the transit time, τ_h, of the last holes produced in the avalanche process to drift back across the drift region.

$$\text{Hole transit time, } \tau_h = \frac{W_2}{v_h}$$

$$\text{Electron transit time, } \tau_e = \frac{W_2}{v_{se}}$$

$$\text{Avalanche to develop in a single hole} = \frac{W_A}{v_h}$$

$$\text{Multiplication avalanche development} = \frac{kMW_A}{v_{se}}$$

Then

$$\tau_{Tr} \cong \frac{W_2}{v_{se}} + \frac{W_2}{v_h} + \frac{W_A}{v_h} + \frac{kMW_A}{v_{se}} \qquad (4\text{-}26)$$

$$\cong \frac{W_2 + kMW_A}{v_{se}} + \frac{W_2 + W_A}{v_h}$$

where $k < 1$. The value of k is a function of the material in question and the electric field. For $W_2 = 25$ μm, $v_{se} = 10^5$ m/s, $v_h = 5 \times 10^4$ m/s, $k = 0.1$, $W_n = 0.5$ μm, and $M = 100$, τ_{Tr} would be calculated as follows:

$$\tau_{Tr} = \left[\frac{(25 + 5)}{10^5} \times 10^{-6} \right] + \left[\frac{25.5 \times 10^{-6}}{5 \times 10^4} \right] \text{ sec}$$

$$= 0.81 \text{ nsec}$$

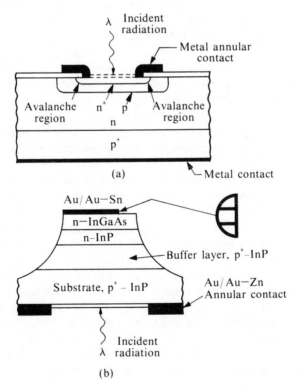

Fig. 4-3. APD diode construction: (a) Diffused reach-through structure, and (b) heterojunction.

Using 0.5 μsec for the -3-dB point, the bandwidth can be calculated using Eq. 4-15 at 560 MHz. The previously calculated bandwidth for a PiN diode of 1.76 GHz is approximately three times larger than that for the APD. This is one of the tradeoffs one must make when choosing an APD over a PiN diode.

APD construction is similar to PiN construction in many respects. Two examples are shown in Fig. 4-3. The first of these, Fig. 4-3(a), is a reach-through structure and the second, Fig. 4-3(b), is a heterojunction APD, which is one of the more recently designed detector types. The construction of more exotic devices will not be presented here because of rapidly changing technology.

LINEAR RECEIVERS

The subject of linear receivers is of great importance to engineers because their performance can be modified by careful selection of detectors, designing their electronics with low-noise parts, and otherwise reducing noise through efficient design. The preamplifiers for most linear receivers can also be used on digital units. In this section, several examples will be discussed. Although many receivers can be used for both single-mode and multimode applications, the next chapter will consider the large differences between the two in coupling efficiency.

The fundamental receiver components are shown in Fig. 4-4. A fiber-optic pigtail (a small section of fiber connected directly to the detector) is fused or coupled to the fiber-optic transmission line. The optical signal is converted to an electrical signal by the detector. The fact that the preamplifier usually converts the detector-signal current to a voltage implies that it must be a trans-impedance amplifier. To give the circuitry large dynamic range, local feedback is added. In low-cost devices, a restriction on minimum link length is often imposed to insure that the detector currents will be sufficiently low to prevent preamplifier saturation. The voltage from the preamplifier is amplified by a driver amplifier, which usually has an automatic gain control (AGC) loop around it. The driver, moveover, is usually routed to some type of demodulation circuitry to extract useful information. All the blocks shown in Fig. 4-4 may not be present, and in some cases none at all. For example, the preamplifier and driver with AGC may be combined and difficult to separate into distinct blocks, or the preamplifier may be included in the AGC loop and have its own local feedback. The point is that Fig. 4-4 is merely a general description of a receiver and may have many variations. It will be presented again prior to the discussion of any particular design to allow the reader to observe the subtle differences. The demodulation and modulation circuitry involved will be addressed in Chap. 8. The issues to be covered here primarily relate to the signal-conditioning circuitry.

The first item that must be modeled is the detector; for the purposes of this

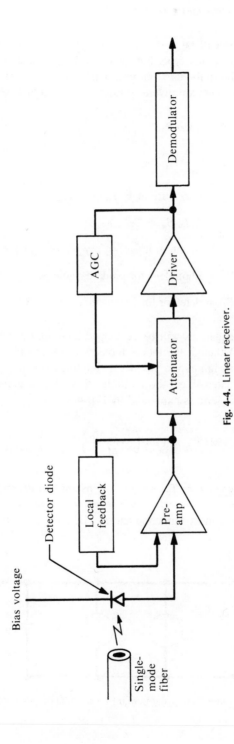

Fig. 4-4. Linear receiver.

book, the detector will be a PiN or APD type. The circuit depicted in Fig. 4-5 can be used for either type, but, as we will observe shortly, the differences between the two lie in the dark current, gain, and noise current. The values for the various components are given by Eqs. 4-27 through 4-33:

$$i_s = \gamma M m P_0 \qquad (4\text{-}27)$$

$$\gamma = \frac{ne\lambda}{hc} \qquad (4\text{-}28)$$

$$i_d = \text{dark current} = I_{du} + M I_{dm} \qquad (4\text{-}29)$$

$$i_{nd}^2 = 2e I_{du} B_n + 2e I_{dm} M^n B_n \qquad (4\text{-}30)$$

$$i_q^2 = \text{quantum noise} = 2e\gamma (P_0 + P_B) B_n M^n \qquad (4\text{-}31)$$

$$C_D = \text{photodetector diode capacitance} \qquad (4\text{-}32)$$

$$R_L = \text{detector load resistance } (R_L < R_{\text{equiv}}) \qquad (4\text{-}33)$$

None of these equations will be developed here until a specific amplifier has been described. Although the model looks very formidable, many of its terms can be combined or ignored to produce a simplified version. For APDs, another term can be evaluated—i.e., the multiplication (M) or gain. The temperature dependence of this term is expressed by Eq. 4-34:

$$M(T) = 1 - \left(\frac{1 - M_0}{M_0}\right)\left(\frac{1}{e^{-\gamma(T - T_0)}}\right) \qquad (4\text{-}34)$$

where $M = M_0$ when $T = T_0$ and γ is a constant that depends on the material.

An examination of this equation will reveal gain increases with temperature—an unusual case for most semiconductors. The gain excursion could cause sat-

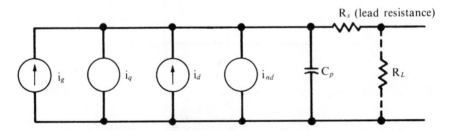

Fig. 4-5. General equivalent circuit for an APD or PIN diode detector.

uration in amplifier states if it is not included as part of the AGC calculations. Table 4-2 depicts some of the general characteristics of detectors.

The performance of a fiber-optic system, whether analog or digital, depends heavily on total SNR, linearity, and transient behavior. Nonlinearities result in harmonic generation, intermodulation, and cross-modulation, all of which degrade system performance; they are pronounced in analog systems in particular. However, the photodiode is sufficiently linear not to pose a problem for high-performance linear systems provided that postdetector amplifiers are sufficiently linear. The SNR of these high-performance systems is dependent on modulation depth, received optical power, optical bandwidth, gain factor of the detector, and noise characteristics of the detector and past-detector amplifier. The receiver SNR is of prime importance. When designing these systems, the following recommendations should be considered:

1. Select a photodetector adequate to meet the system SNR constraints, information bandwidth, data rate, operational wavelength, fiber size, detector package limitations, if any, bias stability, device capacitance, responsivity, and cost limitations.
2. Characterize the photodiode with an adequate equivalent circuit and provide an impedance-matching network between it and the post-detection amplifier. The detector, matching network, and post-detection amplifiers must operate over the entire bandwidth of interest under the bias condition (photodiode).
3. Select an AGC scheme, temperature compensation circuit, and suitable stabilized bias supply to meet system requirements.
4. Minimize circuit parasitics and use adequate grounding techniques to provide the required high-frequency operation.
5. Provide a high degree of decoupling from all power supplies caused by the high gain required in most high-performance systems.
6. Select a transistor that will match post-detector amplifiers in characteristics such as low noise, high gain/bandwidth product, etc.

Since most of these considerations will become apparent during the design discussion, most will be reiterated.

The photodetector has already been modeled in Fig. 4-5. We shall now discuss some of the attributes of these devices that are presented in Table 4-2. Selection of the correct photodetector is imperative for any high-performance single-mode or multimode receiver. Parameters to be considered are bandwidth, noise, quantum efficiency, and operating wavelength. Silicon devices respond only to wavelengths under 1.1 μm. They are fairly low cost and have large bandwidth and low leakage for most APD applications. Germanium APDs have longer wave-

Table 4-2. A Comparison of Photodetectors.

Photodector	Capacitance, Pf	Leakage Current, nA	Excess Noise Factor	Wavelength of Operation, μm	% Quantum Efficiency at 0.8/1.3/1.5 μm
APD, Si	0.5-2	0.2-20	0.3	0.4-1.1	70/—/—
APD, GaAs	1.2-2	10-50	0.3	0.4-0.9	70/—/—
APD, Ge	1.5-2	100-500	0.7	0.9-1.5	—/50/20
APD, Ge (r + *)	1.5-2	100-500	0.7	0.9-1.6	—/50/50
APD, InGaAsP/InP	0.5-1.5	10-50	0.2-0.4	0.9-1.4	—/60/—
APD, InGaAs/InP	0.5-1.5	100	0.4	0.9-1.3	—/60/—
APD, HgCd, Te	1	150	—	0.9-1.3	—/50/0
APD, GaAlAsSb/GaAlAb	1-2	1-40	2.1	1.0-1.4	—/90/—
PN Schottky barrier GaAs	0.1	10-50	—	0.6-0.85	26/—/—
PiN, Ge	10-30	5000	—	0.9-1.5	—/75/—
PiN, Si	0.5	0.2-10	—	0.4-1.1	75/—/—
PiN, InGaAs/InP	0.3-0.5	10	—	0.9-1.6	—/70/70
Phototransistor					
IN GaAs/InP	0.3-0.5	50-100	—	0.9-1.6	—/50/50

length detection capability, but they have much higher leakage currents and are thus somewhat noisier (poorer SNR).

At present, the most promising material for fabricating photodetectors is the InGaAsP alloy system, which is epitaxially grown on InP substrates. The band gaps range from 1.35 eV (InP) to 0.73 eV ($In_{0.53}Ga_{0.47}As$). This range in band gaps, which results in absorption wavelengths below 1.7 μm, is accomplished by varying the alloy composition of the constituent elements, In, Ga, As, and P. These heterostructure alloy systems are advantageous because of their broad spectral sensitivity. Moreover, the InP substrates are transparent to long-wavelength radiation, and only those compositions that are lattice matched to high-purity InP substrates are chosen for epitaxial growth. Detectors can be fabricated so that light absorption occurs well below semiconductor surfaces, a characteristic that minimizes loss caused by surface recombination of photogenerated carriers.

The several companies manufacturing large bandwidth photodetectors include Mitsubishi, Ford Aerospace, and AEG-Telefunken. APDs with 1- to 8-GHz bandwidths operate in the 0.8- to 0.9-μm wavelength region. Tables were not provided for these devices because the technology is rapidly changing and can become obsolete within a year. The reader is urged to do a literature search prior to designing receivers.

Long-wavelength APD and PiN devices can be constructed of silicon up to 1.1 μm, with reduced quantum efficiency. Most of those being fabricated at present are constructed using direct-bandgap III-V semiconductor alloys such as InGaAs, InGaAsP, and AlGaAsSb, which are ternary and quarternary alloys. These devices operate well in the 1.0- to 1.6-μm region with quantum efficiencies of 50 to 90 percent but have rather limited bandwidth—below 1 GHz. Indirect-bandgap Germanium photodiodes have large dark current (leakage current) but are suitable for bandwidths to approximately 10 GHz, with rise times of from 0.3 to 0.08 nsec. Some of these devices respond to frequencies greater than 10 GHz.

Let us now examine some of the preamplifier design techniques. Figure 4-6(a) exhibits a typical transimpedance amplifier circuit with local feedback. Let us first examine the bias circuitry and determine some of the voltages and currents. Neglecting the decoupling circuit, the collector voltage drops can be written as depicted in Eq. 4-35:

$$V_{CC} - I_{C1}R_L - V_{BE} = V_0$$
$$V_0 - V_{BE} \approx R_f I_f$$

(4-35)

Consequently,

$$V_{CC} - I_{C1}R_L - V_{BE} = V_{BE} + R_f I_f$$

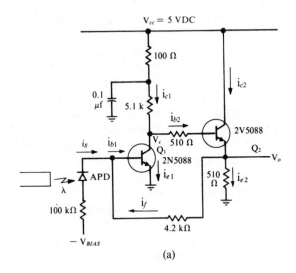

(a)

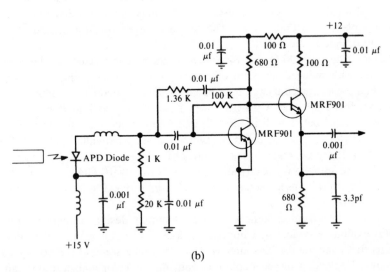

(b)

Fig. 4-6. Common preamplifier circuits: (a) Transimpedance amplifier with linerar feedback (the detector used is an APD); (b) PIN detector preamplifier with local feedback around voltage-amplifier stage only; (c) transimpedance amplifier that can be constructed with or without feedback; (d) equivalent circuit for Fig. 4-6(a), medium frequency model; (e) reduction of equivalent circuit in Fig. 4-6(d); (f) equivalent circuit and first and second reductions for Fig. 4-6(c).

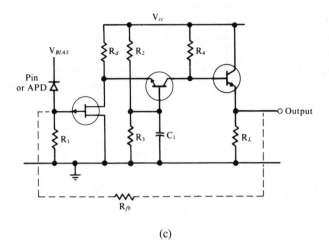

(c)

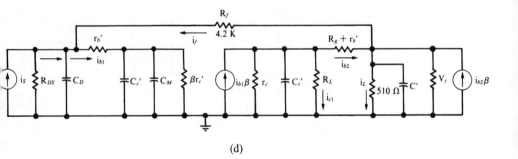

(d)

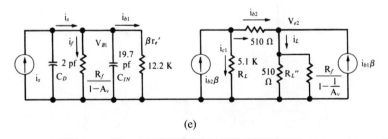

(e)

Fig. 4-6. (*Continued*)

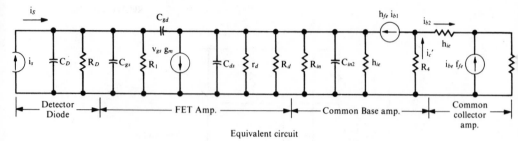

Detector Diode — FET Amp. — Common Base amp. — Common collector amp.

Equivalent circuit

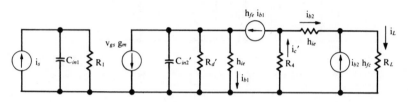

First reduction

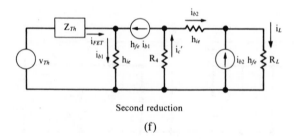

Second reduction

(f)

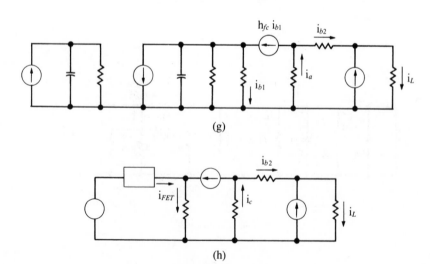

(g)

(h)

Fig. 4-6. (*Continued*)

Assuming that $I_{C1} \approx I_{E1}$ and $R_f I_f \ll I_{C1} R_L$, Eq. 4-36 can then be written to determine the collector and emitter current as follows:

$$\frac{V_{CC} - 2V_{BE}}{R_L} \approx I_{C1} \approx I_{E1} \tag{4-36}$$

By inserting the circuit values from Fig. 4-6(a) into Eq. 4-36 ($V_{BE} = 0.7$ at 25°C) the collector current is 0.71 mA and $V_0 \approx 2.8$ V; i.e., $V_C - V_{BE} = V_0$.

The biasing point of Q_1 is approximately three-fifths of the power supply voltage, and this allows large voltage excursions without saturation ($2 V_{PP}$). I_f is approximately the collector current divided by h_{fe} (350 for 2N5088 transistors), or 2 μA. As one may note, the approximation, $R_f I_f \ll I_{C1} R_L$, is valid. The feedback resistor must supply the leakage current for the APD, which is in the nanoampere region.

The ac circuit will now be examined. The equivalent circuit is shown in Fig. 4-6(b). The base current into the first transistor stage, i_f, is calculated from Eq. 4-37:

$$i_{b1} = i_S - i_f \tag{4-37}$$

where i_{b1} represents the small signal currents at the base of transistor Q_1; i_S, the signal current from the photodiode; and i_f, the feedback current.

Before an analysis can be undertaken let us offer some approximations that will make it less cumbersome. The photodiode shunt resistance R_{PS}, will usually be much larger than the input resistance at the base of Q_1 or the feedback resistance and may therefore be neglected. The total capacitance looking into the base circuit is the same as the base-emitter capacitance, C_e, and the Miller capacitance, C_m; the latter are gain-dependent, as shown in the following equation (neglecting the fact that $r_c \gg R_L$):

$$C_e = C'_e + C_c \left(\frac{R_e}{r'_e} + 1 \right)$$

where:

C_e = emitter capacitance
C_c = collector capacitance
R_e = resistance in the emitter circuit
r'_e = emitter resistance
r_b = base resistance

The value of r'_b at Q_1 input is much smaller than that of $\beta r'_e$, which is 12.2

k. The reduction is shown in Fig. 4-6(d). The current relationships are given by Eqs. 4-38 and 4-39 as well as by Eq. 4-37, as follows:

$$i_L + i_f = i_{b2}\beta + i_{b2} = i_{b2} (\beta + 1) \tag{4-38}$$

$$i_{b1}\beta = i_{c1} + i_{b2} \tag{4-39}$$

Substituting Eq. 4-39 into Eq. 4-38 results in Eq. 4-40:

$$i_L + i_f = (i_{b1}\beta + i_{c1}) (\beta + 1) \approx i_{b1}\beta^2 \tag{4-40}$$

Then, substituting Eq. 4-37 into this result, we have

$$i_L + i_f = (i_S - i_f)\beta^2$$

from which we obtain the current gain equation:

$$A_I = \frac{i_L}{i_S} \approx -\left(\frac{i_f}{i_S}\right) \beta^2 \tag{4-41}$$

The objective is to find the transimpedance gain of the circuit, i.e., v_L/i_S, but this is given by Eq. 4-42 instead:

$$A_T = -\frac{i_L Z_L}{i_S} = -\frac{v_2}{i_S} \tag{4-42}$$

An equivalent circuit of the amplifier is shown in Fig. 4-6(c), with the reduction depicted in Fig. 4-6(d). The parameter calculations for this figure are as follows:

$\beta = 350$ min
$C_c = 0.1$ pf
$C_e = 5$ pf
$C_{in} = C_c' [(5.1 \text{ k}/35) + 1] + C_e$
$\quad \approx (0.1) (147) + 5 = 19.7$ pf
$C' = 10$ pf across the 510-Ω load resistance of Q_2
$r_e' = 25/i_{E1} = 25/0.71 = 35\Omega$ at 25°C

The feedback current, i_f, is given by Eq. 4-43, as follows:

$$i_f = \frac{V_{b1} - V_{e2}}{R_f} \approx -\frac{V_{e2}}{R_f} \tag{4-43}$$

where $V_{e2} \gg V_{b1}$. An alternate form is

$$i_f \approx -\frac{Z_L i_L}{R_f}$$

The current into the base of the gain stage, Q_1, is calculated by using Eq. 4-44, a current divider, as follows:

$$I_{b1}(S) = \frac{Z_{Df}(S) i_S}{Z_{Df}(S) + Z_{in}(S)} \tag{4-44}$$

where S is the LaPlace operator and $Z_i(S)$ is the parallel combination of C_{in} and $\beta r'_e$.

$$i_{b2} = \frac{i_{b1}(S)\beta R_L}{R_L + 501\Omega + R'_L(\beta + 1)} \tag{4-45}$$

$$i_L \approx i_{b2}\beta = \frac{i_{b1}(S)\beta^2 R_L}{R_L + 510\Omega + R'_L(\beta + 1)} \tag{4-46}$$

$$i_L \approx \left(\frac{Z_{fD}(S) i_S}{Z_{fD}(S) + Z_m(S)}\right)\left[\frac{\beta^2 R_L}{R_L + 510\Omega + R'_L(\beta + 1)}\right] \tag{4-47}$$

Equation 4-45 is another current divider, with the current i_{b2} into the base of Q_2. Assuming that $i_C \approx i_e$ for Q_2 and $r_c \gg Z_L$, we can substitute Eqs. 4-44 and 4-45 into Eq. 4-46, with Eq. 4-47 as the result. The transimpedance transfer function is depicted in Eq. 4-48.

$$Z_{fD} = \left(\frac{R_f}{1 - A_v}\right)\left(\frac{1}{1 + \dfrac{SR_c C_D}{1 - A_v}}\right)$$

$$Z_{in} = \frac{\beta r'_e}{1 + S\beta r'_e C_{in}}$$

$$R'_L \parallel \frac{R_f}{1 - \dfrac{1}{A_v}} \approx R'_L$$

$$\frac{v_L}{i_S} \approx R'_L\left[\frac{Z_{fD}(S)}{Z_{fD}(S) + Z_m(S)}\right]\left[\frac{\beta^2 R_L}{R_L + 510\Omega + R'_L(\beta + 1)}\right] \tag{4-48}$$

Evaluating Eq. 4-48 using the equivalent circuit values of Fig. 4-6(d), we have the calculation for the terms shown below:

$$R'_L \left[\frac{\beta^2 R_L}{R_L + 510\Omega + R_L^1(1 + \beta)} \right] = 3.52 \text{ V}/\mu\text{A}$$

$$\frac{Z_{fD}(S)}{Z_{fD}(S) + Z_{in}(S)} \approx \left[\frac{R_f}{\beta r'_e(1 - A_v)} \right] \left[\frac{(1 + \beta r'_e C_{in})S}{S \frac{R_f(C_{in} + C_D)}{1 - A_v} + 1} \right]$$

$$\left[\frac{R_f}{\beta r'_e(1 - A_v)} \right] \left[\frac{1 + \beta r'_e C_{in} S}{1 + SR_f \frac{(C_{in} + C_D)}{1 - A_v}} \right] = 0.00234 \left[\frac{1 + 0.241 \ \mu\text{sec } S}{1 + 0.65 \ \text{nsec } S} \right]$$

$$\frac{v_L}{i_S} = 8.3 \left[\frac{1 + 0.241 \ \mu\text{sec } S}{1 + 0.65 \ \text{nsec } S} \right]$$

For a 1-μW optical-power signal at the receiver input, the output signal will be 8.3 mV. As one may observe, this is a low-level signal, but this particular section of a receiver is a preamplifier. Its main feature is high SNR, which will be discussed later in the chapter. The bandwidth of this preamplifier has a 3-dB rolloff at 24 MHz, and the bandwidth can be adjusted using the voltage gain for values of A_v near 50 or R_f may be decreased. The latter, of course, alters the total gain, i.e., v_L/i_S. The numerator term will increase with frequency. This is not really the situation because R_{DS} and the transistor Q_1 input resistance will shunt the input capacity C_{in} and prevent the numerator from increasing without bound. Therefore, the amplifier will resemble a pass-band amplifier with large bandwidth.

The amplifier shown in Fig. 4-6(b) has a 30- to 300-MHz bandwidth. Adding the inductor in series with the PiN will compensate for the diode capacitance. Note the decoupling on the biasing circuitry; this will keep the RF noise out of the amplifier, which is fairly high gain. The voltage gain stage (first amplifier) is neutralized to improve the MRF 901 performance. The detector has only about 2 pf of capacitance, a common amount. The input resistance of the amplifier is 500 Ω. This circuitry will not be analyzed in detail, but a review question at the end of the chapter will present a series of questions to assist the reader in his own analysis.

Let us examine the equivalent circuit in Fig. 4-6(f), which is the first reduction of Fig. 4-6(e). The parameters in the first reduction are as follows:

$$c_{in1} = c_d + c_{gs} = (2 + 0.2)pf(\text{typical})$$

$$R_1 \ll R_d(c_{gd} \text{ neglected})$$

$$c'_{in2} = c_{ds} + c_{in2} \approx 3pf$$

$$R'_d \approx \frac{R_d R_{in}}{R_d + R_{in}} \left(\text{where } r_d \gg R_d\right)$$

The transfer function for v_{gs} can be written by inspection and the Thievenin equivalent circuit taken for the FET. This will result in the second reduction of Fig. 4-6(f), as follows:

$$v_{Th} = \frac{R_d v_{gs} g_m}{1 + R'_d c'_{in2} S} = \left(\frac{R'_d g_m}{1 + R'_d c'_{in2} S}\right)\left(\frac{R_1}{1 + R_1 c_{in1}}\right) i_s$$

where:

$$v_{gs} = \frac{i_s R_1}{1 + R_1 C_{in1} S}$$

The calculation can be made from this reduction to find i_{b1} by summing the current at the node, with this result:

$$i_{b1} = \frac{v_{Th}}{(2_{Th} + h_{1e})}$$

The calculation to find the relationship between i_{b1} and i_{b2} is as follows:

$$i_{b2} = i'_c - h_{fe} i_{b1}$$

$$\frac{i'_c R_4 - v_0}{h_{1e}} = \frac{i'_c R_4 - i_{b2}(1 + h_{fe})R_L}{h_{1e}} = i_{b2}$$

and

$$i'_c R_4 = i_{b2}\big[h_{ie} + (1 + h_{fe})R_L\big]$$

Then, substituting the value of the i_{b2} current equation into the above expression for $i'_c R_4$, and after rearranging the terms to solve for i'_c, we have the following equation, which expresses i_{b2} as a function of i_{b1}:

$$i_{b2} = \frac{R_4 h_{fe} i_{b1}}{h_{ie} + (1 + h_{fe})R_L - R_4}$$

The equation for the output signal is $i_{b2}(h_{fe} + 1)R_L$. Therefore, the output voltage can now be expressed as a function of i_{b1}, which is a desired result if we are to obtain the total transfer function of the amplifier. Equation 4-49 expresses this relationship as follows:

$$v_0 = \frac{R_4 h_{fe}(h_{fe} + 1)R_L}{h_{ie} + (1 + H_{fe})R_L - R_4} i_{b1} \tag{4-49}$$

Substituting the previous relationship for i_{b1} and v_{Th}, we have Eq. 4-50:

$$v_0 = \frac{v_{Th}}{Z_{Th} + h_{1e}} \left(\frac{R_4 h_{fe} R_L}{h_{ie} + (1 + h_{fe})R_L - R_4} \right) \tag{4-50}$$

where:

$$Z_{Th} = \frac{R'_d}{(1 + C'_{in2} R_d S)} + \frac{R_1}{(1 + R_1 C_{in1} S)} \tag{4-51}$$

Substitute the value of v_{Th} given previously and the desired result is obtained, i.e., the transimpedance transfer function for the amplifier, as follows:

$$\frac{v_0}{i_S} = \left(\frac{R'_d g_m}{1 + R'_d C'_{in2} S} \right) \left(\frac{R_1}{1 + R_1 C_{in1} S} \right) \\ \left(\frac{1}{Z_{Th} + h_{1e}} \right) \left(\frac{R_4 h_{fe} R_L}{h_{ie} + (1 + h_{fe} R_L - R_4)} \right) \tag{4-52}$$

The assumptions were that the common base and emitter followers of the two transistors are identical devices with approximately the same bias conditions. The value for Z_{Th} should be evaluated first with numerical values or approximations based on actual parameters. These will reduce the computation required and relieve the reader of some unnecessary work. Note that the equivalent circuits of Fig. 4-6(f) are taken from the circuit of Fig. 4-6(c), which does not show any decoupling of power supplies. The completed circuit must include decoupling because of the high gain and wide bandwidth sometimes encountered. An additional note: Equation 4-51 will also be reduced when approximations are made to Z_{Th} because many of the parameters are common to both equations. Computation examples appear in the review problems at the end of the chapter.

The feedback version of the amplifier in Fig. 4-6(c) is also presented in the review problems. Replacement of the R_L and R_1 resistors with equivalent feedback elements is necessary only to derive a similar set of equations for v_0/i_S and Z_{Th}.

Some important aspects of microwave biasing techniques are given in the next section. This is an important topic because fiber-optic systems are commonly used in RADAR links because of their low noise and total lack of RF emission and RF pick-up from electromagnetic fields. From the military standpoint, moreover, they are impervious to the EMP (electromagnetic pulses) that occur after a nuclear explosion. These pulses destroy electrical equipment in the surrounding area because of the large currents generated.

MICROWAVE PREAMPLIFIER BIASING

The design of bias circuits for microwave applications of GaAsFET and MES-FET (metallic stability barrier) devices is as important as matching network design. These transistors are important because they have sufficient bandwidth to accommodate single-mode transmission and receiver circuitry. GaAsFETs and MESFETs can be easily fabricated into a combination of fiber-optic and electronically integrated circuitry. Microwave gain for these devices is high, and this performance should not be sacrificed because of poor biasing design. Depending upon the desiderata—i.e., low noise and power; low noise, high gain; Class A high power; or Class B or Class AB high power and high efficiency—the optimum dc operating points can be selected. Low-noise amplifiers require relatively low drain-source voltage V_{DS} and current I_{DS}—usually an I_{DS} of 0.15 I_{DSS} (where I_{DSS} is the drain-to-source current when gate-to-source V_{GS} = 0). This is depicted as point A in Fig. 4-7(g). Larger gain can be obtained from these devices by shifting the biasing point upward to an I_{DSS} of about 0.9 with the same V_{DS} as that depicted in Fig. 4-7(g) at point B. To increase the FET output power and maintain linear class A operation, the operating point should be shifted to point C on the curve, i.e., lower I_{DS} current and higher V_{DS} voltage ($I_{DS} \approx 0.5 I_{DSS}$ and $V_{GS} \approx 8$ V). For high efficiency, the operating point should be shifted to point D ($I_{DS} \approx 0.25 I_{DSS}$ and $V_{DS} \approx 8$ V).

The various biasing configurations to achieve these conditions are shown in Fig. 4-7(a) through (f). The first of these circuits—i.e., Fig. 4-7(a)—depicts a dual supply voltage. To avoid operating regions that could damage the device, the gate voltage, V_G, must be turned on first. This configuration results in the lowest source inductance and lowest noise figure as well as high gain. Figures

4-7(b) and (c) require only a single supply, either positive or negative. Again, the gate must be active before the drain voltage is applied. Therefore, V_S must be turned on first in Fig. 4-7(b) and V_G before V_S in Fig. 4-7(c). These biasing circuits require high-quality microwave bypass capacitors at the source. If this precaution is not observed at higher frequencies, small series source impedance can cause noise-figure degradation and low-frequency instability and these can result in oscillatory bias voltages.

The bias configurations shown in Figs. 4-7(d) and (e) require a single supply, V_D. These circuits avoid the turn-on and analog turn-off problems presented by

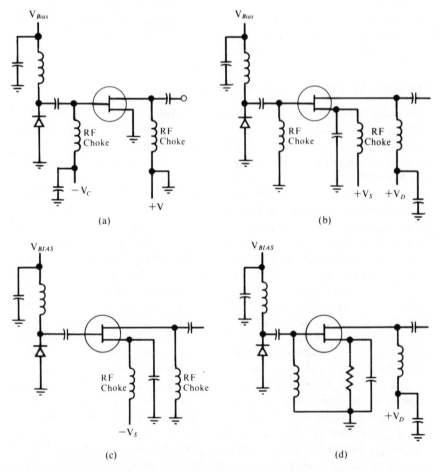

Fig. 4-7. High-frequency MESFET preamplifiers: (a) through (f) different biasing techniques, and (g) biasing points on a typical I-V characteristic for a GaAs FET.

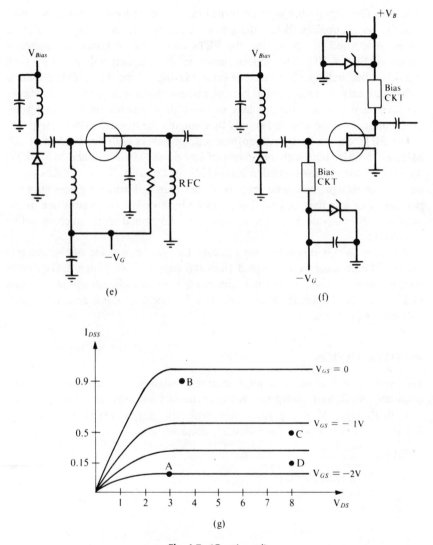

Fig. 4-7. (*Continued*)

the earlier circuitry. Due to the source resistance, these two circuits will be lower in efficiency. The bias point adjustment is a simple matter; one only has to change the value of the source resistance, i.e., use a potentiometer.

Transient protection in the earlier circuits to allow turn-on of one supply before another can be accomplished by adjusting the RC time constants of the power supplies. The supply that is to be turned on first must have a shorter RC time than the other supplies. When the power is turned off, the decay will also be faster. A method of decoupling the FETs and reducing transient damage is shown in Fig. 4-7(f). The zeners shown in the diagram will not only limit transients but will guard against reverse biasing of the FET. Other biasing techniques exist, such as active biasing circuits that use a transistor to supply the necessary voltages, but for lack of space they will not be discussed here. Such voltages can be adjusted simply by changing the transistor operating point.

The FETs can be designed into practical amplifiers with bandwidths of 600 MHz to 15 GHz, but, with the advent of low noise GaAsFETs and MESFETs, the state of the art has exceeded bandwidths of 80 GHz. These devices are of an advance design, with gate lengths of less than 1.0 μm and gate-source capacitances of less than 0.2 pf. They can be built on the same substrate as the detector, all components of the circuit being fabricated using microstrip line techniques.

Note that in this discussion no mention has yet been made of the detector circuits. These must be decoupled from the supplies using high-quality components because most preamplifier circuits are high-gain circuits and stray-signal pick-up can be amplified. These signals will appear as noise and, of course, will reduce performance.

RECEIVER DESIGN

This section will examine a total receiver design—i.e., preamplifier, power amplifier, AGC, and attenuator. Receiver noise issues will also be addressed. The first stage of the preamplifier with the GaAs MESFET—see Fig. 4-8(a)—has a device with the following characteristics:

MESFET:

$$g_m = 15 \text{ m}\mho(\text{millimhos})$$
$$C_{gs} = 0.2 \text{ pf}$$
$$f_T = 12 \text{ GHz}$$
$$I_g = 2 \text{ nA}$$
$$C_{gd} = 0.1 \text{ pf}$$

The APD is fabricated with InGaAs technology and has the following characteristics:

APD:

$$\eta = 0.8$$
$$t_r = 100 \text{ ps}$$

$C_D = 0.5$
$I_{du} = 1$ nA
$I_{dm} = 1$ nA
$K = 0.2$ to 0.5 ionization ratio

The first calculation concerns the bias resistors for the first stage. A good starting point is to make the drain voltage approximately two-thirds the power-supply voltage. If R_D can be made small enough, the input impedance of the previous stage can be ignored. Let the drain current be approximately 6 mA. Then,

$$\frac{V_{cc} - V_0}{I_D} \approx R_D$$

and R_D is 1.3 KΩ. The voltage gain of this stage is calculated using Eq. 4-53. These equations will be more detailed when the ac operation is examined. The dc gain is

$$A_v = -\frac{g_m R'_L}{1 + g_d R'_L} \approx -g_m R'_L \qquad (4\text{-}53)$$

where $g_d R_L \ll 1$.

The drain current was set at 6 mA. The value is determined by Eq. 4-54.

$$I_d = g_m v_{gs} \qquad (4\text{-}54)$$

The bias voltage, v_{gs}, is 0.4 V and the dc gain is approximately 20 if the transistor stage following the FET has sufficiently high input impedance. The biasing may need to be adjusted after the first iteration to adjust some of the device operating points to more favorable positions.

The dc bias voltage is approximately 8 V. The R_{e1} can be calculated at 340 Ω for an emitter current of 1 mA and a V_{be} (voltage drop, base to emitter) of 0.6 V for silicon. The value of V_{C1} of 5 V can be approximated, which makes the value of R_{C1} approximately 5 KΩ. Note that the value of R_{e1} is 340 Ω and that most transistors used will have an h_{fe} of over 100. Therefore, R_D will not be loaded by the biasing circuitry.

Using the same type of reasoning on the biasing circuitry of Q_3, $R_{e2} = 440$ Ω and $R_{C2} = 600$ Ω at $I_E = 10$ mA. The resistor value of Q_4, i.e., R_L, is approximately 540 Ω at $I_E = 10$ mA.

The values of R_g and R_f are 200 kΩ and 2 MΩ, respectively, for a 10-μA biasing current. Since all stages are dc-coupled except for the detector, adjusting one bias voltage will affect the others. When assembling the amplifier, check all stages for cutoff or saturation. Also, since this is only a first pass at design, remember to do a production design. All worst-case values for components must be considered, and potentiometers may require adjustments and selected values of resistors may be necessary in some cases. The preamplifier equivalent circuit is shown in Fig. 4-8(b). All bypass components are assumed to be perfect, and decoupling is assumed to be complete. The equivalent circuit will be discussed first and then the noise issues.

The equivalent V_{Th} and Z_{Th} for the diode and FET can be written by inspection using Thevenin's theorem:

$$v_{Th} = \frac{R_T i_s}{1 + R_T C_T S} = v_{gs}$$

$$Z_{Th} = \frac{R_T}{1 + R'_T C_T S}$$

where, when R_b = bias resistance and R_D = diode resistance,

$$C_T = C_M + C_D = C_{gs} + (1 + g_m R_D) C_{gd} + C_D$$

$$C_M = \text{Miller capacity} = G_{gs} + (1 + g_m R_D) C_{gd}$$

$$R_T = \frac{R'_b R'_{eq}}{R'_b + R'_{eq}}$$

$$R'_b = \frac{R_b R_D}{R_D + R_b}$$

$$R'_{eq} = \frac{R_g R_f / (1 - A_v)}{R_g + \left(\dfrac{R_f}{1 - A_v}\right)} \approx \left(\frac{R_f}{1 - A_v}\right)$$

Here, $R_g (1 - A_v) \gg R_f$.

Next, the relationship between I_{b2} and v_{gs} must be calculated to find the transfer function for the entire amplifier. A current divider can be used to find

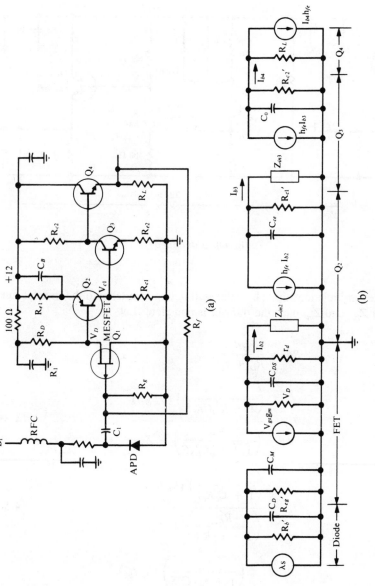

Fig. 4-8. Circuit analysis for a preamplifier with 200-MHz bandwidth: (a) Preamplifier schematic, (b) equivalent circuit, and (c) preamplifier circuit with all coupling capacitors and power supply decoupling shown.

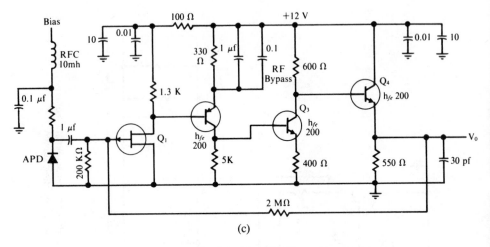

Fig. 4-8. (Continued)

the necessary relationship. The value will not be inserted into the transfer function for Z_{in2} and Z_{inz} until the transfer is complete. Let

$$R'_D = \frac{r_D R_D}{R_D + r_D}$$

$$I_{b2} = \frac{\left(\dfrac{R'_D}{1 + R_D^1 C_{DS}S}\right)(v_{gs}g_m)}{\left(\dfrac{R'_D}{1 + R'_D C_{DS}S}\right) + Z_{in2}} \tag{4-55}$$

$$I_{b3} = \frac{\left(\dfrac{R_{c1}}{1 + C_c R_{c1}}\right)(h_{fe}I_{b2})}{\left(\dfrac{R_{C1}}{1 + C_c R_{c1}}\right) + Z_{in3}} \tag{4-56}$$

$$I_{b4} = \frac{\left(\dfrac{R_{c2}}{1 + R_{c2} C_c S}\right)(h_{fe}I_{b3})}{\left(\dfrac{R_{c2}}{1 + R_c C_c S}\right) + R_L(h_{fe} + 1)} \tag{4-57}$$

$$v_L = I_{b4}(1 + h_{fe})R_L \tag{4-58}$$

All of the relationships are present, and the transfer function can be written using Eqs. 4-55 through 4-58 and the expression for v_{Th}. Equation 4-59 expresses the relationship between v_L and i_S, i.e., the transfer function:

$$\frac{v_L}{i_S} = \left[\frac{R_T}{1 + R_T C_T S}\right]\left[\frac{R_D' g_m}{R_D' + (1 + R_D' C_{DS} S)Z_{in2}}\right]\left[\frac{R_{c1}' h_{fe}}{R_{c1}' + (1 + C_c R_c S)Z_{in3}}\right]$$

$$\times \left[\frac{R_{c2} h_{fe}(1 + h_{fe})R_L}{R_{c2} + R_L(h_{fe} + 1)(1 + R_{c2} C_c S)}\right] \tag{4-59}$$

Note that a quick examination of the equation indicates that the units check, i.e., volts/amp. The final result will be in millivolts per microampere, which is the desired unit of gain.

Now let us evaluate the impedances, Z_{in2} and Z_{in3}. Both terms are derived by the same method; therefore, only Z_{in2} will be considered and Z_{in3} will be identical except for a change of variables. At room ambient,

$$r_{e2}' = \frac{26 \text{ mV}}{I_e}$$

$$R_{in2} = (1 + h_{fe})r_{e2}'$$

$$R_{in3} = (1 + h_{fe})(r_e' + R_{e2})$$

$$C_{in2} = C_{be} = C_{cc}\left(\frac{R_{c1}'}{r_e'} + 1\right)$$

Therefore,

$$Z_{in2} = \frac{R_{in2}}{1 + R_{in2} C_{in2} S} \tag{4-60}$$

$$Z_{in3} = \frac{R_{in3}}{1 + R_{in3} C_{in3} S} \tag{4-61}$$

Many values can be calculated using the previous values for biasing resistors. If the receiver is to operate at frequencies of 10 to 20 MHz, then the capacitances and some of the resistances shunting the transistors may be ignored. The equivalent circuit model is dependent on device parameters and frequency of operation.

Let us begin our numerical evaluation of the circuit using all the parameters known thus far:

$$C_T = 0.2 \text{ pf} + (1 + 19.5)(0.1) \text{ pf} + 0.5 \text{ pf}$$

$$= 2.7 \text{ pf}$$

$$A_v \approx -(R'_D g_m)\left(\frac{R_{c1}}{r'_{e1}}\right)\left(\frac{R_{c2}}{R'_{e2}}\right) = -(19.5)\left(\frac{5.1 \text{ k}}{26}\right)\left(\frac{620}{430}\right)$$

$$\approx -5,500$$

$$R_{eq} = 360 \ \Omega$$

The input resistance to the amplifier with feedback as seen by the detector is 360 Ω, i.e., R_T. The time constant established by the detector diode and FET input impedance is $R_T C_T$; therefore, the bandwidth at the FET input is as follows:

$$f_T = \frac{1}{2\pi R_T C_T}$$

$$= 163 \text{ MHz (3-dB bandwidth)}$$

The bandwidth can be increased through a reduction in the numerical value of C_T or R_T. If the gain of the FET is made smaller—for example, if R_D is made $\frac{1}{2}$ its value—its Miller capacity will be smaller, thus increasing bandwidth. Reduction of the feedback resistor will reduce R_T, thus also increasing bandwidth. Since there are several possibilities, the designer has a great deal of latitude to optimize his/her design for some particular parameter.

Let us now determine the general equation for gain (Eq. 4-60) with all the parameters of Z_{in2} and Z_{in3} accounted for. Rearranging the terms of Eq. 4-59 as shown in Eq. 4-62, the low-frequency gain term can be factored out, as shown in Eq. 4-64.

$$\frac{v_L}{i_S} = \left(\frac{R_T}{1 + R_T C_T S}\right)$$

$$\times \left[\frac{R'_D g_m (1 + R_{in2} C_{in2} S)}{(R'_D + R_{in2}) + R_{in2} R'_D (C_{in2} + C_{DS}) S}\right]$$

$$\times \left[\frac{R'_{c1} h_{fe} (1 + R_{in3} C_{in3} S)}{(R'_{c3} + R_{in2}) + (C_{in3} C_c) R_{in3} R'_{c1} S}\right]$$

$$\times \left[\frac{R_{c2} h_{fe} (1 + h_{fe}) R_L}{R_{c2} + R_L (h_{fe} + 1)(1 + R_{c2} C_c S)}\right] \tag{4-62}$$

First, however, some of the other terms can be neglected if the time constants are sufficiently small. Thus,

$$A_{Lf} = \frac{R_T R'_D g_m R'_{c1} h_{fe}^2 (1 + h_{fe}) R_{c2} R_L}{(R'_D + R_{in2})(R'_{c1} + R_{in3})[R_{c2} + R_L(1 + h_{fe})]} \qquad (4\text{-}63)$$

Evaluation of Eq. 4-62 will give the values of gain before any roll occurs as a result of higher frequencies. Equation 4-64, an alternative form of Eq. 4-62, depicts the transfer function with all time constants included (if biasing resistors are selected to minimize interstage loading on the amplifiers, several of the parallel combinations of resistances can be approximated by the smaller of the two values):

$$\frac{v_L}{i_S} = \frac{A_{Lf}(1 + R_{in2}C_{in2}S)(1 + R_{in3}C_{in3}S)}{(1 + R'_{in2}C'_{in2}S)(1 + R'_{in3}C'_{in3}S)(1 + R'_L C_c S)(1 + R_T C_T S)}$$

$$(4\text{-}64)$$

where:

$$R'_D = R'_{in2} = \frac{R_{in2}R'_D}{R_D + R_{in2}}$$

$$C'_{in2} = C_{in2} + C_{DS} \qquad (4\text{-}65)$$

$$R'_{in3} = \frac{R_{in3}R'_{c1}}{R_{in3} + R'_{c1}}$$

$$C'_{in3} = C_{in3} + C_c \qquad (4\text{-}66)$$

$$R_L = \frac{R_{c2}R_L(h_{fe} + 1)}{R_{c2} + R_L(h_{fe} + 1)} \qquad (4\text{-}67)$$

The equations are evaluated for the preamplifier in Figs. 4-8(a) and (b), with the following result:

$$\frac{v_L(S)}{i_S(S)} = \frac{(1.0)(1 + 20S)(1 + 176S)}{(1 + 5.4S)(1 + 10S)(1 + 180S)(1 + 0.9S)} \text{ mV}/\mu\text{A}$$

The S here is the LaPlace operator, and all numerical values preceding it are in nanoseconds. The transfer function requires some additional roll-off capacitance because peaking occurs as a result of the numerator zeros (two S terms in parentheses). The amplifier can then become unstable and oscillate.

It will be left as an exercise for the reader to calculate the values of C_{in2}, C_{in3}, R_{in2}, and R_{in3}, or any combination of them, or add roll-off capacity to reduce peaking. The values in Eq. 4-64 are calculated using the following transistor parameters:

$h_{fe} = 200 \, Q_2, Q_3, Q_4$ (no loading due to Q_2)
$R_D = 1.3$ k
$R'_D = 1$ k
$g_m = 15$ millimhos
$R_{c2} = 600 \, \Omega$
$R_{c2} \approx R'_{c2}$
$R_{c1} = 5$ kΩ
$R_{c1} \approx R'_{c1}$
$R_{in2} = 5.2$ kΩ
$R_{in3} = 88$ kΩ
$C_{be} = 2$ pf
$C_{ce} = 0.01$ pf
$C_0 = 30$ pf

The FET and detector diode parameters were given previously.

A schematic of the final preamplifier with all components is shown in Fig. 4-8(c). This completes the preamplifier section, which has a bandwidth of approximately 30 MHz. The passband can easily be extended by a reduction of C_{in2} capacitance or a decrease in R_{in2} resistance, or a combination of both. The diagram is complete with decoupling and bypass capacitors. Note that the emitter bypass capacitance requires an extra good RF capacitor to insure adequate emitter-resistor bypass.

PREAMPLIFIER NOISE

Preamplifier noise is an important topic in both analog and digital systems. The noise will be separated into two groups: one related to optical components and the other to receiver thermal noise and leakage currents. Equation 4-68 represents the signal-to-noise ratio of the signal at the FET output, Q_1:

$$\text{SNR} = \frac{i_S^2}{[i_q^2 + i_G^2 + i_B^2 + i_M^2 + i_p^2] + [I_T^2 + i_D^2 + i_L^2 + i_F^2]} \quad (4\text{-}68)$$

$\qquad\qquad\qquad$ *Optical noise terms* $\qquad\qquad$ *Thermal and leakage terms* where the optical noise terms are as follows:

$$i_S^2 = \text{mean square of the signal} = \frac{1}{2}\left(\eta\frac{eGmP_0}{h\nu}\right)^2$$

$$i_q^2 = \text{quantum noise} = \frac{2e^2\eta P_0 G^2 F_d B_n}{h\nu}, \text{ where } F_d = G^K$$

$$i_B^2 = \text{LEDs only beat noise} = \left[\frac{2(e\eta P_0)^2 B_n}{(h\nu)^2 Jw}\right]\left[1 - \frac{B}{2w}\right]$$

i_M^2 and i_P^2 = modal and partition noise, respectively (usually neglected or empirically measured)

$$i_G^2 = \text{incoherent background radiation} = \frac{2e^2\eta P_G G^2 F_d B_n}{h\nu}$$

and the thermal and leakage terms are as follows:

$$i_T^2 = \text{thermal noise} = \frac{4KTB_n F_T}{R_{eq}}$$

$$i_D^2 = \text{dark current noise} = 2eI_d G^2 F_d B_n$$

$$i_L^2 = \text{leakage current noise} = 2eI_L B_n$$

$$i_F^2 = \text{channel noise (FET only)} = \frac{3.73(\pi C)^2 B_n^3 KT}{g_m}$$

The variables in the terms of these equations are as follows:

η = quantum efficiency of the APD
P_0 = average received optical power
m = modulation index
P_G = average incident background radiation
B_n = noise bandwidth
G = APD gain
F_d = noise figure associated with random avalanche process
F_T = noise figure of the amplifier (FET)
λ_w = spectral width of the source
J = number of spatial modes received
W = spectral width of source
K = Boltzmann's constant (1.380622×10^{-23} J°K)
T = temperature in degrees Kelvin
N_m = number of spatial modes received
C_T = equivalent input capacity
g_m = transconductance gain of the FET

Let us evaluate each of the noise and signal components, with the SNR as an end result, using Eq. 4-68 and substituting the appropriate values. The first calculation will be responsivity of the APD without gain ($G = 1$), or

$$\gamma = \frac{\eta e}{h\gamma} = \frac{\eta e \lambda}{hc}$$

$$= \frac{0.8 \times 1.6 \times 10^{-19} \times 1300}{6.625 \times 10^{-34} \times 3 \times 10^8}$$

$$= 0.837 \text{ A/W (or } \mu A/\mu W)$$

where $\eta = 0.8$, $e = 1.6 \times 10^{-19}$ C, $\lambda = 1300$ nm, $c = 3 \times 10^8$ m/sec, and $h = 6.625 \times 10^{-34}$ J sec.

The mean square signal is calculated as follows:

$$i_S^2 = \tfrac{1}{2}(0.837 \times 20 \times 0.5 \times 0.1 \ \mu W)^2$$

$$= 0.35 \ \mu A^2$$

where $m = 0.5$, $G = 20$, and $P_0 = 0.1 \ \mu W$ (-40 dBm). Note that $m = 50$ percent, which is the usual limit before the signal becomes distorted; i.e., it insures linearity.

The noise calculations for i_q^2, i_T^2, i_L^2, i_D^2, and i_F^2 (the others will be neglected) are as follows:

$$i_q^2 = 2(0.837) \times 1.6 \times 10^{-19} \times 0.1 \times 10^{-6} \times (20)^2 \times 4.47$$

$$\times 30 \times 10^6$$

$$= 14.37 \times 10^{-16} \ A^2 \ (\text{Quantum noise})$$

$$i_T^2 = \frac{4(4.1 \times 10^{-21}) \times 30 \times 10^6 \times 2}{0.360 \text{ K}}$$

$$= 0.273 \times 10^{-14} \ (\text{Thermal noise})$$

$$i_D^2 = 2 \times 1.6 \times 10^{-19} \times 2 \times 10^{-9} \times 4.60 \times 4.47 \times 30 \times 10^6$$

$$= 31.6 \times 10^{-18} \ A^2 \ (\text{Dark current noise})$$

$$i_L^2 = 2 \times 1.6 \times 10^{-19} \times 1 \times 10^{-9} \times 30 \times 10^6$$

$$= 9.6 \times 10^{-21} \text{ (Leakage noise)}$$

$$i_F^2 = \frac{3.73\left(\pi \times 2.7 \times 10^{-12}\right)^2 \left(30 \times 10^6\right)^3 \times 4.1 \times 10^{-21}}{15 \times 10^{-3}}$$

$$= 1.98 \times 10^{-18} \text{ (FET noise)}$$

Total noise $= -i_N^2 = 0.42 \times 10^{-14}$

Using Eq. 4-68 and the results of the noise and signal calculations, we then have

$$SNR = 83.3$$

Using the more common form,

$$SNR' = 10 \log SNR$$

$$= 19.2 \text{ dB}$$

For analog applications, this is considered a fairly good value. As the signal progresses through the other stages—i.e., Q_2, Q_3, and Q_4—it will degrade somewhat. However, for video transmission using a carrier, the carrier-to-noise ratio (CNR) must be 46 dB or greater for adequate video recovery at the receiver. Therefore, the received signal must be increased and the noise level reduced. Note that the quantum noise term can be decreased considerably if a PiN detector is used. This is one of the reasons why they are popular for use in low-noise amplifiers.

Thermal noise will be the dominant component in most low-noise amplifiers. Depending on the coding scheme, the signal-to-noise ratio may be adequate to produce a BER of 10^{-8} after the reduction brought about by noise contributions from the Q_2, Q_3, and Q_4 stages. For optimally designed amplifiers to produce low-noise figures, F_T must be equal to SNR_{in}/SNR_{out}, or, expressed in dB, NF $= 10 \log F_T$. For the reader interested in optimizing noise, see Motchenbacker and Fitchen,[2] an excellent text on the subject.

Most of our discussion has assumed that the optical sources are lasers, but for short data lengths (less than 10 km), LEDs may be considered as the sources. LEDs may be characterized as an emitting continuum over a large spatial area, with the emitters having a low temporal coherence. The beat noise term will be present, but it can be neglected. Coupling losses are also much larger in LEDs, but many of the newer LEDs have optical power outputs approaching 1 mW. The edge emitter is fast becoming the best of both worlds; i.e., it is a LED but has a directional output similar to that of a laser.

Often it is necessary to use lasers because of long transmission distances, large bandwidths, or both. Multimode lasers produce a spatial granular speckle pattern at waveguide discontinuities that appears as a source of noise. In a star connection these become even more pronounced. Single-mode fiber, however, will eliminate modal noise, but in this case the signal is composed of two polarization states and any polarization selective device can cause a complete loss of signal.

Modal noise in single-mode fiber can be eliminated by using a true single-mode waveguide. The measured cutoff wavelength of higher-order modes usually occurs at a shorter wavelength on a long waveguide because of the higher loss of high-order modes. Therefore, a long waveguide may appear to operate in single mode even though it supports the next mode, or modes—with high loss. Since the two-mode case is the worst case with regard to modal noise, the waveguide must be chosen to propagate single mode over its short section.

Conventional single-mode fiber supports two orthogonal polarized modes, which propagate at slightly different phase velocities (commonly known as *birefringence*). This condition results from imperfections in the waveguide and causes polarization fluctuations. Frequency fluctuation also causes this condition. Consequently, a source with polarization selective loss—such as coupling, integrated optical components, polarization-selective power dividers, bend loss, and misalignment at waveguide joints—will modulate the transmitted power, resulting in polarization modal noise.

Polarization modal noise is reduced by the use of multilongitudinal-mode lasers because each spectral line produces a different state of polarization, and the mean optical signal becomes depolarized. However, reduction of coherence time of the individual longitudinal modes is of negligible advantage because the interpolarization delays are so short.

If light can be launched into only one polarization mode, polarization modal noise will disappear. The use of highly birefringent—i.e., polarization maintained—waveguide reduces coupling between polarization modes, but this condition will make precise orientation of joints imperative. Single-polarization waveguide is an ideal solution; however, most waveguide at present supports the second polarization with higher loss.

It is possible to calculate the minimum detectable power for a PiN or APD, and this is commonly called the *quantum limit*. If a receiver is desired that has a sensitivity greater than this value, it will not be possible to design it.

The development of the argument is as follows: The probability that no electron-hole pairs will exist in the absence of any noise can be expressed using Poisson's distribution equation, as follows:

$$P(n) = \frac{\Lambda^n e^{-\Lambda}}{n!} \qquad (4\text{-}68)$$

Evaluated at $n = 0$ where $\Lambda = E_P/hf$,

$$P(0) = e^{-\Lambda}$$

$$P = 10^{-9} \text{ for a BER of } 10^{-9}$$

Then, since $e^{-21} = 10^{-9}$,

$$\frac{E_P}{hf} = 21 \qquad (4\text{-}69)$$

For an optical signal with a 50-percent duty cycle, 21 photons must be produced 50 percent of the time. Therefore, dividing both sides of Eq. 4-69 by 2, we have the energy required in the pulse:

$$10.5hf = E_P'$$

where $E_P' = E_P/2$.

Then, the minimum power can be expressed as follows:

$$P_{\min} = 10.5hf \qquad (4\text{-}70)$$

Evaluating Eq. 4-70 for the preamplifier, we have

$$P_{\min} = 10.5 \, (2 \times 10^{-19}) \, (167 \times 10^6)$$

$$= 1.75 \times 10^{-10} \text{ W or } -67.6 \text{ dBm}$$

The preamplifier input was 0.1 μW, or -40 dBm, and the amplifier will produce a signal that will just meet the SNR provided amplifiers Q_2, Q_3, and Q_4 have very low noise. Amplifiers can usually be optimized to produce sensitivities within 12 dB of the quantum limit; therefore, another design iteration could produce fruitful results. This exercise is left to the reader, providing an opportunity to make use of some of the noise reduction techniques previously discussed.

An examination of the block diagram for the receiver in Fig. 4-4 reveals that the next item to consider is the AGC controlled attenuation. The design details will not be presented here for want of space. The diagram in Fig. 4-9(a) is a MOSFET attenuator with the AGC applied to the upper gate. The AGC voltage control induces a variable resistance that attenuates the signal in accordance with the voltage setting. This particular attenuator has a 100 MHz bandwidth. Its control voltage range is from 0 to −5.2 V. Its sections can be cascaded to increase attenuation at the expense of an increase in noise level. The design of its circuits can be found in Motchenbacker and Fitchen[2] and Ha.[3] The equivalent resistance in the drain circuit is a variable resistor in series with the 50-Ω load. As the voltage on the AGC input changes, the drain resistance will also change. The output of each attenuator stage is between a voltage divider composed of a 50-Ω resistor and a voltage-controlled (AGC) resistor.

The main amplifier or power amplifier is shown in Fig. 4-9(b) with a comparator at the output. This is a digital receiver. The analog case is a similar version; just add another amplifier stage. A tap is performed between the first and second comparator to supply an input to the AGC amplifier.

The main amplifier has a minimum input of 1 mV and output of 100 mV at this input level. The comparator circuit consists of three MECL MC 10216 line receivers with a rise time of 3 to 4 nanoseconds. These devices operate from a single supply, which helps reduce the parts count. Local feedback (5-kΩ resistor) from the comparator output to input applied at the low-pass filter (100-pf capacitor, 75-k resistor) will compensate for baseline wander caused by long strings

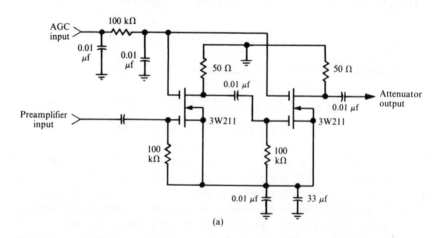

Fig. 4-9. Other receiver components: (a) Voltage-controlled attenuator, (b) power amplifier and comparator circuit, (c) AGC amplifier that drives the attenuator, and (d) power supply for the APD bias.

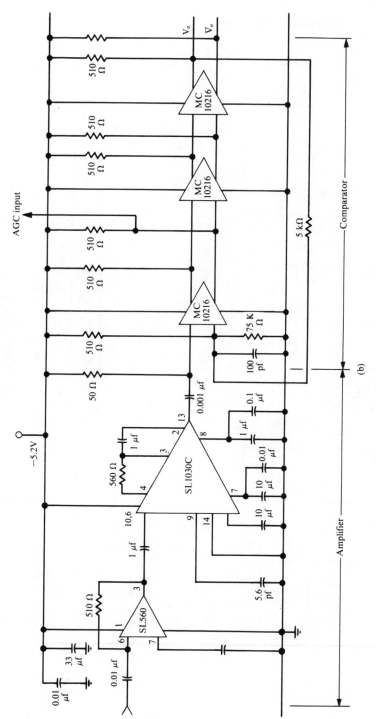

Fig. 4-9. (*Continued*)

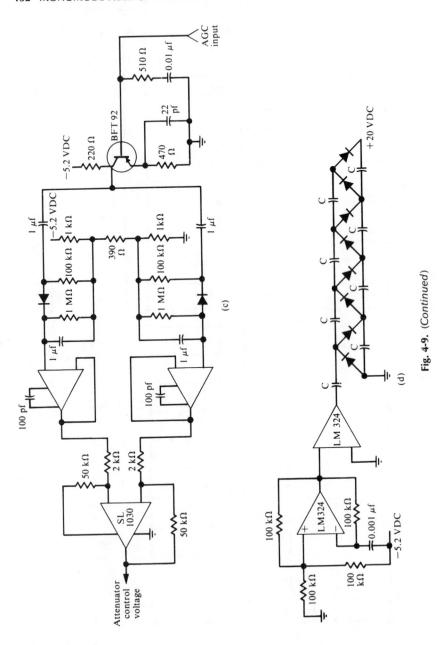

Fig. 4-9. (*Continued*)

of zeros or ones in the data stream. Improvements in this design can be accomplished through the use of GaAs logic. At the time this book was being prepared, GaAs technology was still in the research laboratories and very expensive. These devices operate in the hundreds of megahertz region.

The AGC circuit shown in Fig. 4-9(c) will produce the necessary control voltage to prevent overload of the main amplifier circuits. The first stage of the AGC amplifier BFT 92 will increase the AGC voltage to approximately 1 V_{PP} prior to detection. The detection diodes are fast switching and detect and filter both positive and negative peaks. The voltage followers at the detector outputs will maintain the voltage impressed across the detector capacitors until they are refreshed. The input impedance to the voltage followers is very high, and the capacitors and diodes must have extremely low leakage to prevent charge bleed-off in the capacitors. This AGC circuit will operate with balanced or unbalanced waveforms; i.e., it operates independently of mark-to-space ratios.

The bias supply, shown in Fig. 4-9(d), will multiply the power supply voltage up to 20 V for back-biasing the APD. It will produce 100 μA at 20 V. This is only a suggested circuit. A secondary can be provided on the power transformer that will produce biasing voltage with better regulation. Remember that changes in the back-bias will cause variations in the diode capacitance changes, C_T, and in bandwidth, and these may or may not affect the amplifier, depending on the application.

REVIEW PROBLEMS

1. For a InGaAsP/InP detector with bandgap energy from 0.73 to 1.35 eV, what is the threshold wavelength?
2. Given the wavelength 1.6 μm as the threshold wavelength: (a) What is the bandgap energy? (b) Which materials or material best fits this application?
3. Find the responsivity of the diode detector in Problem 2 using Table 4-2.
4. What is the transit time for a detector with a drift velocity of 10^5 m/sec and a drift region of 50 μm? What must the bias voltage be if the electric field is 10^6 V/m?
5. What is the modulation frequency in Problem 4?
6. For the responsivity calculated in Problem 3, bandwidth of 1.2 GHz, and a signal level of 0.1 μW, what is the SNR of the detector? Is it adequate to produce a BER of 10^{-9}? How may the SNR be improved (assuming a PiN diode)?
7. What is the minimum detectable power for an avalanche diode with a noise factor of 3, bandwidth of 100 MHz, G of 10, responsivity of 0.7, and dark current of 10 nA?
8. Redesign the preamplifier circuit of Fig. 4-9(c) to operate at +5 VDC.
9. Reduce the noise of the FET amplifier stage to approach the quantum limit to within 15 dB or less.
10. Devise another voltage-controlled attenuator but draw only the circuit.

11. Use a 3N211 as the FET input stage to the preamplifier of Fig. 4-9(c), inject local feedback from Q_4 to the upper gate of the 3N211.

12. Try to produce variation of the circuit in Problem 11 with a common-base amplifier.

REFERENCES

1. Gower, J. *Optical Communication Systems*. New York: Prentice Hall International, Inc., 1984, pp. 383–388.
2. Motchenbacker, C. D., and Fitchen, F. C. *Low-Noise Electronic Design*. New York: John Wiley & Sons, 1973.
3. Ha, Tri T. *Solid-State Microwave Amplifier Design*. New York: John Wiley & Sons, 1981.

5 | WAVEGUIDE TERMINATIONS AND SPLICES

MISALIGNMENT LOSSES

Core diameters of single-mode waveguide range from 4 to 10 μm; to keep connector insertion losses low, therefore, alignment must be in the submicron region. Moreover, lateral offsets must be kept below 1 μm and angular misalignment less than 0.5 degrees to keep these losses to acceptable levels. For a typical single-mode optical waveguide, the insertion loss can be 0.5 dB for the previously stated tolerances.

Another problem with single-mode systems—one that was of less concern with multimode systems—is reflection from connectors. Such reflections induce system degradation of laser transmitters. Index matching fluids and close waveguide face-to-face contact will reduce this type of loss.

Let us first examine the equation that governs spot size:

$$\alpha_r = -20 \log \left(\frac{2r_1 r_2}{r_1^2 + r_2^2} \right) \text{ dB} \qquad (5\text{-}1)$$

where r_1 is the radius of the transmitter pigtail end and r_2, the spot radius of the projected image on the transmission fiber.

The minimum loss for this expression occurs when $r_1 = r_2$. Note that this expression is not concerned with actual core tolerances. For single-mode sources, the core diameter will not be completely filled with light.

The discussion here will separately analyze each tolerance causing loss; the total loss is the sum of the losses provided each loss is less than approximately 1 dB. Multimode loss analysis is more complex because of the mode mixing, leaky modes, and other effects that must be considered.

The next item to consider is end separation. This is based on the assumption that lateral and angular misalignments are zero. The equation describing this behavior is as follows:

$$\alpha_{sp} = -20 \log \left[\frac{1 + 4z^2}{(1 + 2z^2)^2 + z^2} \right]^{1/2} \qquad (5\text{-}2)$$

where

$$z = x/k\eta_2 r_0^2 \qquad (5\text{-}3)$$

135

Here, x = separation, k = free-space propagation constant of light, η_2 = cladding refractive index, and r_0 = mode spot radius of the LP mode.

For variations in z due to x only, a plot showing attenuation is depicted in Fig. 5-1. The value of r_0 is approximated using the following equation established by Marcuse,[1] which has an accuracy of within 1 percent for the restruction, $0.8 \leq \lambda/\lambda_c \leq 2.4$, where λ_c is the cutoff wavelength:

$$r_0 = \left[0.65 + 0.43 \left(\frac{\lambda}{\lambda_c} \right)^{3/2} + 0.0149 \left(\frac{\lambda}{\lambda_c} \right)^{6} \right] a \qquad (5\text{-}4)$$

where a is the radius of the fiber.

The curve in Fig. 5-1 is generated for λ_c = 1180 μm and λ = 1300 nm. Equation 5-5 depicts the expression for V:

$$V^2 = a^2 k^2 (\eta_1^2 - \eta_2^2) \qquad (5\text{-}5)$$

where $V < 3$ and a = core radius.

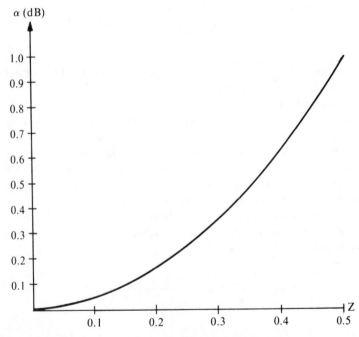

Fig. 5-1. Plot of Z versus α for Δn = 0.555% and λ/λ_c = 1.14.

The value of z can further be depicted in terms of $\Delta\eta$. First, by substituting the values of λ and λ_c into Eq. 5-4, we have the result shown below:

$$r_0 = [0.65 + 0.52 + 0.033] \, a = 1.2a \tag{5-6}$$

Equation 5-5 can be evaluated for k as shown below:

$$V^2 = a^2k^2(\eta_1 - \eta_2)(\eta_1 + \eta_2)$$
$$V^2 \approx 2a^2k^2(\Delta\eta) \, \eta_2$$

Thus,

$$k = \frac{V}{a \, \sqrt{2\Delta\eta\eta_2}} \tag{5-7}$$

By substituting Eqs. 5-6 and 5-3 into Eq. 5-2, z can now be evaluated in terms of the values found on a typical specification sheet, as shown in Eq. 5-8:

$$z = \frac{x \, \sqrt{2\Delta\eta\eta_2}}{V\eta_2 1.44a} \tag{5-8}$$

For $V = 2.405$,

$$z = 0.29 \left(\frac{x}{a}\right) \frac{\sqrt{2\Delta\eta}}{\eta_2}$$

Evaluating 5-8 as a function of separation to core radius and separation as the difference in refractive index between this core and clad is a more meaningful approach for the designer. Problems 3 and 4 in the review questions address these issues.

Let us now consider the radial misalignment and offset loss given in Eq. 5-9:

$$\alpha_{\theta d} = 4.34 \left[\left(\frac{d}{r_0}\right)^2 + \left(\frac{\pi\eta_2 r_0\theta}{\lambda}\right)^2 \right] \tag{5-9}$$

where d = lateral offset and θ = radial misalignment. For the condition stated for the evaluation of Eq. 5-6, this equation normalized in terms of the offset and the core radius is expressed as follows:

$$\alpha_{\theta d} = 4.34 \left[\frac{k_r^2 a^2}{K_1^2 a^2} + \left(\frac{\pi \eta_2 K_1 a \theta}{\lambda} \right)^2 \right]$$

where $k_r = d/a$ and $k_r < 1$. Then,

$$\alpha_{\theta d} = 4.34 \left(\frac{k_r^2}{K_1^2} + \pi K_1^2 A_c \left(\frac{\theta}{\lambda} \right)^2 \right] \tag{5-10}$$

Equation 5-10 has some interesting ramifications. It indicates how loss is dependent on core area A_c and the ratio of wavelength and angular misalignment. As one might intuitively deduce, as wavelength increases, larger angular misalignments can be tolerated. The constant K_1 must be evaluated for ratios of λ/λ_c using Eq. 5-4. At a wavelength of 1300 nm, Eq. 5-10 can be evaluated as shown below:

$$\alpha_{\theta d} = 4.34 [0.69 k_r^2 + (8.15 \times 10^6 A_c \theta^2)] \tag{5-11}$$

The total loss of the joint is given in Eq. 5-12:

$$\alpha = \alpha_r + \alpha_{sp} + \alpha_{\theta d} \tag{5-12}$$

The losses described for a joint are the same for a connector or splice. In splices, however, the misalignments are usually smaller because positioning is optimized and the joint is permanent, whereas in connectors the joint can be recoupled many times and each time the loss will vary slightly because of mating variations. The variations may be 0.05 dB. Manufacturers' tolerances must be considered when calculating joint losses. Therefore, the loss equation can be better characterized by including the reconnection loss that results in additions to all three of the other losses. These additional terms are shown in Eq. 5-13:

$$\alpha = (\alpha_r + \Delta\alpha_r) + (\alpha_{sp} + \Delta\alpha_{sp}) + (\alpha_{\theta d} + \Delta\alpha_{\theta d}) \tag{5-13}$$

When liquid couplants are used, the additional terms in this equation are small enough to be ignored. For calculating the worst-case coupling loss equation, manufacturers' specifications usually give both the worst-case loss and the reconnecting loss so that α is the sum of the worst-case loss plus some value $\Delta\alpha_m$, which represents the reconnection loss. For example, if the manufacturer assigns the connector a loss of 0.5 dB and the variation in loss upon mating is 0.05 dB, then the worst-case total loss is 0.55 dB. The network discussion in Chap. 8 will consider worst-case loss budgets and produce the worst case when all tolerances are considered that increase loss.

Splices are a very important issue to consider because long-haul (i.e., 100- to 150-km) networks may use as many as 100 to 150 of them. As one can visualize, a network with a loss of only 0.1 dB/splice can be equivalent to 1 kilometer of waveguide with a loss of 0.1 dB/km, which is not unusual for single-mode cable. Therefore, precision positioning is absolutely imperative when making single-mode splices. Connector loss is larger, but often only two are necessary. For the link with 150 kilometers of waveguide and 0.1 dB/splice, the loss is 15 dB, and the connector loss will usually add only 10 percent at most to that figure for a total loss of 16.5 dB end-to-end.

CONNECTORS

The first section of this chapter dealt primarily with the loss mechanism of connectors and splices. The connectors for single-mode waveguide termination will now be addressed. Often devices such as couplers, switches, multiplexers, etc., are spliced into a system to minimize losses. Splices make system troubleshooting rather difficult, however, and for local area network applications, they are not an adequate solution. Long-haul networks such as telephone trunks may require no more than a single connector at each end of the network.

Connectors must be easily connected and have a high degree of positioning accuracy if they are to be successful in the field. To keep loss down to approximately 1 dB at a 1300-mm transmission wavelength, the accuracy required is a face-to-face separation of less than 30 nm, an angular alignment of less than 1 degree, and a lateral offset of less than 1.5 μm. Geometric variations in core dimensions or core eccentricity is between 0.2 to 0.5 μm. Many connectors exhibit losses as low as $\frac{1}{2}$ dB, with mating and unmating loss of 0.2 dB or less after 500 to 600 cycles. Temperature excursions between $-25°C$ and $80°C$ will cause an additional loss of less than 0.2 dB. All of these loss mechanisms must be considered when making an estimate of the link loss budget. The loss budget will be addressed in Chap. 8 when system aspects and applications are considered.

Assembly of connectors will not be considered in any detail because the technology is rapidly changing, and today's connector may be tomorrow's obsolete part. The generalities of connector loss mechanisms will not change, but accuracy will no doubt improve as manufacturing techniques improve. Loss for terminations will eventually be reduced to $\frac{1}{4}$ dB. As one may recall, Swiss watchmakers have inscribed the Lord's prayer on the head of a pin. This feat required the same type of precision machining needed for machine precision holes in connectors.

Connectors using ferrule-type alignment can be constructed as either cylindrical or conical types, as shown in Figs. 5-2(a) and 5-2(b). These techniques for common-mode fiber connectors will be examined and their application to single-mode fiber then considered.

The hole required in the ferrule to accommodate normal 125-μm waveguide with a waveguide tolerance of ± 3 μm is 128 μm. Lateral misalignment can reach a maximum of 2.5 μm. If all the misalignments for a 125-μm (outside-

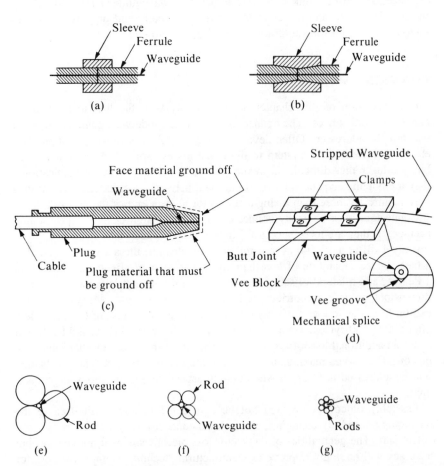

Fig. 5-2. Mechanical splice techniques: (a) Cylindrical splice with ferrule, (b) sleeve with tapered ferrules, (c) plug with a machined surface fit for single-mode connector, (d) vee block and butt splice, (e) three-rod splice, (f) four-rod splice, and (g) six-rod splice.

diameter) waveguide are added up as shown below, the worst-case misalignment will be 11 μm:

Misalignment	Worst case
Hole in ferrule	2.5 μm
Waveguide tolerance	0.5 μm
Ferrule-to-sleeve	8.0 μm
Total	11.0 μm

The mean lateral misalignment for the cores is approximately 4 μm. For a multimode, 50-μm core waveguide, the loss is approximately 0.2 dB and the worst case (11-μm misalignment), 1 dB. For the single-mode waveguide, however, a 4-μm misalignment represents a loss of 5 or 6 dB, and the worst-case misalignment will result in a loss of all optical power across the connector. The core diameter may be only 10 μm.

The conical-rung connector shown in Fig. 5-2(b) has superior alignment characteristics. These conical ferrules can be constructed with a slide fit into a biconical sleeve. This type of connector also prevents the wear exhibited by the cylindrical types, which eventually become loose and offer poor loss characteristics. To improve on the tolerances, the conical ferrule can be ground to produce a closer tolerance fit, as shown in Fig. 5-2(c). The normal misalignments and angular misalignment must be considered when using this connector system. Connector losses with this system are less than 0.5 dB for single-mode waveguide. Since the variations due to mating are within 0.1 dB, these connectors exhibit a worst-case loss of 0.6 dB after 500 cycles of mating and unmating.

Fresnel reflections are another source of loss that can become substantial if the waveguide ends are not butted tightly—as large as 0.3 to 0.6 dB. These losses represent interference generated at the end of the waveguide and are similar to the reflections in any other electromagnetic transmission medium. Certain gels are manufactured that will improve the coupling between butted joints. A common problem with these matching gels is that contaminants get into the joint and cause reflections or loss of signal. These couplants must not exhibit any chemical breakdown that causes deterioration of the joint. The connectors must also be mechanically sound so that they carry the stress rather than the waveguide itself.

As one may observe, large numbers of connectors in a system are equivalent to several kilometers of optical waveguide. In telephone systems, for example, patch panels are necessary to provide the necessary connectivity between inside plant (electronics) and outside plant (transmission lines). The two plant facilities are administered and maintained by different organizations. When connections are changed only rarely, splices can be substituted for connectors. Splices exhibit much lower loss than demountable connector terminations—usually about a magnitude less—but they require equipment and trained service personnel to install them.

SPLICES

Splices are usually of three types. The first is the fusion splice, which is characterized by very low loss and restores the strength of the joint to approximately 70 percent of the original waveguide. The two ends to be spliced are heated to a viscous liquid state and forced to form a continuous medium. The fusion joint must not distort the waveguide, but a slight distortion is unavoidable. This produces some loss, usually less than 0.1 dB. The faces of the waveguides to be joined must be cleaned so that neither has any protrusions and must be perpendicular to the core axis to a tolerance of less than 1 degree, which is easily obtainable. CO_2 lasers have been successfully used as heat sources since they produce large quantities of highly localized heat. Arcs, oxyacetylene, and hot wires have also been used as heat sources, but localized laser heating produces the highest mechanical strength. The splice joint exhibits many of the losses caused by misalignment with the exception of separation, but such losses are much smaller and thus account for the overall low loss.

The second type of splice is the vee-groove splice without fusion. These splices are characterized by higher loss than fusion splices (0.2 to 0.3 dB). The joint is composed of the faces of the two waveguides fitted into a vee groove and butted, as shown in Fig. 5-2(d). Vee grooves that are precision machined yield low misalignment loss. Separation loss will constitute one of the dominant losses, but couplant gels can be used to reduce this component. The joint must be strain relieved to prevent any movement of the waveguides. Since it is extremely fragile, strength members should be added to prevent joint movement. Variations of vee-groove joints include three-, four-, and six-rod splices, as shown in Figs. 5-2(e), (f), and (g). Figure 5-2(e) has three contact points that hold the waveguide captive; the other two, four and six, respectively. Since these rods can be ground to within 0.2 μm rather easily, a high degree of precision alignment is possible. The rods can be either precision-ground steel or glass. Note that the size of the splice for the six-rod splice is rather small compared to that for the three-rod version.

The last type of splice to consider is the mechanical splice. After assembly, this type of splice usually requires a cam arrangement for final alignment with the splice under power. Such splices often have low loss similar to that of fusion types.

Fusion splices require technical people with high skill levels since field splicing with fusion techniques under adverse conditions is rather difficult. Mechanical splices, on the other hand, are more adaptable to field implementation. The reader should do a literature search before deciding on a splicing technique because of the rapidly changing technology. The mechanical configurations may change to improve accuracy, but the methodology will involve one of the three types just discussed. For short-distance links, such as those used in local-area networks, mechanical splices are adequate because only one or two are usually necessary. When long-haul transmission is implemented, large numbers of splices are necessary; therefore, losses must be kept as small as possible. For example, 100 kilometers of optical cable will require a splice for every 1 or 2 kilometers. One hundred splices will result in 10 dB of loss, or more, for mechanical splices and about 1 or 2 dB for fusion splices. We will deal with these issues during the discussion of a loss budget for systems in Chap. 8.

COMMON COMMERCIAL CONNECTOR DESIGNS

Figure 5-3(a) is the schematic of a biconic connector showing the conic alignment structure. A connector of this type manufactured by Dorran Photonics Inc. has a loss of less than 1 dB.

This loss estimate is calculated using statistical techniques. The three interfacing surfaces to consider for the connector shown in Figure 5-3(a) are (1) the core/taper offset of plug 1, (2) the core/taper offset of plug 2, and (3) the biconical sleeve offset. Typical values for mean and standard deviations for the offsets between adjacent waveguides are 0.5 and 0.2 μm, respectively. Using a Monte Carlo simulation technique, the mean and standard deviations for waveguide-to-waveguide offset are calculated to be 0.9 and 0.4 μm, respectively. Calculation of the loss at 1300 nm will produce a mean deviation of 0.3 dB and a standard deviation of 0.1 dB. Actual measured values are 0.25 dB mean deviation and 0.12 dB standard deviation. Spring-loaded connectors with faces polished to a flatness less than 0.25 μm produce these results. The flatness produces a return loss with an average value of 30 dB.

It should be noted that high-precision fabrication techniques are required as well as a high-quality inspection process. These connectors will most likely need to be matched pairs for losses lower than 1 dB, and this condition will drive up costs considerably. For the reader interested in pursuing the analysis of these connectors further, see Ref. 2.

The connector for single-mode waveguide produced by the Fiber Optic Communications Development Division of NEC is shown in Fig. 5-3(b). The insertion loss and repeatability loss due to mating and unmating are 0.5 dB and 0.2 dB, respectively. The measurements were made at 1300 nm wavelengths. This connector is equipped with a triangular alignment sleeve, a method similar

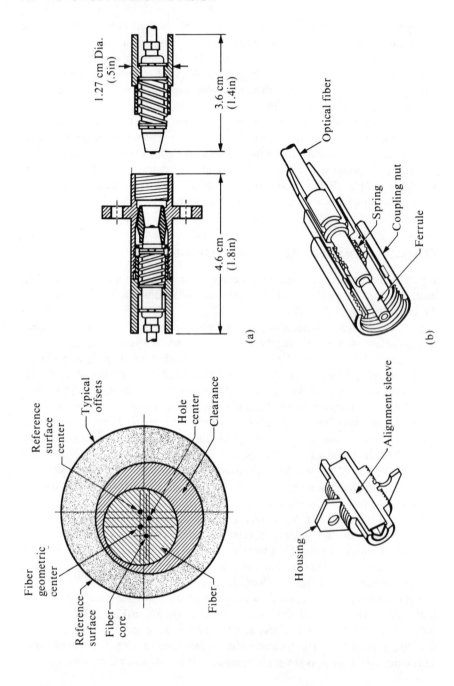

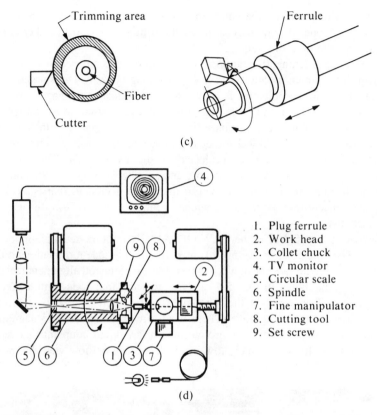

1. Plug ferrule
2. Work head
3. Collet chuck
4. TV monitor
5. Circular scale
6. Spindle
7. Fine manipulator
8. Cutting tool
9. Set screw

Fig. 5-3. Typical single-mode connectors: (a) Biconic connector with typical offsets, (b) NEC connector structure, (c) ferrule trimming technique, and (d) ferrule cutting machine.

to using a three rod captivation technique. The fact that the sleeves are press-formed makes mass production of the device feasible.

For worst-case loss, the components can be approximated as follows: 0.3-dB Fresnel loss, 0.1-dB intrinsic loss (due to mismatch in waveguide parameters), 0.3-dB loss from misalignment, and 0.3-dB loss from tolerances in assembly. The tolerance contribution is estimated based on present machining technology.

A high degree of core-to-core alignment accuracy is possible because the ferrule is machined to the desired diameter with the core center as the reference. Therefore, both core and cladding concentricity error and cladding outside-diameter deviations are compensated to produce a cylindrical core-centered surface. The alignment sleeve is smaller than the outside diameter of the ferrule and balances elasticity with rigidity. The two ferruled ends are spring-loaded to maintain a good butt joint and keep lateral misalignment low.

The trimming technique for the ferrule is depicted in Fig. 5-3(c). The ferrule is held stationary while the cutter moves around it to cut it to a near perfect cylindrical shape. As one may surmise, the trimming system is rather complex because of the precision required.

The cutting machine, diagrammed in Fig. 5-3(d), is composed of a microscope system, TV camera/monitor, and x-y positioning system. The spindle interior is equipped with an objective lens system. The optical core image is produced by illuminating the optical core and projected and focused on a circular scale. This scale is set inside the spindle in a position concentric with the spindle rotation axis. Once the optical waveguide image has been focused on the circular scale, both are magnified and projected on the TV monitor. The positioning device is then adjusted so that the core and spindle rotation axis are concentric and the ferrule can thus be trimmed to maximum accuracy, that is, within an average concentricity of 0.1 μm, which produces losses of 0.5 dB at 1.3-μm wavelengths.

Shown in Fig. 5-4(a) is the TRW OPTALIGN multimode/single-mode connector. The connector in Fig.5-4(b), used for single-mode terminations only, is fabricated with the same design principles. It uses a four-rod alignment technique similar to that shown in Fig. 5-4(c). Note that the four rods form a precision vee groove. The bends in the lengthwise section force the two ends of the waveguide into the vee groove. These bends are approximately 15 degrees. To reduce total loss, the bent vee-groove tube has a liquid couplant that reduces separation loss to an extremely low level, depending on how well the index of

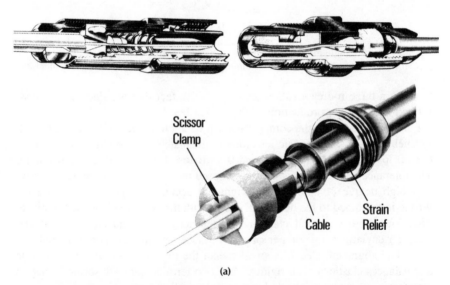

(a)

Fig. 5-4. OPTALIGN connectors: (a) TRW OPTALIGN multimode/single-mode connector, (b) OPTALIGN single-mode connector, (c) single-mode four-rod alignment technique, (d) matching couplant reservoir, and (e) mechanical dimensions of single-mode connector.

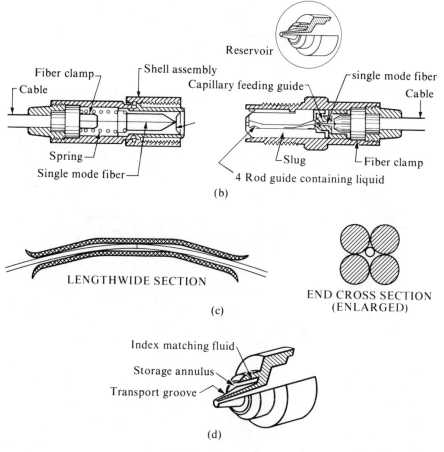

Fig. 5-4. (Continued)

refraction of the two waveguides and the couplant match. Typical loss for this connector waveguide with a 4.4-μm core and 80-μm clad is 1 dB, but only the worst-case loss given on the manufacturers specification sheets should be used when designing. If the worst-case figure is unavailable, double the typical value to insure that the system operates correctly.

The liquid couplant virtually eliminates Fresnel-reflection loss and angular-displacement loss during connector mating. The most critical precision part in the connector must be the four-rod glass tube, which is fabricated from precision glass rods that are fused. These can easily be made to within 0.5-μm tolerances. Since the end faces of the waveguide require neither polishing nor the use of epoxy, assembly does not require higher-level skills. Another feature of this type of connector is its ability to be reused if the waveguide is damaged during

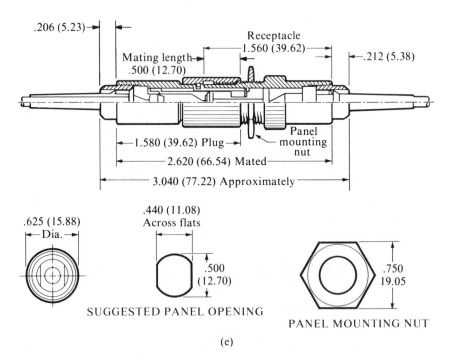

.206 (5.23)

Receptacle
1.560 (39.62)

.212 (5.38)

Mating length
.500 (12.70)

1.580 (39.62) Plug

Panel
mounting
nut

2.620 (66.54) Mated

3.040 (77.22) Approximately

.625 (15.88)
Dia.

.440 (11.08)
Across flats

.500
(12.70)

SUGGESTED PANEL OPENING

.750
19.05

PANEL MOUNTING NUT

(e)

Fig. 5-4. (*Continued*)

assembly. Figure 5-4(a) indicates that it is cylindrical in shape and consists of a plug and receptacle. The receptacle end contains a waveguide aligned in a vee-groove guide that is mounted in a plastic plug. When the plug and receptacle are coupled to make a connection, the plug pushes the piston back against a coil spring, exposing the waveguide that enters the alignment guide. Prior to assembly, the ends of the waveguides are cleaved to give them a surface perpendicular to the core axis.

This connector is also equipped with a reservoir filled with the matching couplant, as shown in Fig. 5-4(d). Fluid from the reservoir is fed to the transport groove from the annular cavity through capillary action and travels up the guide through the flared ends.

Figure 5-4(e) shows various mechanical dimensions of the connector. The maximum loss is 1.5 dB, 1.0 dB being typical. Mating and unmating of 200 cycles caused a 0.2-dB increase in connector insertion loss. Temperature excursions between $-55°C$ and $+85°C$ produce a loss increase of 0.1 dB after a number of temperature cycles (approximately five).

One of the problems with connectors that use liquid couplants is the possibility of contamination of the matching fluid. To prevent this problem, the connectors must always have a dust cap installed. Also, after frequent instances of mating and unmating, the reservoir may become depleted of couplants, and this, of course, will cause a large rise in insertion loss. For controlled environments such as offices, no serious problems should occur if proper precautions are taken, but in harsh environments such as steel mills, foundries, etc., connector loss may increase after only a few matings and unmatings even with precautions.

Lensed connectors are another possible technique for terminating waveguides. Expansion of the beam diameter makes offset not as stringent but angular alignment becomes critical. Angular alignment is easier to control than the loss components that contribute to offset. One of the considerable advantages of lensed connectors is that filters, beam splitters, etc., may be inserted in the expanded beam to perform more complex functions. Positioning of waveguides and optical elements is also not an easy task. Since dust has less affect on connector performance with lens-expanded beams, they may be used in dusty or industrial environments. Typical loss for these connectors is less than 1 dB. They do not exhibit separation loss, which is a source of harmonic distortion in connectors that do not use index matching fluid.

Other connectors will appear on the market in the near future. This section will give the reader a sampling of what is available and how present technology compares with older established multimode connectors. The captivation method involves a vee-groovelike alignment whether the vee grooves are single or multiple. Vee grooves are usually easy to machine with high tolerances.

COMMERCIALLY AVAILABLE SPLICES

Previously, this discussion has addressed techniques for splicing, such as fusion, mechanical (including adjustment for minimum loss), and vee groove. Here we will deal with the specific commercial splices available. In particular, only mechanical splices will be covered here.

The OPTASPLICE is a splice based on the four-rod guide technology of Fig. 5-4(c). This splice is shown in more detail in Fig. 5-5(a). Based on practical experience, such splices can be aligned under power, and fast-drying cement can be used to assemble them in 10 to 15 minutes. Both ends of the waveguides to be joined are inserted in the four-rod guide, and optical power is passed through the joint. Prior to inserting the waveguide ends, the guide is filled with an epoxy that is sensitive to ultraviolet light. An optical power meter or OTDR is connected at the end of waveguide opposite the power source. The two ends in the splice guide are adjusted until the peak power is observed, as measured by the power meter, or until the loss is minimized, as observed on the OTDR. The splice is held rigidly, and a flash unit exposes the joint to ultraviolet light. This joint will then set in minutes. Some loss will occur after the epoxy is set because fast drying usually induces shrinkage; slight movements may therefore

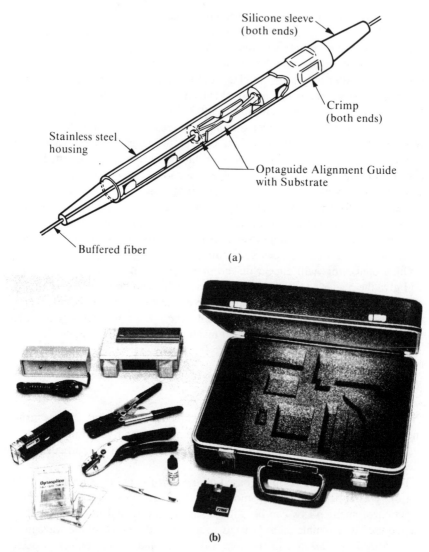

Fig. 5-5. TRW four-rod splice: (a) TRW OPTASPLICE, and (b) OPTASPLICE module SK-3 splice kit.

occur in the joint, causing a slight loss increase. Before attempting to make these types of splices, consult the manufacturer for any new types of epoxy available.

Some splice techniques require no epoxy or fusing but a purely mechanical means of assembly. One such device, the Elastomer splice produced by GTE, has a captivation technique similar to that of the NEC connector, i.e., it uses a triangular alignment sleeve. This splice allows mechanical alignment after it has been assembled using a cam arrangement. It also comes in nonadjustable types. Dry splices of this type have losses of approximately 1 dB; if liquid couplant is used, the loss is approximately 20 percent lower. Only the OPTASPLICE is described here in any detail.

The splicing kit shown in Fig. 5-5(b) consists of the following: splice kit, NONIK stripper, hand-held cleaving tool, Norland optical adhesive number 61, holding fixture, UV flash lamp, and sleeve crimping tool. The splice kit includes the splice alignment guide, two rubber sleeves, and stainless steel tubular housing, all of which are shown in Fig. 5-5(a).

Splices are installed in the following manner: If the waveguide end is frayed, cut off just enough to remove the frayed portion. Slip the silicon rubber sleeves and metal housing over the waveguides to be spliced. Strip an inch of buffer from the end of each waveguide to be joined, using either mechanical or chemical means (chemical preferable). The ends of the waveguide are then cleaved to produce smooth end faces that are perpendicular to the core axis. These cleaved ends should be inspected for chipped ends or cracks; should any appear, the stripping and cleaving should be redone.

A microscope is useful for a high magnification means of inspection. Adhesive is applied to the end of each waveguide (Norland No. 61), and the ends are stepped into the alignment guide. The fixture provided is used to retain the waveguide alignment. Optical power may be passed through the joint, with a meter monitoring the other end. Adjust the waveguides in the fixture for peak power at the meter. The UV flash is then applied over the glass guide to induce adhesive curing. It takes approximately 5 minutes, but the splice shouldn't be removed from the fixture for 10 minutes lest environmental factors retard curing time.

After the adhesive is cured, the splice is removed from the fixture and the rubber sleeves moved along the cable and slipped into the tabs provided on the guide. The next step is to slide the metal housing over the splice and rubber sleeves. This housing is then crimped at both ends with the tool provided. For harsh environments, the splice can be given additional treatment, but at this point it has higher mechanical strength than the fiber-optic cable itself.

These splices have a rather small form factor, being $\frac{1}{8}$ inch in diameter and $2\frac{1}{2}$ inches long. This makes them ideal for use where large numbers of them are needed in a confined space. Often they can be mounted in organizers, and cable with multiple waveguide can be strain relieved by the organizer.

These splices are particularly useful when field installations are necessary. In field installations, an order wire may be provided to assist in the installation. Active alignment is not required but does reduce loss slightly. Reference 3 provides further information on other types of splices.

For short-haul data links where loss in single-mode fiber-optic cable splices can be tolerated—i.e., installations with unrepeated distances of 10 to 15 km—in-line connectors can be used. The advantage of these splices is that they can be easily disconnected for troubleshooting purposes. Often they exhibit losses as low as those of many of the mechanical splices. They are inherently more expensive than fusion or mechanical splices, however, because they require two connector plugs and a sleeve. The cost is usually 10 to 20 times that of other splicing techniques.

For installation on a pigtail such as in transmitters, receivers, or couplers, in-line connectors are used if the joint is not fused. They thus have a great deal of utility. Single-mode transmitters often come with pigtails rather than connectors because, for long-haul operations, fusion splices are frequently made along the entire link to keep losses very low.

One such connector is manufactured by Interoptics. Only its performance will be discussed here, in the hope it will give the reader some insight into what is available in the marketplace. The loss of this connector dry is less than 1 dB; with a turnable ferrule, however, which is provided, the loss can be as small as 0.2 dB. Repeatability with an unmatched pair is 0.2 dB and with a matched pair, 0.15. This device will accommodate 4-μm/80-μm, 6-μm/125-μm, 8-μm/125-μm, and 9-μm/125-μm waveguides, these being core/clad dimensions. The loss figure of 1 dB dry was measured at 1300 nm, with a 8-μm core/125-μm clad waveguide (untuned). The ferrule provides 12-position tuning. The assembly of the termination can be accomplished in less than 15 minutes.

In-line connectors are ideal where a LAN is installed in an office environment. Extending or moving the network cables is a simple matter of disconnecting the joint and adding new sections or rerouting the cable (if a move is required), that is, providing the optical power budget and time delay budget (both of which are discussed in Chap. 8) are adequate to prevent system performance degradation. One of the problems with these joints is the distortion of analog optical signals. However, most of the LANs discussed in Chap. 8 will be concerned with digital baseband signals. These signals can also be distorted, but the distortion seldom presents a serious problem for applications of this kind.

One application that deals with analog signal distortion is that of single-mode fiber optics to broadband cable plants. This application has a tremendous potential in the CATV industry. Although the main cable is broadband coaxial, some aspects of single-mode replacement, drops to subscribers, or combinations of the two, will be addressed in Chap. 8. As one might surmise, connectors and splices will have a major impact on the implementation of any proposed plant. They are of paramount importance simply because of the large number needed to implement a design. Large losses result in larger numbers of repeaters. Since the latter are both costly and a potential source of network failure, the result is poorer reliability. Imagine the reaction of subscribers tuned into the

Super Bowl football game if the cable plant goes down during a long run in the last minute of the game when the score is tied. When dealing with the public, such matters may not seem worth getting excited about, but people can and do respond to poor reliability in CATV, cars, appliances, etc.

At the time this book was being written, multiple-waveguide connectors were expensive and difficult to obtain. Alignment problems multiply because of the tolerances involved. The telephone industry uses connectors that resemble splices, i.e., they are not made to mate and unmate a limitless number of times. The local area network industry demand is for one- and two-fiber connectors, whereas many of the larger networks requiring backbone design require four-, six-, and eight-fiber connectors. Fortunately, a great many of the applications requiring two-, four-, six-, and eight-waveguide terminations are for short-haul cable plants. The losses of these connectors will be larger than those of the single-position type, usually over 1 dB.

Let us examine some of the tolerances based on what present technology can produce. Multiple-position connectors will not be discussed on a theoretical basis only. The presentation here aims to give the reader some insight into what performance may be expected from these connectors. Tolerances may be selected, of course, so that a worst-case connector will never be fabricated, but selection—i.e., matching parts to reduce total error—is an expensive means of fabrication; in fact, the technique borders on producing hand-crafted connectors.

A dual-waveguide connector will be the only type considered. Problem 10 in the Review Questions addresses larger numbers of waveguides. If the tolerances of the two pins and the spacing between them are all worst-case, the tolerances can cause offsets that are twice those of a single connector (as much as 1 μm). Also to be considered is angular displacement; e.g., the two pins bend slightly when pushed into the connector. As one may observe, these tolerance variations make it difficult to achieve an insertion loss of less than 1 to $1\frac{1}{2}$ dB.

To analyze a multiple-position (pin) connector, one must examine the geometry of the configuration, as in Problem 10. Note that the error between the pins that form diagonals across the connector constitute the worst case. The diagonals can have tolerance errors large enough to cause losses exceeding 2 dB. Because of space limitations, such analyses will be left to the reader. They are not trivial by any means; a knowledge of minimum matching tolerances for connector patterns is required. Also, most thermal coefficients are sufficiently different to cause loss as a result of large temperature excursions.

Lenses in multiple-position connectors can reduce stringent alignment requirements. Since very little information was available at the time of this writing, it will be necessary for the reader to research the topics more fully. Until the technology has fully matured—most likely within the next 8 to 10 years—rapid changes in termination components will continue to occur. New connector machining techniques, longer wavelength operation, and perhaps materials research will produce the necessary breakthroughs to reduce the cost of components.

Longer wavelength operation in the 3-μm region will allow larger waveguide cores to operate single-mode. This would alleviate close tolerances required at 1300 nm, which will provide lower loss connectors at reasonable cost.

REVIEW PROBLEMS

1. Derive the expression for minimum loss in Eq. 5-1.
2. Plot Eq. 5-1 for r_1/r_2 vs α_r, with the values of $r_1 = \pm 50$ percent of r_2.
3. Plot Eq. 5-8 as a function of $x = m$ percent of a, $\Delta n = 0.55$ percent, and $n_2 = 1.53$.
4. Plot Eq. 5-8 as a family of curves, with (x/a) having the values, 0.2, 0.5, 0.8, 1, 2, and 3, for refractive index variations from $\Delta = 0.1$ to 1 percent.
5. What are your conclusions about the plots in problems 3 and 4?
6. Calculate the total connector loss for various values of offset assuming that $\theta = 0$; then assume that $k_r = 0$ and vary θ. How does each assumption affect the loss?
7. Using a connector fabricated with a three-rod captivation technique, what is the maximum tolerance allowed to keep offset below 0.2 μm for each of the rods, assuming that the ferrule is perfectly centered?
8. If the rods of the three-rod connector are replaced with precision ball bearings, what must their tolerance be if offset is to be kept below 0.2 μm?
9. Devise a connector with four precision ball bearings to captivate the ferrule in question 8.
10. What are the losses for a four-waveguide connector, i.e., four-position?

REFERENCES

1. Marcuse, D. *Bell System Technical Journal* 56(5):703–718 (1977).
2. Fiber Optic Couplers, Connectors, and Splice Technology. *SPIE Proceedings*, vol. 479, May 1–2, 1984.
3. Baker, D. *Fiber Optic Design and Applications*. Reston, VA: Reston Publishing Co., 1985.

6 | INTEGRATED OPTIC AND NETWORK COMPONENTS

This chapter is dedicated to fiber-optic component use in integrated optic and network implementations. The optical components that may be mounted on a substrate to form subsystems are switches, couplers, multiplexers, filters, optical flip-flops (formed from switches), wavelength division multiplexers, angular division multiplexers, and many other devices that combine optical and electronic parts. This chapter is dedicated to single-mode components only, which is, in all probability, the direction that fiber-optic technology is taking.

The material here will give the reader some insight into the complexities of these components. The physics involved will be covered in a rather superficial manner for the most part because new substrates and rapid changes in technology would require that the text be rewritten before publication. The functionality of these devices, however, is of paramount importance and upgrades in their performance can be expected in the future.

FIBER-OPTIC SWITCHES AND COUPLERS (ACTIVE)

Because of the need to perform logic functions, perhaps one of the most important components is the fiber-optic switch. Switches can be categorized into several classes depending on the control mechanism. The type used in multimode and single-mode technology—which has limited use because of its slow switching speed—is the electromechanical switch. Its switching speeds are similar to those of electrical relays. As in these latter devices, a mass must be moved with the use of electrical energy. The mechanical circuit forms analogs of R, L, and C components in electrical systems with large values that result in large time constants.

The next class of switch operates at high speeds because no moving mass is required. Electric, magnetic, or ultrasonic signals alter the properties of the waveguide materials that block or divert the optical signals. This type of switch will be examined in the greatest detail.

A third category of switches is similar to the second. The optical properties of the waveguide are not altered, but the switching is performed by a secondary material, for example, a liquid crystal performing a blocking action on optical signals.

The last category will not be covered because it is trivial compared to the other three, consisting of manually operated switches initiated by springs, humans, or levers. The reason this type of switch is mentioned is that many machines—copy machines, for example—perform mechanical logic functions but do not use microprocessors because their logic is so simple. Instead, they use plastic, nylon, and stamped metal parts to keep down cost. Also, very small fluidic networks may be used in some applications. Therefore, this third class of switch may be used in a host of appliances, office machines, and similar applications. From the author's experience with copy machines, a microprocessor board that costs approximately $300 would not be used because a chain drive fitted with cams operating microswitches and a mechanical counter proves more cost effective.

Let us examine some of the switches in the first category, several of which are shown in Fig. 6-1. First we must develop a method for dealing with these switches in a mathematical form. Those shown in Fig. 6-1(a) and (b) are solenoid-actuated devices. The solenoid may have a movable core that acts against spring tension to produce movement. The differential equation for the mechanical movement in switch 6-1(a) can be written as follows:

$$M \frac{d^2x}{dt^2} + c_f \frac{dx}{dt} + k_s x = F(t) \qquad (6\text{-}1)$$

The first term is force due to the mass, M; the second term, where c_f is the friction coefficient of damping, results from viscous damping; and the third term, where k_s is the spring modulus, is the result of spring tension. The system is actuated by a forcing function, $F(t)$, which will be a step function for these switches, i.e., they will be off or on. The equation for a simple RLC circuit is given by Eq. 6-2, which uses Kirchhoff's second law, i.e., the sum of the voltage drops around a loop are zero:

$$L \frac{di}{dt} + \frac{1}{C} \int_0^t i \, dt + iR = v(t) \qquad (6\text{-}2)$$

If Eq. 6-2 is differentiated and the terms rearranged, we have Eq. 6-3, which is analogous to Eq. 6-1:

$$L \frac{d^2i}{dt} + R \frac{di}{dt} + \frac{i}{C} = \frac{dv}{dt} \qquad (6\text{-}3)$$

Here mass M is equivalent to inductance L, c_f is analogous to R, and k_s is similar

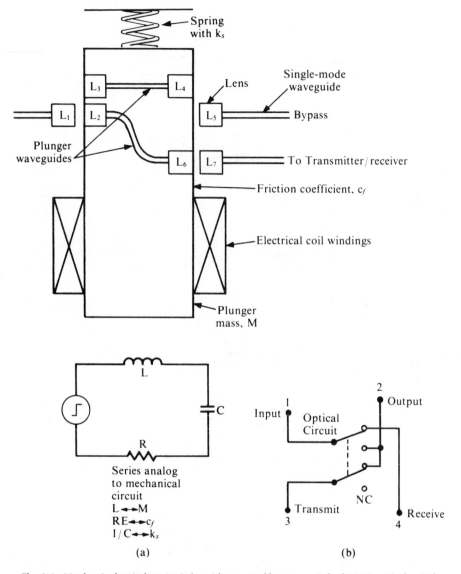

Fig. 6-1. Mechanical switches: (a) Solenoid-operated bypass switch, (b) 2×2 optical switch circuit, (c) galvanometer switch, (d) prism switch, and (e) prism switch operator.

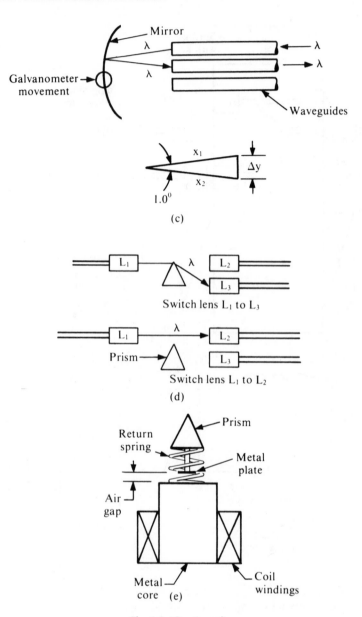

Fig. 6-1 (*Continued*)

to i/C. If Eq. 6-2 is written using the Laplacian operator, S, the result will be the unit step-forcing function in Eq. 6-4:

$$LSI(S) + \frac{I(S)}{SC} + I(S)R = \frac{V}{S} \tag{6-4}$$

Then,

$$I(S) = \frac{V/L}{S^2 + \left(\dfrac{R}{L}\right)S + \dfrac{1}{LC}}$$

Writing the denominator in a more familiar form for electrical engineers, we have Eq. 6-5.

$$I(S) = \frac{V/L}{S^2 + 2\delta\omega_n S + \omega_n^2} = \frac{V/L}{(S^2 + \delta\omega)^2 + \omega_n^2(1 - \delta)} \tag{6-5}$$

The roots of the characteristic equation (the denominator) are as follows:

$$S = -\frac{R}{2L} \pm \sqrt{\frac{R^2}{4L^2} - \frac{1}{LC}}$$

where:

$$\omega_n = \frac{1}{\sqrt{LC}}$$

$$\delta = \frac{R}{2}\sqrt{\frac{C}{L}}$$

Substituting mechanical variables, M, C_f, and k_s, we have

$$\omega_n = \sqrt{\frac{k_s}{M}}$$

$$\delta = \left(\frac{C_f}{2}\right)\left(\frac{1}{\sqrt{k_s M}}\right)$$

The time response to the system will depend on the numerical values that affect the damping ratio, δ. The four cases for the damping ratio are shown in Table 6-1. Note that, under certain conditions, the solenoid plunger will oscillate; correct values for k_s, M, and C are therefore essential for good performance. Note also that the switch is equipped with lenses L_1 through L_7 to expand the beam; this makes alignment of the plunger waveguides less critical.

The switch shown in Fig. 6-1(a) functions in the following manner: The plunger is forced against the spring to the position shown in the diagram by electrical energy. The channel waveguides connect a receiver or transmitter into the optical circuit. If an electrical failure occurs, the spring will force the plunger into a bypass condition.

The switch shown in Fig. 6-1(b) is useful as a bypass switch in optical networks because it will bypass both a transmitter and a receiver; consequently, only one device is required as compared to Fig. 6-1(a), where two devices are required. Of course, the two functions shown in Fig. 6-1(a) can be implemented in a single plunger assembly.

Figure 6-1(b) represents a 2×2 switch function. It may be implemented using prisms, movable fibers, moving mirrors, electric field switching, etc. Functionality is of prime importance because these devices are used in network rings for bypassing defective nodes; their utility will be discussed in Chap. 8.

The final mechanical switch to be discussed is the galvanometer switch shown in Fig. 6-1(c). A mirror mounted on a galvanometer movement is positioned by impressing a voltage across the winding that will deflect the mirror as shown in Fig. 6-1(c). The distance, Δy, between core centerlines of the waveguides being switched is determined by using Eq. 6-6, as follows:

$$x = \frac{\Delta y \sin 89.5 \text{ deg}}{\sin 1 \text{ deg}} \tag{6-6}$$

The parameters in Fig. 6-1(c) are as follows:

$x_1 = x_2 = K_\phi \Delta y$
$K_\phi = 57.3$

Table 6-1. Damping Ratio Cases.

$\delta < 1$ roots (underdamped case): S_1; $S_2 = -\delta\omega_n \pm j\omega_n \sqrt{1 - \delta^2}$

$\delta = 1$ roots (critically damped case): S_1; $S_2 = -\omega_n$

$\delta > 1$ roots (overdamped case): S_1; $S_2 = -\delta\omega_n \pm \omega_n \sqrt{\delta^2 - 1}$

$\delta = 0$ roots (no damping; device is oscillatory): S_1; $S_2 = \pm j\delta\omega_n$

$\Delta y = N_a a$ (in terms of radius a)
N_a = number of core radii

Angular displacement of the beam of 1 degree will cause loss of 1 to 2 dB. Therefore, the galvanometer must not undergo an angular displacement of greater than 1 degree. The optical path, x, is 11.46 mm for a distance, Δy, of 400 μm. If the waveguides are arranged in an angular pattern to reduce angular displacement, x_1 and x_2 will be shorter, thus reducing loss. These techniques are usually used on multimode switches.

The mechanical equations for rotational motion of the galvanometer are as follows:

$$T = J \frac{d^c \theta}{dt^2} + f_\theta \frac{d\theta}{dt} + \theta S_\theta \qquad (6\text{-}7)$$

where:

T = Torque
J = moment of inertia
f_θ = friction
S = stiffness retarding spring of meter movement

Electrical circuit analog:

$$I = C \left(\frac{dv}{dt} \right) + \frac{1}{L} \int_0^1 v \, dt + vG$$

$$\frac{dI}{dt} = C \left(\frac{d^2 v}{dt^2} \right) + \frac{v}{L} + G \left(\frac{dv}{dt} \right) \qquad (6\text{-}8)$$

where C is equivalent to J, f_θ is equivalent to G, and S_θ is equivalent to $1/L$. If the Laplacian operator is used, then

$$V(S) = \frac{I/C}{S^2 + \left(\dfrac{G}{C} \right) S + \dfrac{1}{LC}}$$

Using the same technique, finding the roots to the characteristic equation, and substituting the mechanical variables, we have

$$\omega_n = \sqrt{\frac{S_\theta}{J}}$$

$$\delta = \left(\frac{f_\theta}{2}\right)\left(\frac{1}{\sqrt{S_\theta J}}\right)$$

The restrictions on δ are the same as those previously discussed. This mechanical circuit is analogous to an RLC parallel circuit.

For the mechanical circuits described, an electrical time constant is also present, but it is usually small compared to the mechanical time constant, which will be the dominant one. As an exercise for deriving the total transfer function with electrical and mechanical variables, see review problem 5.

Figure 6-1(d) is a diagram of a prism switch operator. The actuator may be a solenoid with a plunger, or the prism may be moved by a magnetic field much as in relay actuators. A metal plate with a return string may be fitted to the bottom of the prism, as shown in Fig. 6-1(e). The electromagnetic coil will attract the metal plate on the prism, thus moving it the distance of the air gap. This distance will be approximately 150 to 200 μm, which is extremely small. The time to actuate the switch will be in the msec range because of the small mass.

This introduction to mechanical switches will give the reader some idea of what may be found in the marketplace. There is a great deal of room for innovation in mechanical switching devices. Many of the switches are built to be similar to relays, with similar types of operators and actuators. Design of the mechanical switch network may therefore be performed using relay logic techniques, and consequently, symmetrics are useful for these designs. (See Refs. 1 and 2 for further information on symmetrics.)

One of the problems that limits the use of switches is crosstalk. Most mechanical switches have crosstalk of 60 to 80 dB down. For example, if the optical signal passing through the switch is 1 mW (0 dBm), the crosstalk will be -60 dBm, which is related to power. Crosstalk not only reduces detection thresholds but makes eavesdropping on the data stream possible.

Since the insertion loss for these switches is from 1.5 to 3 dB, several switches in series in a network can incur heavy loss and cause the network to malfunction. It often happens in failsafe situations that one or perhaps two will be allowed to bypass optical power around a node before failure of the network occurs. In Chap. 8, these types of network will be studied in more detail. It will suffice here to make the reader aware of some of the problems encountered when designing with mechanical switches. At the time this book was being written, these devices were extremely expensive—on the order of 80 times that of an electrical relay.

The next category of switches to be discussed is useful for integrated optic applications, as well as switching single-mode optical signals. These switches lend themselves well to fabrication with monolithic integrated-circuit fabricating techniques. Monolithic technology makes holding to tight tolerances on dimensions possible, thus facilitating fabrication reproducibility, reducing cost, and increasing precision.

One of the most important classes of single-mode solid-state switches is the $\Delta\beta$ reversal switch. This particular device will allow wavelength tuning that is light polarization insensitive. This is an important feature in single-mode fiber-optic transmission since two states of polarization normally propagate, which allows for electrical compensation of the switch and thus relaxes manufacturing tolerances. Actually, some of the more recently designed devices have additional electrodes in cascade to further compensate for fabrication errors.

A schematic of a $\Delta\beta$ reversal switch is shown in Fig. 6-2(a), and a cutaway

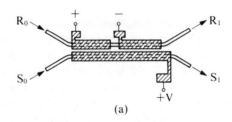

(a)

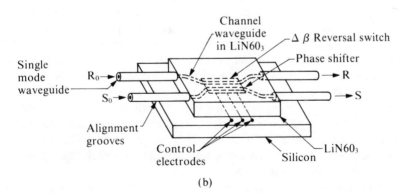

(b)

Fig. 6-2. Electronic switches: (a) $\Delta\beta$ reversal switch, and (b) single-mode switch ($\Delta\beta$ reversal).

view is depicted in Fig. 6-2(b). This switch functions as a four-port switch, with the state definitions and characteristic parameters shown below:

State Definitions

(x) $R_0 = 1, S_0, R = 0, S = 1$
(=) $R_0 = 1, S_0, R = 1, S = 0$

Characteristic Parameters

Interaction length $= L$
Coupling coefficient $= k$
Conversion length $= \ell = \pi/2k$
Propagation constant mismatch $= \Delta\beta = \beta_1 - \beta_2$

Phase shifter

$$\Delta n = 1/2 \, n^3 r_{13} E_3; \, \Delta\phi = \frac{2\pi}{\lambda_0} \Delta nD$$

Let us examine how the state definitions affect the function of switch. The switch is connected via pigtails to the outside world. A Z-cut LiNbO$_3$ crystal has two single-mode channels of thickness d with a gap, G, between them; these form the switch. Titanium (Ti) is diffused in the LiNbO$_3$ to form the channels. Electrodes divide the coupler (or $\Delta\beta$ switch) into equal-length sections in which voltages of equal or opposite sign can electro-optically control (Pockels effect) the phase-propagation constants of each of the waveguide channels (the plane of polarization may be rotated by the application of a voltage to the electro-optic crystal). With equal-magnitude and opposite-polarity voltages applied to alternating sections of the guides, a propagation constant mismatch, $(\beta_1 - \beta_2) = \Delta\beta$, is induced in half the switch and $-\Delta\beta$ in the other half.

The coupling between these channels and energy transfer can be characterized by parameters L, $\Delta\beta$, and k. Parameter k is the measure of interchannel coupling strength. If $\Delta\beta = 0$, then the distance, ℓ, over which 100-percent power transfer occurs is given by: $\ell = \pi/2k$. In single-mode switch design, the conditions, $\Delta\beta = 0$ and $\ell = \pi/2k$, are controlled. For optimum operation, the value $n\ell$ should be maintained, where n is an odd-integer multiple of the 100-percent transfer length, ℓ. To ease fabrication tolerances in the switch, the $\Delta\beta$-reversal switch configuration was developed.

The coefficient r_{13} is extracted from the electro-optic tensor for the particular crystal, i.e., trigonal with the uniaxial symmetry with the extraordinary z axis. The coefficient in the tensor for both LiNbO$_3$ and GaAs can be represented as shown in Fig. 6-3:

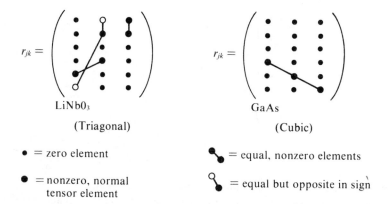

$$r_{jk} = \begin{pmatrix} \bullet & \circ & \bullet \\ \bullet & \bullet & \bullet \\ \bullet & \bullet & \bullet \\ \bullet & \bullet & \bullet \\ \circ & \bullet & \bullet \end{pmatrix}$$

$$r_{jk} = \begin{pmatrix} \bullet & \bullet & \bullet \\ \bullet & \bullet & \bullet \\ \bullet & \bullet & \bullet \\ \bullet & \bullet & \bullet \\ \bullet & \bullet & \bullet \\ \bullet & \bullet & \bullet \end{pmatrix}$$

LiNbO₃ GaAs

(Triagonal) (Cubic)

• = zero element   = equal, nonzero elements

● = nonzero, normal tensor element   = equal but opposite in sign

Fig. 6-3. Coefficients extracted from the tensor for LiNbO₃ and GaAs.

The values in the tensors are shown below:

$$r_{jk} = \begin{pmatrix} 0 & -r_{22} & r_{13} \\ 0 & r_{22} & r_{13} \\ 0 & 0 & r_{33} \\ 0 & r_{51} & 0 \\ r_{51} & 0 & 0 \\ -r_{22} & 0 & 0 \end{pmatrix} \qquad r_{jk} = \begin{pmatrix} 0 & 0 & 0 \\ 0 & 0 & 0 \\ 0 & 0 & 0 \\ r_{14} & 0 & 0 \\ 0 & r_{14} & 0 \\ 0 & 0 & r_{14} \end{pmatrix}$$

LiNbO₃ evaluated   GaAs evaluated

The changes in index of refraction along the various axes are calculated below, with the results shown in Eqs. 6-8, 6-9, and 6-10.

Evaluating the relationship, $\Delta(1/n^2)\, i = \Sigma_{j=1}^{3}\, r_{ij} E_j$, on each row of the tensor, we have:

$$\Delta \left(\frac{1}{n^2}\right)_x = -r_{22}E_x + r_{13}E_y$$

$$\Delta n_x \approx -\tfrac{1}{2} n_0^3(-r_{22}E_x + r_{13}E_y) \qquad (6\text{-}9)$$

$$\Delta \left(\frac{1}{n^2}\right)_y = (r_{22} E_y + r_{13} E)$$

$$\Delta n_y \approx -\frac{1}{2} n_0^3 (r_{22} E_y + r_{13} E_z) \tag{6-10}$$

$$\Delta \left(\frac{1}{n^2}\right)_z = r_{33} E_z$$

$$\Delta n_z \approx -\frac{1}{2} n_0^3 (r_{33} E_z) \tag{6-11}$$

where $x = 1$, $y = 2$, $z = 3$, and $n_0 = $ index of refraction.

The values of the coefficients are different at different wavelengths and frequencies, as shown in Table 6-2 for two wavelengths.

Let us examine the $\Delta\beta$ switch geometry. Figure 6-4 depicts the diffusion of titanium metal in the LiNbO$_3$ to form a waveguide, with the geometry and all dimensions as shown. Note, in Table 6-2, that a change of Δn_3 in LiNbO$_3$ (one of the commercial devices with the strongest electro-optic effect) results in a change of less than two parts in 10^4 for a field strength of 10^4 V/cm. The small dimensions will keep voltage levels low by at least a couple orders of magnitude. Before this subject can be pursued, the switching relationships to the geometry must be examined.

The coupling relationships are given by Eqs. 6-12 and 6-13:

$$\frac{dR}{dZ} - j\delta_1 R = jkS \tag{6-12}$$

$$\frac{dS}{dZ} + j\delta S = jkR \tag{6-13}$$

where $\delta \equiv (\beta_1 - \beta_2)/2 = \Delta\beta/2$.

Table 6-2. Coefficients for Two Wavelengths.

r_{22}	r_{13}	r_{33}	r_{51}	λ	6-8	6-9	6-10	Units
				LiNbO$_3$				
3.4	8.6	33.8	28	550 nm	112	37	328	10^{-6} μm/V
6.8	10	32.2	—	633 nm				
3.4	—	30.8	—	High freq., 633 nm				
				GaAs				
—	—	—	1.6	10.6 μm			59	10^{-6} μm/V

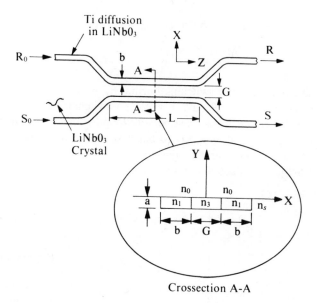

Fig. 6-4. Switch geometry.

The solution of Eqs. 6-12 and 6-13 gives the matrix in Eq. 6-14:

$$\begin{bmatrix} R \\ S \end{bmatrix} = \begin{bmatrix} A_1 & -j\beta_1 \\ -j\beta_1^* & A_1^* \end{bmatrix} \begin{bmatrix} R_0 \\ S_0 \end{bmatrix} \tag{6-14}$$

where:

$$A_1 = \cos\left[z\sqrt{k^2 + \delta^2}\right] + j\delta \sin\left[z\sqrt{k^2 + \delta^2}\right]/\sqrt{(k^2 + \delta^2)}$$
$$A_1^* = \cos\left[z\sqrt{k^2 + \delta^2}\right] - j\delta \sin\left[z\sqrt{k^2 + \delta^2}\right]/\sqrt{k^2 + \delta^2}$$
$$\beta_1 = k \sin\left[z(k^2 + \delta^2)^{1/2}\right]/\sqrt{k^2 + \delta^2}$$
$$\beta_1^* = \beta_1$$

Evaluating Eq. 6-14 for 100 percent of the optical power coupled between the two channels, $\Delta\beta = 0$ and $L\pi/2k$, where $L = n\ell$, we have

$$\begin{bmatrix} R \\ S \end{bmatrix} = \begin{bmatrix} 0 & -j \sin\dfrac{\pi}{2} \\ -j \sin\dfrac{\pi}{2} & 0 \end{bmatrix} \begin{bmatrix} R_0 \\ S_0 \end{bmatrix} \tag{6-14a}$$

Switched state: $R = -jS_0 \qquad R_0 = 1; \quad S_0 = 0; \quad R = 0; \quad S = 1$

$\qquad\qquad S = -jR_0$

For the situation when no switching occurs, $\beta_1 \approx 0$ since $k \approx 0$ and $\delta \ll \pi/2$. Note that when $R = R_0$ and $S = S_0$, no switching occurs and $L \to \infty$; i.e., the two adjacent channels must be very long to couple any signal.

The switch can be used as a 50/50 beam splitter if $\beta_1^2 = 0.5$ and $\delta = 0$; then, $\sin^2 k\ell$ for $L > \ell/2$ can be adjusted.

Equation 6-14a is evaluated as follows:

$$
\begin{bmatrix} R \\ S \end{bmatrix} \approx \begin{bmatrix} \cos\dfrac{\pi}{4} & -j\sin\dfrac{\pi}{4} \\[2ex] -j\sin\dfrac{\pi}{4} & \cos\dfrac{\pi}{4} \end{bmatrix} \begin{bmatrix} R_0 \\ S_0 \end{bmatrix}
$$

$R_0 = 1; \quad S_0 = 0; \quad R = 0.7; \quad S = 0.7$

This equation depicts amplitudes; for power units, the terms are squared—i.e., $P_S \cong 0.5$, $P_R \cong 0.5$, or a 50/50 power split. The difference in indices between the channels are related to propagation constants β_1 and β_2 by Eq. 6-15:

$$
\Delta\beta = \beta_1 - \beta_2 = (4\pi/\lambda_0)\,\Delta n_{\text{ind}} \tag{6-15}
$$

This equation shows the necessary relationship between $\Delta\beta$ and Δn_{ind} to control the switch. Also, one must note that cross-talk is possible between adjacent channels if the two β_1 terms are not zero when the switch is not active. A two-electrode switch can be constructed in cascade. The matrix equation is derived as shown below, leading to Eq. 6-16:

$$
\begin{bmatrix} R \\ S \end{bmatrix} = \begin{bmatrix} A_2 & -jB_2 \\ -jB_2^* & A_2^* \end{bmatrix} \begin{bmatrix} R_0 \\ S_0 \end{bmatrix}
$$

$$
A_2 = A_1^* A_1 - 2B_1^2
$$

$$
B_2 = 2A_1^* B_1
$$

$$\begin{bmatrix} R \\ S \end{bmatrix} = \begin{bmatrix} A_1^* A_1 - 2B_1^2 & -j\,2A_1^* B_1 \\ -j\,2A_1 B_1^* & A_1^* A_1 - 2B_1^2 \end{bmatrix} \begin{bmatrix} R_0 \\ S_0 \end{bmatrix}$$

$$\begin{bmatrix} R \\ S \end{bmatrix} = \underbrace{\begin{bmatrix} -A_1^* & -jB_1 \\ -jB_1^* & A_1 \end{bmatrix}}_{\substack{-\delta \\ \text{section}}} \underbrace{\begin{bmatrix} A_1 & -jB_1 \\ -jB_1^* & A_1^* \end{bmatrix}}_{\substack{+\delta \\ \text{section}}} \begin{bmatrix} R_0 \\ S_0 \end{bmatrix} \qquad (6\text{-}16)$$

If all the power is initially in one guide, initial conditions R_0, $S_0 = 1$, 0, and crosstalk is assumed to be zero by setting $A_2 \equiv B_2 \equiv 0$. Each of these conditions establishes a correspondence between the normalized parameters, L/ℓ and $\Delta\beta/L$. This is plotted in Fig. 6-5(a) for a single electrode switch, and in 6-5(b) for two, in 6-5(c) for three, in 6-5(d) for four, and in 6-5(e) for six electrode switches. For the condition when L/ℓ is between 1 and 3, there are $\Delta\beta$ values (a function of the applied voltage or electric field under the electrodes) that will drive the device into both parallel and crossed curves. Therefore, device length is a noncritical parameter. These curves also reveal that it is smaller values of $\Delta\beta$ that produce these effects. The electric field required to produce the switching is lower, and therefore switching voltages are also lower. Furthermore, longer devices and multiple electrodes yield lower voltage operation. A trade-off exists between device length, operating voltage, and number of sections in the switch.

Figure 6-5(a) shows how critical the dimension of the switch becomes when only single sections are used. Figure 6-5(b) allows a multiple of 1 or 2 or 3, but the $\Delta\beta$ is relatively unchanged, thereby implying that the switch operates in a way similar to that of a single-section type but with a less stringent length requirement. The three-section switch in Fig. 6-5(c) allows both a loose tolerance in length and lower $\Delta\beta$, which is related to Δn_{ind}. The latter characteristic means that the switch length tolerance is relaxed and the voltage supplying the field can be reduced by at least one-half for $1 < L/\ell < 5$ and one-fourth for $2 < L/\ell < 4$. As one may observe from Figs. 6-5(b) and (e), more sections added to a switch relax either the L/ℓ tolerance, $\Delta\beta$ switching requirement, or both. For a four-section switch, at 800-nm wavelength, the switching voltages were 11 V for cross coupling—i.e., when $R_0 = 1$, $S_0 = 0$, $R = 0$, and $S = 1$—and 22 V for $R_0 = 1$, $S_0 = 0$, $R = 1$, and $S = 0$, with the latter being the straight through state.

An undesirable effect that must now be alluded to is Δk, or the variation in the switch coupling coefficient which has a symmetrical variation in the refractive index. Equation 6-17 expresses the functional relationship, with K_0 being the coupling coefficient with $\Delta n_{ind} = 0$:

$$f_k(\Delta n_{ind}) = K_0 + f_{\Delta k}(\Delta n_{ind}) \tag{6-17}$$

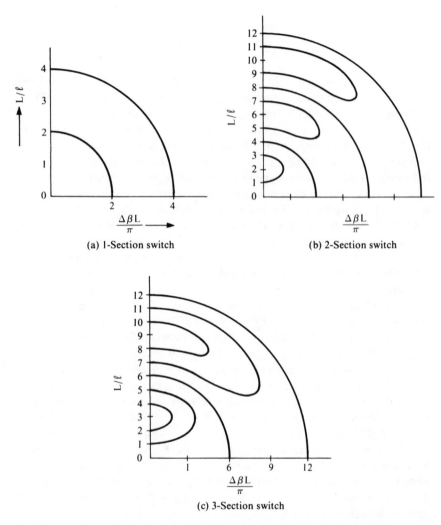

Fig. 6-5. Switches with single and multiple electrode sections: (a) One section; (b) two sections, (c) three sections, (d) four sections, and (e) six sections.

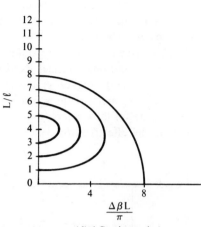

(d) 4-Section switch

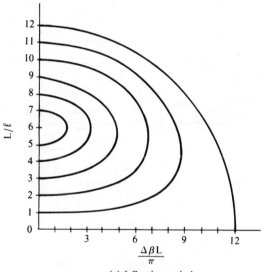

(e) 6-Section switch

Fig. 6-5. (*Continued*)

A series of curves showing the relationship of variations in the waveguide of the switch and distances between them as a function of the coupling coefficient is shown in Fig. 6-6. The top curve shows the result for a variation in gap index profile with the channel indices constant, i.e., no variations. The middle set of curves show symmetrical variations in the channel indices with the gap index constant. The push-pull case reflects the highest efficiency, i.e., where the Δn_c profile $= -\Delta n_g$ and Δn_c and Δn_g are variations in channel and gap indices, respectively. The waveguide data for these curves are $\lambda_0 = 0.88$ μm; $n_s = 2.2$ (substrate); $n_{ce} = 2.203$ (channel effective; this takes into account the vertical

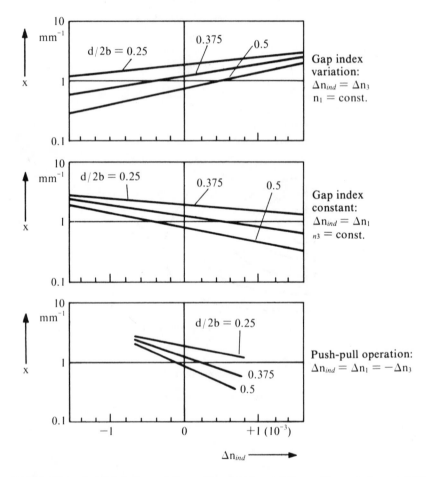

Fig. 6-6. Plots of switch index variation with gap index variation, with gap index constant, and for push-pull operation ($b = 5$ μm).

index profile); and n_{ge} (gap effective; this also takes into account the vertical profile).

The curves indicate that k is almost an exponential function of Δn_{ind}. This relationship is expressed in Eq. 6-18 below:

$$k \approx K_0 e^{\delta \alpha} = K_0 \left[1 + \delta \alpha + \frac{1}{2} (\delta \alpha)^2 + - \frac{(\delta \alpha)^{n-1}}{n!} \right] \qquad (6\text{-}18)$$

where:

$$\alpha = \frac{\Delta \beta L_0}{\pi} = \frac{4 \Delta n_{ind} L_0}{\lambda_0} \qquad (6\text{-}19)$$

and

$$k = K_0 + \Delta k$$

Then,

$$\Delta k \approx K_0 \left[\delta \alpha + \frac{1}{2} (\delta \alpha)^2 + \cdots \frac{(\delta \alpha)^n}{(n-1)!} \right]$$

The equation

$$\frac{\Delta k}{K_0} = \frac{2 \Delta k L_0}{\pi}$$

is derived from Eq. 6-17 for $\Delta n_{ind} = 0$, which is the normalized coupling coefficient, k_n; it is shown below using only the first two terms of the exponential expansion:

$$k_n = \frac{\Delta k}{K_0} = \left(\frac{4 \Delta k L_0}{\Delta n_{ind} \lambda_0} \right) (\Delta n_{ind}) \approx \delta \alpha$$

Then,

$$\delta = \left(\frac{dk}{dn_{ind}} \right) \frac{\lambda_0}{4\phi} \qquad (6\text{-}20)$$

where the value ϕ is the amount of coupling that has taken place to and from the interaction region. The value of both δ and α can be calculated using Eqs. 6-19 and 6-20.

A series of curves relating coupling length, wavelength, and Δn are depicted in Fig. 6-7. The dimensions for the waveguide, shown in the upper right-hand corner, are in micrometers. Also shown are the refractive index of the substrate ($n_{sb} = 2.2$) and of the guide ($n_g = 2.2 + \Delta n$). This can be deduced intuitively since the dimension becomes closer to the operating wavelength.

The curves shown in Fig. 6-8 depict the relationship between the coupling coefficients of Ti-diffused $LiNbO_3$ waveguide at wavelengths of 1.06 μm, 0.83 μm, and 0.633 μm as a function of waveguide separation. The waveguides plotted in the figure are constructed for single-mode operation, with the guide dimensions for switch geometry (Fig. 6-4) as indicated in Table 6-3.

If a sinusoidal signal is injected in the R_0 port of a coupler, an output is

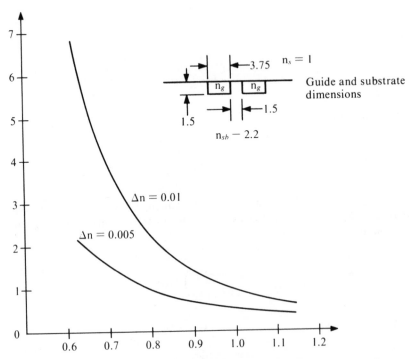

Fig. 6-7. Coupling length versus wavelength for uniform rectangular waveguides (superstrate and substrate have indices of refraction of 1 and 2.2, respectively, and the waveguide index of refraction is 2.2 + Δn).

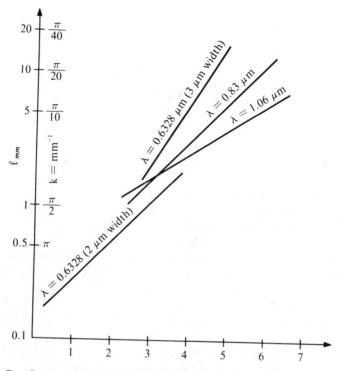

Fig. 6-8. Coupling coefficients of Ti-diffused LiNbO$_3$ waveguide at 1.06-, 0.83-, and 0.633-μm wavelengths as a function of waveguide separation; each waveguide is operated single mode because of the fabrication.

Waveguide thickness		b width
0.6332 m	300 Å	3 m, 2 m
0.830 m	400 Å	4 m
1.06 m	460 Å	5 m

Table 6-3. Channel and Thickness Dimensions.

λ	b	Thickness of Ti
633 nm	3 μm	300 Å (30 nm)
830 nm	4 μm	400 Å (40 nm)
1060 nm	5 μm	460 Å (46 nm)

obtained at R and S equivalent to $\cos^2 (k\ell + \phi)$ at R and $\sin^2 (kL + \phi)$ at S. Then the coupling coefficient may be calculated using Eq. 6-21. Since

$$\tan^2 (kL + \phi) = \frac{\sin^2 (kL + \phi)}{\cos^2 (kL + \phi)} = \text{Ratio}$$

then,

$$kL + \phi = \tan^{-1} (\text{Ratio})^{1/2} \qquad (6\text{-}21)$$

where k is the coupling coefficient and L, the interaction length. The coupling length ℓ as reiterated here is $L = n\ell$ and $\ell k = \pi/2$. Then, $L = n\pi/2k$. The two values used for these calculations—i.e., $\sin^2 (Lk + \phi)$ and $\cos^2 (Lk + \phi)$—are intensity outputs, not amplitudes.

The losses in the coupler or switch, depending on how it is used, are similar to any other optical loss mechanism, such as separation, lateral misalignment, or angular misalignment. There are others as well, including index of refraction mismatch and loss due to mask making accuracy when doping the channels. The waveguide loss is approximately 1 dB/cm. Of course, this situation may change in the future.

Tolerance and bend losses occur between angled and nonangled sections; these losses can be 1 to 2 dB. The total loss per switch or coupler can be as large as 5 to 7 dB, depending on how precise the coupler is made. One of the losses included in this figure is that due to electrode placement.

Electrodes fabricated directly on the waveguides necessary for Z-cut crystals of LiNbO$_3$ produce a loss of approximately 1 dB/mm for TE mode and 10 dB/mm for TM mode. Therefore, it is necessary to isolate the electrodes from the LiNbO$_3$ surface with a low-loss dielectric for low-loss switches; both TM and TE mode losses can be reduced. The switching voltage for the TM mode is approximately one-third that of the TE mode corresponding to the r_{33}/r_{13} ratio.

The buffer layer between the LiNbO$_3$ and electrodes results in a different electronic characteristic for the switch because each dielectric has different leakage resistances; this is equivalent to three different dielectrics between the plates of a capacitor. With dc voltage, a voltage division will occur across the three leakage resistances. If the LiNbO$_3$ resistance is lower than that of the other two layers, the switching voltages would have to be higher to produce the necessary field across the LiNbO$_3$ to get the necessary index of refraction change. The reverse situation is also possible, i.e., larger leakage resistance in the LiNbO$_3$ and smaller in the buffers. Low resistance TiO$_2$ film is sometimes used as a buffer. SiO$_2$ would also be a natural choice because it is used in fabrication of these circuits, but its leakage resistance is much larger than that of TiO_2. Note that adding the buffer layer reduces the total capacitance between the electrodes because the buffer is in series with the $LiNbO_3$ crystal.

The $\Delta\beta$ switches or couplers have some of the most interesting properties because they can be altered by applying control voltages to the electrodes. The next section will deal with fixed-ratio couplers and star couplers. These devices are passive because they have no control element. The tap ratio, or number of ports, is decided upon, and the device is manufactured with built-in power split ratios.

COUPLERS (FIXED)

When dealing with single-mode technology, it is very difficult to fabricate couplers larger than 2 × 2, i.e., two inputs and two outputs. During the writing of this text, for example, no commercial product was available. Readers acquainted with multimode technology, however, will recall that star couplers with multiple inputs and outputs are readily available. Phase of the optical waveform may not be considered, but single-mode couplers are phase sensitive because of the single wavelength; consequently, dimensions are critical and the cladding must be fused together within correct geometric limitations.

These single-mode fixed couplers have less excess loss than multimode devices. Some of the latter exhibit 0.1 to 0.5 dB of excess loss, whereas their single-mode counterparts show 0.05 to 0.5 dB. Fabrication of couplers takes many forms. Beam-splitting devices, which use mirrors or gratings on the beam emanating from the waveguide end and whose alignment is rather critical, have not been sufficiently successful to enjoy widespread use. Another technique, evanescent coupling, has been used on most of the commercial couplers to date. It brings two waveguides into close proximity so that the core of one can intercept the evanescent field of the other. The coupling is dependent on the distance between the cores, the index of refraction of the material separating the cores, and the interaction length where the coupling takes place. This should all sound familiar because the coupling is similar to that of the electronic controlled device of the previous section, except that the index of refraction of the material is not altered.

In evanescent wave coupling, optical energy is transferred from one optical fiber to another as a result of field overlap between the cores. Since evanescent fields decay exponentially, coupling can be accomplished only if the waveguides are brought into close proximity. A typical 4- to 10-micrometer core encased in 80- to 125-μm clad must be brought within close proximity to another waveguide with the same characteristics to form a directional coupler. The evanescent field must allow for overlap, which is usually accomplished through etching, polishing, or fusion.

The etching technique requires that the waveguides be stripped of all coatings to the cladding and then twisted together and immersed in hydrofluoric acid. When the cladding etches away, the cores will come into close proximity, thus forming a coupler. These devices are sensitive to vibration and temperature, and, in general, lack any measure of ruggedness. Because of this fragile nature, they are very rarely used for commercial products.

A second, mechanical means of forming a coupler is to remove half of each waveguide to be joined and embedding them in a glass plate over the coupling length. The advantages of embedding waveguide in glass plates is that polishing can be fairly closely controlled and cracking in the cores (due to the extremely small size) during the material removal stage prevented. After polishing, the plates with the waveguides are brought into contact by inserting a drop of index matching fluid between them. The plates can then be adjusted for maximum coupling (or any other coupling ratio desired) and cemented together. Such couplers can be made to be adjustable but would be very difficult to fabricate in any quantity. Any device dealing with index matching fluids can eventually exhibit problems. Temperature coefficients for the glass, matching fluid, cement, and waveguide are usually different, and the cement and index matching fluid may interact or exhibit some aging properties. Moreover, the fluid between the plates and cement usually cannot tolerate mechanical stress such as vibration and shock loads.

The last technique for coupling is commonly known as "fused biconical taper coupling," or FBT. Fabrication is similar to that for multimode couplers in some respects. The two waveguides used are first fused at 1500°C and then stretched while heat is still being applied. The cores become smaller in cross section, thus causing the evanescent electric field to spread out further from the core. The coupling can then take place when the two cores are farther apart than they are in the other two cases. Stretching is continued during heating until the desired coupling ratio is reached. Such couplers are fairly rugged and easier to fabricate than the previous two. Most present-day couplers are fabricated using FBT techniques or a variation of them.

The effects of evanescent field coupling can be observed using Fig. 6-9. As the normalized frequency is made smaller, spreading occurs in the field. Equation 6-22 describes the relationship between the normalized frequency parameter, V; the core radius, a; the wavelength, λ; the core index of refraction, n_c; and the clad index of refraction, n_{CL}:

$$V = \frac{2\pi a}{\lambda} \left(n_{CL}^2 - n_C^2 \right)^{1/2} \tag{6-21}$$

Note that the V value is not 5 for the single-mode case; it is given here merely to illustrate the effect. Of course, the coupler can be made with a V of 5 if the

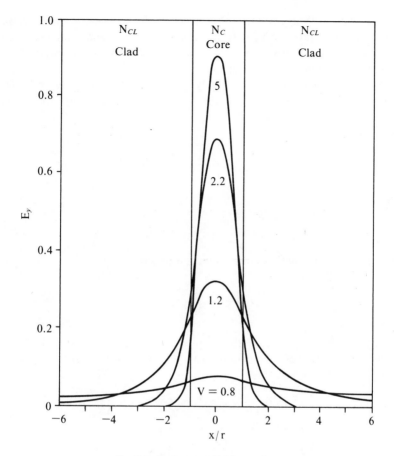

Fig. 6-9. Evanescent field coupling.

attendant loss can be tolerated at the interface to the main cable to which it is to be attached. As the cross-section becomes smaller during the stretching of the core, V is also affected; it too becomes smaller. The coupling length increases until the core medium becomes insignificant—for a V value less than about 1. The clad will no longer be infinite but begin to act as the guiding structure, and the medium surrounding the clad will act like the clad. In effect, the structure resembles a three-layer waveguide.

Equation 6-23 is the expression for the coupling coefficient, C, as given by McIntyre and Snyder[3]:

$$C(V) = \frac{(n_C^2 - n_{CL}^2)^{1/2} V^2 K_0(w_d/a)}{n_C a V^3 K_1^2(w)} \qquad (6\text{-}23)$$

where V and w are eigenvalues of the mode in the core and cladding, respectively; K_0 and K_1 are modified Bessel functions; and a and d are core radius and separation, respectively. The coupling is a function of V along the length because of the taper in the waveguide.

If waveguide 1 is excited with unity power initially, then Eqs. 6-24 and 6-25 are the power outputs for waveguides 1 and 2, respectively [for the nontapered case, $\overline{C}(z)$ is a constant]:

For waveguide 1:

$$P_1(z) = \cos^2\left[\overline{C}(z)\right] \tag{6-24}$$

For waveguide 2:

$$P_2(z) = \sin^2\left[\overline{C}(z)\right] \tag{6-25}$$

where the coupling coefficient as a function of z is

$$\overline{C}(z) = \frac{1}{z} \int_0^z \overline{C}(z')\, dz'$$

As the wavelength is varied, the coupled power/throughput power varies between ± 5 to 7 dB every $\Delta\lambda$ period ($\Delta\lambda = 12$ nm), as follows:

$$f\left(\frac{P_c}{P_t}\right) \text{dB} = 10 \log A(\lambda) \cos 2\pi\left(\frac{\lambda_c - \lambda}{\Delta\lambda}\right)$$

$$f\left(\frac{P_c}{P_t}\right) \text{dB} \approx 10 \log A \cos \frac{2\pi}{12}(\lambda_c - \lambda) \tag{6-26}$$

For short variations in wavelength, the amplitude is approximately constant. The value λ_0, the cutoff wavelength, and λ are in nanometers. If the coupling ratio varies with wavelength, then it is possible to use these devices as wavelength multiplexers/demultiplexers.

One of the pressing issues of this discussion is performance. Obviously, one of the single most important parameters is excess loss, which has been reduced from 2 to 3 dB to less than 0.05 dB. The latter figure cannot be achieved with beam splitters because of the losses in the optics. FBT devices are required. In multimode couplers, polarization becomes an issue. To prevent polarization sensitivity in FBT couplers, the waveguides should not be twisted. On the other hand, if polarization sensitivity is desired, as it is in sensor design, couplers may be used for phase detection devices; these will be discussed in Chap. 11. Other effects that govern coupler characteristics are index of refraction, geometric irregularities in separation and distance between cores, etc. These have less effect on coupling, however, than issues previously discussed.

Figure 6-10 is a diagram of a typical coupler, while Eqs. 6-27, 6-28, and 6-29 describe the various performance parameters that relate to it. The coupling ratio is expressed, in dB, in Eq. 6-27:

$$\text{Coupling ratio} = 10 \log \left(\frac{P_2}{P_1 + P_2} \right) \qquad (6\text{-}27)$$

The excess loss is neglected in this calculation because of the extremely low values encountered. In fact, it may be neglected in most cases, i.e., from 0.1 to 0.05 percent. Equation 6-28 may be used to calculate it, in dB, as required:

$$\text{Excess loss} = 10 \log \left(\frac{P_1 + P_2}{P_{\text{in}}} \right) \qquad (6\text{-}28)$$

Directivity, as given in dB by Eq. 6-29, is very important because large values sometimes result in power directed toward sources:

$$\text{Directivity} = 10 \log \left(\frac{P_3}{P_{\text{in}}} \right) \qquad (6\text{-}29)$$

Equation 6-29 also represents crosstalk in instances where security may become a problem.

General application of directional FBT couplers will be presented later in the text. A brief description of a few of them will be given here with more elaborate applications presented later.

Interferometers are implemented with these devices. Sensors with extremely high sensitivity, they are used in gyros, temperature sensors, rainfall indicators,

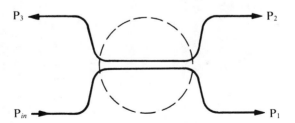

Fig. 6-10. Single-mode coupler.

TABLE 6-4. Single-Mode Coupler Manufacturers.

Company	Construction Method	Wavelengths	Operating Temperature	Polarization Sensitivity (360°)	Insertion Loss
Cabloptia Cortalloid, Switz.	Polished	N/A	N/A	N/A	<0.5 dB
Canadian Inst. & Res. LTD Mississauga, Ont.	Polished	633 nm 833 nm 1300 nm	0 to 120°F	2%	<.04 dB
Canstar Scarborough, Ont.	Fused biconical taper (FBT)	850 nm 1300 nm 1500 nm	−50° to 70°C	±2%	<1 dB to <1.50 dB
CLTO Bezons, France	Fused	1300 nm	−20° to 60°C	N/A	<0.5 dB to <1.0 dB
GEC Res. Lab Wembley, U.K.	FBT and polished	850 nm 1300 nm	−10° to 60°C	N/A	<0.1 dB
General Optronics Edison, N.J.	FBT	850 nm 1300 nm	N/A	N/A	<1.5 dB
Gould, Inc. Glen Burnie, Md.	FBT	633 nm 820 nm 1300 nm 1550 nm	−55° to 125°C	±0.5%	<0.1 dB to 1.0 dB

Hitachi Cable America Inc. Tokyo, Japan	Polished	1300 nm	−10° to 50°C	±2%	<1 dB
ITT Roanoke, Va.	FBT	633 nm 850 nm 130 nm	−20° to 90°C	N/A	<0.75 dB
JDS Optics Inc. Ottawa, Ontario	Bulk optics	1300 nm	N/A	N/A	<1.5 dB
Kaptron Palo Alto, Calif.	Bulk optics	All	−40° to 65°C	N/A	<1.5 dB
Opto-Electronics Oakville, Ont.	FBT	850 nm 1060 nm 1300 nm 1500 nm	−25° to 55°C	N/A	<1 dB
Phalo Optical Sys. Manchester, N.H.	FBT	1300 nm	−50° to 80°C	N/A	<0.5 to <1.5 dB
NEC Electronics Mt. View, Calif.	N/A	1300 nm	−10° to 40°C	N/A	<0.8 dB to <1.5 dB
STC Components London, England	FBT	1300 nm	−10° to 60°C	N/A	<0.5 dB
York Tech LTD Hampshire, England	FBT	850 nm 1300 nm	−30° to 70°C	N/A	<1 dB

strain gauges, etc. In communications, they are used as taps, multiplexers in integrated optics, etc.

Several commercial products are presented in Table 6-4, which lists manufacturers involved in single-mode technology. The reader should use this table as a guide when designing; improvements in various parameters will no doubt be achieved. Note that the products of very few of these vendors have excess loss greater than 1 dB. When a single coupler is needed—e.g., in sensor circuits—large excess loss may not be of any consequence; in fact, the cost advantage may warrant the use of couplers with large excess loss. For communication circuits where large excess loss reduces performance, however, lower loss components will be required. Since temperature of operation may be a factor, discretion must be used; i.e., large temperature excursions are not necessary for office environments, but military or outdoor installations may require wide temperature excursion devices.

WAVELENGTH MULTIPLEXING/DEMULTIPLEXING

The technique of wavelength multiplexing/demultiplexing is commonly referred to as "wavelength division multiplexing," or WDM. This technique of spectral division is similar in many respects to frequency division multiplexing (FDM) in electronics communication. Instead of assigning a frequency to each channel or group of channels as in FDM, a wavelength is assigned in WDM. In the previous section, it was briefly mentioned that single-mode biconical FBT couplers could be used for wavelength multiplexing/demultiplexing by adjusting the coupling coefficient. This technique allows only two wavelengths at a time to be multiplexed. Another method of multiplexing that allows the use of multiple wavelengths is shown in Fig. 6-11(a).

The input waveguides are fitted with GRIN lenses that are used to expand and collimate the laser or LED inputs. The input light beams are focused on a larger GRIN lens and focused on the output waveguide. The inputs may all be single mode, but the output waveguide will resemble multimode operation because of the multiple wavelengths and associated modes for each. The GRIN lens (to be discussed later in this section) is wavelength-sensitive—i.e., the image focused on the exiting surface of the lens is dependent on wavelength. Wavelengths outside the GRIN lens wavelength of operation will either be noncollimated or the focal point of the particular wavelengths will be inside or outside the lens rather than on the rear surface. Therefore, WDM with this scheme is limited to closely spaced wavelengths.

The WDM device described will also result in mode coupling and other properties similar to those of multimode waveguides. Since these couplers result in multiple reflection paths, insertion loss will increase with the number of chan-

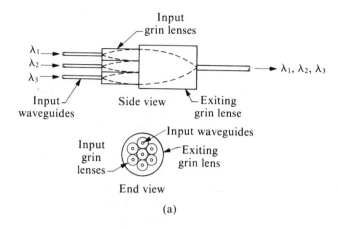

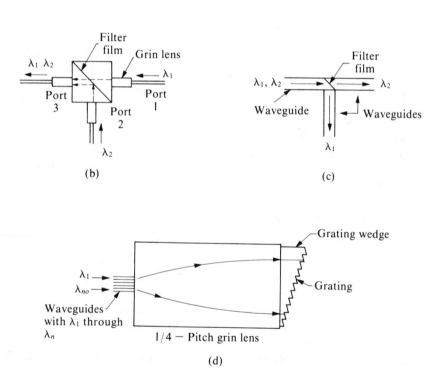

Fig. 6-11. Wavelength division multiplexers and demultiplexers: (a) WDM grin lens type; (b) to demultiplex $\lambda_1\lambda_2$, reverse the input into port 3, and wavelengths λ_1 and λ_2 will be outputted at ports 2 and 1, respectively; (c) multiplexer/demultiplexer without GRIN lenses; and (d) GRIN lens and grating multiplexer/demultiplexer.

nels. In the following chapter on integrated circuits, a form of WDM is presented that does not make use of GRIN lenses.

Demultiplexers are constructed with a multitude of techniques. The couplers discussed in previous sections are one method of separating wavelengths, i.e., by adjusting the coupling coefficient, which is a function of wavelength.

A mirrored surface with a wavelength selective film is another method of demultiplexing, as shown in Fig. 6-11(b). This device, which functions equally well as a multiplexer, is fitted with GRIN lenses to reduce alignment difficulties. The technique shown in Figure 6-11(c), which doesn't make use of GRIN lenses, will have lower insertion loss provided alignment can be accomplished to a precision dependent on the waveguide. When long wavelengths [1200 to 1550 nm, or 3 μm (for some of the new long-wave technology)] are implemented, alignment may eventually cease to be a problem.

Another method of WDM multiplexing/demultiplexing is shown in Fig. 6-11(d), which is implemented with a GRIN lens and diffraction. The loss attributed to the grating is approximately 1 dB or less. For multiplexing operations, the incoming light is collimated with GRIN lenses on inputs 1 to $n -$ 1. The light is then focused by the $\frac{1}{4}$-pitch GRIN lens onto output λ_{no}. All of the inputs will be focused on the output because of the properties of the grating. If the device functions as a demultiplexer, the incoming light (which can represent a trunk input) consists of λ_1, λ_2 . . . λ_{n-1}. The incoming light from λ_{no} is collimated and dispersed by the grating to the outputs, depending on the particular wavelength, λ_1, λ_2 . . . λ_{n-1}.

Other techniques not discussed in detail are dispersive devices, concave grating, polarization, and modal multiplexing/demultiplexing techniques. These dispersive WDM couplers use a prism, diffraction grating, or holographic optical element to perform angular dispersion. These methods allow linear dispersion in the focal plane of the coupler, which can then be modified to enhance channel density and sensitivity.

The first of these devices, the prism, allows angular separation. Prisms require collimating lenses for both the launch and collection of optical signals. A WDM fashioned with these devices can be made fairly small. Some of the inherent negative qualities include insertion loss and crosstalk, which are a function of lens quality, waveguide dimensions, and waveguide location on the object and image planes. Also, dispersive effects are heavily dependent on prism dimensions and material. Spectral width of the WDM varies as λ^{-3} because of the nonlinear nature of prism dispersion. Therefore, significant channel separation cannot be easily accomplished.

The next dispersive device is the diffraction grating WDM coupler. These devices offer higher dispersive power for given size than prism couplers. Since the dispersion is linearly proportional to wavelength, channel spacing can be distributed over a sizable spectral width. Three types of gratings are useful for

WDM multiplexing/demultiplexing: the halographic element (HOE), plane grating, and concave grating.

The HOE implementation has eliminated the need for an auxiliary collimating lens and focusing optics at the expense of only moderate diffraction efficiency, which translates into higher excess loss. Many of the materials required to fabricate these devices are not readily available for the 1100 nm to 16 nm spectral range. This is, of course, an important consideration because of the increase in long wavelength activity (large research and development effort). These gratings exhibit losses in the 2- to 4-dB range without considering other losses attributed to the optics and terminations.

The most versatile grating is the plane, blazed diffraction grating, which is available at fairly low cost with a large variety of ruling constants and blaze wavelengths. These gratings are fairly easily fabricated with a high degree of precision. As previously stated, efficiency is fairly high—approximately 80 to 95 percent of the incident. Since optical power is diffracted into the first order, insertion loss attributed to the diffraction grating will be less than, or equal to, 1 dB. The grating design to implement WDM multiplexer/demultiplexers is dependent on constraints such as the number of channels, maximum insertion loss, source availability, grating availability and cost, source stability with temperature, ease of alignment, crosstalk (minimum isolation between channels), mechanical, ruggedness, etc.

As one may observe, a large number of factors must be considered when designing any multiplexer/demultiplexer, not just grating devices. Since design of a GRIN lens-grating device has been previously covered, there is no need to consider it again here.

Concave gratings are the last of the dispersive devices to be addressed. These devices, which were not very well perfected at the time this book was being written, do not require auxiliary optics such as collimating lenses and focusing and diffraction elements. The gratings are difficult to rule with required precision and exhibit low efficiency because of the rapidly changing blaze angle across the device's curved surface. Therefore, excess loss is also fairly large.

Let us now examine another multiplexing/demultiplexing technique that uses polarization. It is closely akin to the phase in electronic systems. The ability of optical waveguide to propagate at distinct polarization states is the principle used to separate phase into channels. The channels are constructed of orthogonal polarization states with the use of couplers (polarization-dependent). The systems parameters that must be determined are the number of channels, transmission loss for the desired link, and optical isolation between adjacent channels, all of which are interrelated. For example, if phase spacing between channels is not adequate, isolation will be poor, and large crosstalk components will be present. If the total transmission loss is too large, there will be problems in the detection of phase.

A knowledge of all phase-sensitive devices is necessary if the system is to perform adequately. Each component must be examined to determine its phase effects. As some examples, consider the optical source, detector, waveguide, couplers, and connectors. The source wavelengths are identical, but each channel has a separate source and detector. Waveguide polarization properties are critical parameters, ones useful in determining if a link design is viable or not.

Lasers and LEDs are the usual source of interest in fiber optics. LEDs are polarization independent, but their output can be polarized through the use of dichroic filters, which consist of birefringent materials that absorb transverse electric (TE) and transverse magnetic (TM) fields differently to produce linear polarized light. Typically, up to 50 percent of the light is lost through this conversion, but some of the newer LEDs have optical power outputs approaching those of the laser. These systems were not very feasible a few years ago, but they are today.

Detector diodes are also polarization-independent. Therefore, channel isolation can be provided by demultiplexing couplers, polarization elements, or both. Further loss is incurred at the detection end of the channel.

As was observed in the discussion of semiconductor device physics in Chap. 3, lasers are polarization-dependent devices. The TE mode is typically 20 dB larger than the TM mode. The degree of polarization can also be expressed using I_{TE} and I_{TM} of the two fields, as indicated by Eq. 6-30.

$$\% \text{ polarization} = \left(\frac{I_{TE} - I_{TM}}{I_{TE} + I_{TM}}\right) \times 100 \qquad (6\text{-}30)$$

For lasers with 1-mW power output and 20 dB of difference between I_{TE} and I_{TM}, the % polarization is 90 percent; the I_{TE} can then be as large as 0.9 mW and the I_{TM}, 0.1 mW. Various percentages of I_{TE} and I_{TM} comprise the channel.

Waveguides used in polarized-modulated systems are useful for our purposes in single-mode networks or systems. Polarization-maintaining waveguides have been successfully used in single-mode systems for several kilometers. If polarization-maintaining waveguide is not used, the orthogonal fields begin to equalize over distance. Any splices, connectors, couplers, etc., that produce discontinuities will distort the polarized signals. By constructing the waveguide with an elliptical core, it will become polarized. Another technique, called "stress-induced birefringence," involves a circular core and elliptical clad.

The extinction ratio is defined by Eq. 6-31, as follows:

$$\text{Extinction ratio} = 10 \log \left(\frac{P_{\max}}{P_{\min}}\right) \qquad (6\text{-}31)$$

Extinction ratios for various waveguides range from 14 to 30 dB, assuming that the T_M and T_E are orthogonal. When dealing with fiber-optic sensors, the polarization-maintained waveguide will be discussed in more detail.

Figure 6-12 is a simplified link using polarization multiplexing/demultiplexing. Coupler 1 will reflect the T_{X1} beam into the single-mode waveguide due to the nature of the polarized sensitive filter. T_{X2} on the other hand will pass through the filter unobstructed. The two polarized beams will travel down the waveguide to the receive end where the reverse process takes place. T_{X1} is again reflected because of the filter, and T_{X2} passes through unobstructed. If the inputs to coupler 1 and outputs of coupler 2 are fitted with GRIN lenses, alignment can be made less critical. The waveguide is assumed to be adequate to maintain polarization integrity. The joints to the coupler may require a filter to correct for polarization changes even if the couplers are fusion-spliced via pigtails to the waveguide. Insertion loss will usually be large as a result of the absorption of light waves that are not in the polarization mode. The waveguide itself will corrupt the polarization to some extent. Some of the energy will be lost, but a component will appear in the adjacent channel as crosstalk. Crosstalk minimum value should be at least 20 dB for commercial-grade applications and even greater for the military.

In the networking section of the text in Chap. 8, some small-scale networks will be presented. Some of these techniques are addressed during the discussions of fiber-optic sensors where polarization has several applications. Since no local-area networks are implemented using this technique, not much space has been given to it here.

A superficial introduction to modal multilexing is warranted to give the reader some insight into what is being done in research and development. Due to lack of space, only the highlights will be presented. Single-mode channels are multiplexed on multimode fiber-optic waveguides, and this results in a certain amount

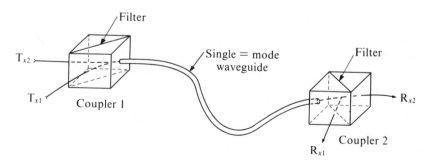

Fig. 6-12. Polarization multiplexer/demultiplexer link.

of mode mixing. Mode mixing, in turn, results in corruption of the initial signals and crosstalk. The number of modes supported in multimode waveguide can reach the hundreds or thousands. Equation 6-32 expresses the relationship between the number of modes and waveguide physical parameters:

$$N_M = \tfrac{1}{2} v^2 \qquad (6\text{-}32)$$

where

$$v = \left(\frac{2\pi a}{\lambda}\right) NA$$

NA = numerical aperture = $(n_1^2 - n_2^2)^{1/2}$

An alternate form of Eq. 6-32 is as follows:

$$N_m = 2 \frac{\pi^2}{\lambda^2} a^2 NA^2$$

If the alternate form of equation 6-32 is used to calculate the number of modes for a common 50-μm/125-μm communication waveguide with the following characteristics—core diameter $a = 50$ μm; stepped index, NA = 0.14; and λ = 850 nm—the waveguide will support 334 modes. This does not imply that 334 channels are available to the network designer, however. Mode mixing presents a problem when using this technique, and only a limited amount of success has been achieved to date. Transmission distances are limited to a few hundred meters, and only three to four channels have been implemented. This may become a viable technique, however, when integrated optics increases in growth and short distances cease to be of any consequence. Also, fiber-optic circuitry interconnects may eventually use some form of this technique. The reader should always keep aware of the changes in technology.

GRIN lenses have already been mentioned, but their characteristics have not been discussed. These lenses have some rather useful properties for fiber optics. The quarter-pitch lens is the most commonly used device in fiber optics. A quarter-pitch lens is depicted in Fig. 6-13(a) and a half-pitch lens, in Fig. 6-13(b). Note that the half-pitch device results in a reversed image or focus whereas the quarter-pitch device results in an expanded beam and is collimated. Because of the periodic nature of a GRIN lens, it is possible to obtain collimated, inverted focus, inverted collimated, and upright focus in sequence. The quarter-pitch devices are used when a beam must be expanded to prevent high alignment losses. The lens is able to take the large beam width and reduce it back to the original size.

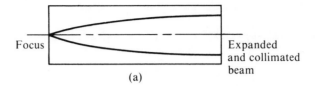

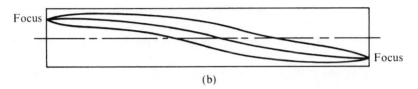

Fig. 6-13. GRIN lenses: (a) Quarter-pitch (0.25P) GRIN lens, and (b) half-pitch (0.5P) GRIN lens.

The lenses have an index of refraction that varies with the radial distance from the axis. From Ferat's principle, the refractive index distribution for meridianal rays (those rays that propagate in the plane containing the lens axis) is given by Eq. 6-33.

$$n^2(r) = N_0^2 \sec h \, (ar) \tag{6-33}$$

where:

n_0 = the refractive index on the lens axis
$n(r)$ = the refractive index as a function of lens radius
α = the quadratic gradient constant

Equation 6-33 can be expanded into a series form, which is expressed in Eq. 6-34:

$$n^2(r) = n_0^2 \left[1 - (ar)^2 + \frac{2}{3}(ar)^4 - \frac{17}{45}(ar)^2 + \cdots \right] \tag{6-34}$$

Helical rays that spiral down the lens at a fixed radius can be described using Fermat's principle for helical rays as expressed by equation 6-35.

$$n^2(r) = n_0^2 \left[1 - (ar)^2 \right]^{-1} \tag{6-35}$$

Expanding this equation into a power series, we have Eq. 6-36:

$$n^2(r) = n_0^2 \left[1 - (ar)^2 + (ar)^4 - (ar)^6 + \cdots (-1)^{n+1} (ar)^{2n-2} \right]$$

$$(6\text{-}36)$$

Note that Eqs. 6-34 and 6-36 differ in the higher order terms; it should be clear, therefore, that the no distribution profile can simultaneously satisfy both meridional and helical rays.

The GRIN lens profile distribution equation is written as shown in Eq. 6-37:

$$n^2(r) = n_0^2 \left[1 - (ar)^2 + h_4(ar)^4 - h_6(ar)^6 + \cdots \right] \qquad (6\text{-}37)$$

The values h_4 and h_6 plus all other like constants are referred to as "aberration constants." These terms become important at the periphery of the lens. They are normally neglected, and Eq. 6-37 may be written as shown in Eq. 6-38:

$$n^2(r) \approx n_0^2 \left[1 - (ar)^2 \right] \qquad (6\text{-}38)$$

The GRIN lens is dependent on wavelength because of α and the refractive index. The sinusoidal patterns for a quarter-pitch and a half-pitch lens are given by Eqs. 6-39 and 6-40, respectively:

$$Z = \frac{\pi}{2\alpha\lambda} \qquad (6\text{-}39)$$

$$Z = \frac{\pi}{\alpha\lambda} \qquad (6\text{-}40)$$

The GRIN lens index profile must be controlled with sufficient precision to permit acceptable imaging characteristics. Since the optical and mechanical axes do not always coincide, alignment must be such as to overcome this difficulty.

For an accurate GRIN lens design used on a WDM coupler, the relationships between wavelength and launch and collection positions must be known. The launch position is the transmitter end, and the collection position is the receive end. Kobayashi and Seki have derived the equation for this relationship, as follows:

$$x_1 = \frac{n_1}{n_0^\alpha} \left[\theta - \sin^{-1} \left\{ \frac{\lambda}{n_1 \Lambda} - \sin \left(\theta - \frac{n_0}{n_1} \alpha x_0 \right) \right\} \right] \qquad (6\text{-}41)$$

where:

x_0 = transmitter position
x_1 = receiver position
n_0 = refractive index of glass wedge
n_1 = refractive index of grating
λ = wavelength
θ = grating angle
α = gradient constant of lens
Λ = period of grating

Equation 6-41 also defines the design relationship for the lens/grating multiplexer shown in Fig. 6-11(d). With a little mathematical manipulation, other relations can also be derived. For example, differentiating λ with respect to x_1 implicitly and solving for the derivative will result in the linear dispersion of the lens/grating configuration as shown in Eq. 6-44. First the terms of equation 6-41 are rearranged as shown in Eq. 6-42 and sine of both sides taken as shown in Eq. 6-43:

$$\theta - x_1 \frac{n_0^\alpha}{n_1} = \sin^{-1}\left\{\frac{\lambda}{n_1\Lambda} - \sin\left(\theta - \frac{n_0}{n_1}\alpha x_0\right)\right\} \qquad (6\text{-}42)$$

$$\sin\left(\theta - x_1 \frac{n_0^\alpha}{n_1}\right) = \frac{\lambda}{n_1\Lambda} - \sin\left(\theta - \frac{n_0}{n_1}\alpha x_0\right) \qquad (6\text{-}43)$$

Equation 6-43 is then differentiated and Eq. 6-44 can be derived by rearranging terms. First,

$$-\left(\frac{n_0^\alpha}{n_1}\right)\cos\left(\theta - x_1\frac{n_0^\alpha}{n_1}\right)\left(\frac{dx_1}{d\lambda}\right) = \frac{1}{n_1\Lambda}$$

then,

$$\frac{d\lambda}{dx_1} = -n_0^\alpha\Lambda\cos\left(\theta - x_1\frac{n_0^\alpha}{n_1}\right) \qquad (6\text{-}44)$$

For the situation when the cosine term is near unity, which is true for most practical cases, the magnitude for linear dispersion results in Eq. 6-45:

$$\left|\frac{d\lambda}{dx_1}\right| = n_0^\alpha\Lambda \qquad (6\text{-}45)$$

In a configuration similar to 6-11(b), constructed with a linear array of equal diameter fibers, the channel spacing can be derived by multiplying linear dispersion Eq. 6-45 and D, the distance between centers of adjacent waveguides, as follows:

$$\delta\lambda = D \left| \frac{d\lambda}{dx_1} \right| = Dn_0^\alpha \Lambda$$

Other items may be calculated using Eq. 6-41, such as solving for the grating period or any other variables.

Some of the items presented in this chapter will be used in the next, which deals with integrated optics. The couplers, however, are all designed without lenses because the integrated circuit would then be impossible to fabricate or align. Some of the devices to be presented can also be used as discrete units, such as surface acoustic wave (SAW) devices. Coupling to waveguides fabricated on substrate material is a bit more difficult.

REVIEW PROBLEMS

1. Calculate the damping ratio for a solenoid operator with $M = 1$ gm, $K_s = 1$ dyne/cm, $c_f = 1$ dyne/cm/sec.
2. For a critically damped solenoid operator with $c_f = 0.05$ dyne/cm/sec and $M = 1.2$ gm, what is time constant R/L in the mechanical form?
3. In problem 2, what is the value of K_s?
4. The galvanometer switch can be replaced with a digital equivalent. Draw a sketch of such a device.
5. Derive the electromechanical equation for a galvanometer circuit where L_f is the field inductance, R_f is the field resistance, and K_T is the torque constant in mA/in.-oz, i.e., $T = KI_f$ and v_f is the field voltage.
6. Let $J = 0.01$ gm-cm²; $f_\theta = 0.02$ gm-cm/rad/sec; $S_\theta = 0.01$ gm-cm/rad; $L_f = 300$ μH; $R = 10$ Ω; and $V = 2$ V. What are the mechanical and electrical time constants? Does the system oscillate? If so, what is required to correct the instability?
7. Evaluate Eq. 6-14 for a 40/60 power split. Remember that the equation depicts amplitude of the outputs.
8. Given the values, 1 μW at R and 1.1 μW at S, find the coupling coefficient for $L = \ell$ and $L = 2\ell$.
9. Design an optimized coupler using all the previous material presented to operate at 1.06 μm.
10. Show the effects that radius reduction has on $C(V)$ of Eq. 6-23. Plot the variation for a/a_0 for $1, \frac{1}{2}, \frac{1}{4}, \frac{1}{8}$, and $\frac{1}{10}$, where a_0 is the initial radius.
11. Plot Eq. 6-23 for wavelength variation between 800 and 1550 nm with V and W constant.

12. What is the cutoff wavelength required to devise a coupler that will separate 820 nm from 848 nm?

13. Is it possible to separate the wavelengths of problem 12 if the spectral width of each source is 26 nm?

14. How many modes are supported in a waveguide with the following characteristics: $NA = 0.21$, core diameter $= 100$ μm, and $\lambda = 1$ μm?

REFERENCES

1. Benscher, H. J.; Budlong, A. H.; Haverty, M. B.; and Waldbaum, G. *Electronic Switching Theory and Circuits*. New York: Van Nostrand Reinhold, pp. 300–361, 1971.
2. Baker, D. G. *Fiber Optic Design and Applications*. Reston Publishing, pp. 232–239, 1985.
3. McIntyre, P. D.; and Snyder, A. W. Power Transfer between non-parallel and tapered optical fibers. *Journal of Optics Society of America* **64**:285–285 (1974).

7 | INTEGRATED OPTICS

This chapter addresses more than just optics issues. Drive circuitry and integrated circuitry are also presented in terms of gallium arsenide and silicon custom VLSI technology. Common MOS structures are covered first because they have been around for several years and can possibly be upgraded. Gallium arsenide (GaAs), on the other hand, is still rather expensive. Its attributes, however, outweigh the extra cost if high switching speeds are required. An introduction to both technologies as applied to integrated optics is warranted if only because of the immaturity of each.

For future military purposes, GaAs-integrated circuits will be almost exclusively used because of their high tolerance to various types of radiation. GaAs devices are useful in space and wherever nuclear radiation is present, either at a low level or during a full-fledged nuclear event. Many devices are presently available in discrete form, but as soon as the fabrication difficulties are solved, larger integration of circuitry can be expected.

MOS integrated circuits are rather common and suffice for any commercial or military application for which high switching speed or radiation hardness is of no consequence. The only devices, for example, that require high switching speeds in a serial link with moderate transmission speed may be the output and input shift registers for the transmitter and receiver, respectively. Data, on the other hand, may be processed at much lower switching speeds. Some of the techniques involved will be examined in detail within this chapter. Circuits of this sort are considered to be hybrid circuits.

To examine silicon and GaAs technology, a comparison of the transient time required in each is necessary. Equation 7-1 is an expression for the transient time in a MOS transistor (FET):

$$\tau = \frac{L}{\mu_m \epsilon} \qquad (7\text{-}1)$$

where τ, L, μ_m, and ϵ are the transit time, channel length, electron mobility, and electric field, respectively. If L and ϵ are constant—i.e., assuming that all the fabrication techniques used for MOS can be used for GaAs—silicon $\mu_m = 6 \times 10^6$ cm/sec and GaAs $\mu m = 1.4 \times 10^7$ to 5×10^7 cm/sec. Thus, GaAs is 2.3 to 8.3 times faster than silicon. Most silicon devices are rather hard pressed to operate at 200 MHz, whereas GaAs devices operate easily at 2 GHz.

Note that the performance ratio is 10 to 1, whereas a straight mobility comparison is approximately 8 to 1. The reason for this discrepancy is that the GaAs material is a better insulator. Therefore, the ϵ field may also be increased, thus further reducing the transit time. Another effect is the reduction of the parasitic capacitance between devices. Consequently, devices may be put closer together; i.e., transistor density is increased. The density increase implies that more functions can be packed into an IC.

Radiation is another item that should be examined because radiation other than nuclear will damage ICs. Silicon can withstand only 10^3 to 10^4 rads, whereas GaAs can withstand 10^7 to 10^8 rads. This, of course, is its appeal to any military application.

Temperature is another factor that makes the appeal of GaAs even greater. The wider energy band gap of GaAs allows these ICs to operate at 300° to 400°C, whereas silicon has a working temperature range of 200° to 250°C. The extra margin will allow manufacturers to make commercial and military grade GaAs parts that are approximately the same. Temperature may not become a test issue, although performance can differ.

This chapter will present components that can be integrated into either optical or combination electrical/optical circuits. After each has been considered, a functional electrical/optical circuit will be examined. In chap. 6, some of the circuits could be either integrated or constructed as a stand-alone device. Here, some of the same philosophy applies, but only integrated-circuit approaches will be examined.

INTEGRATED OPTICS COMPONENTS

The first items to consider are the waveguides used to interconnect components. These interconnecting paths cannot have cylindrical geometries because of fabrication difficulties; i.e., normal IC construction techniques cannot be used. Therefore, all fabrication will be considered to have an IC-like substrate, making use of the various techniques such as sputtering, defusion, ion implant, etc., that are usually employed.

Let us examine some of the waveguide structures common to integrated optics as shown by Fig. 7-1. The first four structures in the figure represent strip-waveguide techniques. The light is confined to the material with the largest refractive index; i.e., $n_f > n_s, n_c, n_i$. The first waveguide is fabricated by starting with a GaAs substrate that is not shown to scale in the figure. The thickness is determined by the need for mechanical strength. It is commonly called the "bulk," that is, bulk GaAs. For a typical "bulk," the substrate is usually part of a wafer 250 μm thick. The guide material is then deposited on the substrate. A third layer, the photoresist, is later removed. A mask is made of the waveguide layout desired, on the substrate. The mask is placed over the photoresist and

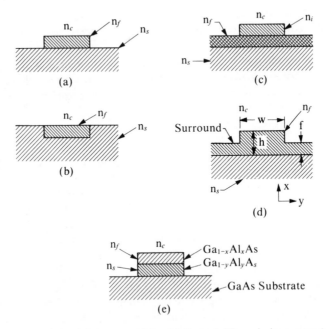

Fig. 7-1. Several strip waveguides: (a) through (d), and planar type (e).

exposed to ultraviolet light. The exposed areas are etched away to the substrate, leaving a waveguide coated with photo-resist; the latter is then removed. This procedure produces the desired effect, i.e., the one shown in Fig. 7-1(a).

The waveguide shown in Fig. 7-1(b) is formed in the substrate material using either the masked diffusion or ion implant method. Each of these methods allows the thickness to be precisely controlled. The third strip waveguide technique, shown in Fig. 7-1(c), requires a deposition of lower index material on a planar waveguide. This waveguide may also be fabricated like that in Fig. 7-1(a), except that three layers are deposited and a mask is used to form the pattern for the lower index material. The ridge guide, as it is commonly called, of Fig. 7-1(d) can be fabricated like the guide of Fig. 7-1(a) as well, except that all of the guide material is not etched away. The last waveguide is the planar waveguide of Fig. 7-1(e), which can also be fabricated like that in Fig. 7-1(a).

The waveguides in Fig. 7-1 have well-defined boundaries, as shown. Edge roughness is a particular problem. Loaded guides and [see Fig. 7-1(e)] and ridge guides reduce this problem somewhat. The thinner portion of ridge guides operates below cut-off, and light propagates only in the thicker region. This technique allows only part of the propagating waveform to strike a vertical wall, thus elevating the roughness problem somewhat. In the loaded waveguide, the same principle holds true except that no vertical wall exists. Another variation of the ridge waveguide is the mirror ridge, which resembles two ridge waveguides back-to-back, with the cross section that would be achieved were Fig. 7-1(d) superimposed on Fig. 7-1(b).

Electromagnetic-mode treatment of a dielectric slab will be presented using Maxwell's equations. This is necessary because ray analysis does not lend itself to single-mode analysis. In the slab waveguide shown in Fig. 7-2, $n_1 > n_2, n_3$. Maxwell's equations are expressed as follows:

$$\overline{\nabla} \times \overline{H} = n_j^2\, e_0 \frac{d\overline{E}}{dt} \qquad (7\text{-}2)$$

$$\overline{\nabla} \times \overline{E} = -\mu_0 \frac{d\overline{H}}{dt} \qquad (7\text{-}3)$$

$$\overline{\nabla} \cdot \overline{E} = 0 \qquad (7\text{-}4)$$

$$\overline{\nabla} \cdot \overline{H} = 0 \qquad (7\text{-}5)$$

where:

$\overline{E} \propto e^{-2wt}$

$j = 1, 2, 3$

μ_0 = magnetic permeability of free space (homogeneous)

Taking the $\overline{\text{CURL}}$ of each side of Eq. 7-3, we have the expression shown in Eq. 7-6:

$$\overline{\text{CURL}}\,(\overline{\nabla} \times \overline{E}) = -\mu_0 \overline{\text{CURL}} \left(\frac{d\overline{H}}{dt}\right) \qquad (7\text{-}6)$$

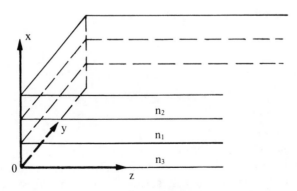

Fig. 7-2. Slab waveguide.

If the identity of Eq. 7-7, as follows,

$$\overline{\nabla} \times (\overline{\nabla} \times \overline{A}) = \overline{\nabla}(\overline{\nabla} \cdot \overline{A}) - \nabla^2 \overline{A} \tag{7-7}$$

where $\overline{A}$ is a general vector, is applied to the expression in Eq. 7-6, then

$$\overline{\nabla}(\overline{\nabla} \cdot \overline{E}) - \nabla^2 \overline{E} = -\mu_0 \frac{d(\overline{\nabla} \times \overline{H})}{dt}$$

Making the appropriate substitutions into Eqs. 7-3 and 7-4, we have

$$\nabla^2 \overline{E} = \mu_0 e_0 n_j^2 \frac{d^2 \overline{E}}{dt^2} \tag{7-8}$$

Let us now substitute the general vector, $\overline{r}$, for space variables x, y, z, and make $E = \overline{E}(\overline{r}, t)$. Then, since μ_0 and e_0 are constant in a vacuum, Eq. 7-8 can be rewritten as Eq. 7-9:

$$\nabla^2 \overline{E}(\overline{r}, t) = \frac{n_j^2}{c^2} \frac{d^2 \overline{E}(\overline{r}, t)}{dt^2} \tag{7-9}$$

The solution to the above equation is of the form:

$$\overline{E}(\overline{r}, t) = \overline{E}(\overline{r}) \, e^{i[wt - \phi(\overline{r})]}$$

Substituting this solution into Eq. 7-9 yields

$$\nabla^2 \overline{E}(\overline{r}) \, e^{i[wt - \phi(\overline{r})]} = \left(\frac{i^2 \omega^2 n_j^2}{c^2}\right) e^{i[wt - \phi(\overline{r})]}$$

so that

$$\nabla^2 \overline{E}(\overline{r}) + k^2 n_j^2 \, E(\overline{r}) = 0 \tag{7-10}$$

where $k^2 \equiv \omega^2/c^2$.

If $\phi(\overline{r}) = \beta Z$, then Eq. 7-10 has no explicit Z dependence other than that of the propagation constant, β. The equations can be written for the three layers of the slab directly if the assumption is made that y extends in both positive and negative directions to infinity and that the modes are uniform throughout the y direction. The equations are as follows:

$$\text{Region 1:} \quad \frac{\partial^2 \overline{E}_1}{\partial x^2} - q^2 \overline{E}_1 = 0 \tag{7-11}$$

$$\text{Region 2:} \quad \frac{\partial^2 \overline{E}_2}{\partial x^2} - p^2 \overline{E}_2 = 0 \tag{7-12}$$

$$\text{Region 3:} \quad \frac{\partial^2 \overline{E}_3}{\partial x^2} - h^2 \overline{E}_3 = 0 \tag{7-13}$$

where:

$$q^2 = n_1^2 k^2 - \beta^2$$
$$p^2 = \beta^2 - n_2^2 k^2$$
$$h^2 = \beta^2 - n_3^2 k^2$$

The observation can quickly be made from Eqs. 7-11, 7-12, and 7-13 that $\beta > kn_1^2$ will result in all exponential solutions to the equations and that for the case when $kn_1 > \beta > kn_3$, kn_2, Eq. 7-12 will have a sinusoidal solution. The previous assumption that y is infinite is a fairly good approximation when the width of the waveguide is much larger than the thickness, i.e., when there is a high aspect ratio.

For this planar waveguide, some function of thickness in the x-direction dimension and Δn, the difference in refractive indices, must be found. This information is necessary to determine these parameters for single-mode operation. We begin with the wave equation and the solution to it of Eqs. 7-14 and 7-15, respectively.

Transverse Electric (TE) Guided Mode Derivation (see Fig. 7-3):

$$\nabla^2 E_y = \left(\frac{n_j^2}{c^2} \right) \left(\frac{\partial^2 E_y}{\partial t^2} \right) \tag{7-14}$$

$$E_z(x, Z, t) = \epsilon_y(x) \, e^{i(\omega t - \beta Z)} \tag{7-15}$$

where:

$$\epsilon_y = A e^{-hx}$$

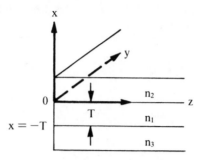

Fig. 7-3. Field diagram of slab waveguide.

for the n_2 region ($0 \leq x \leq \infty$),

$$\epsilon_y = A \cos (qx) + B \sin (qx)$$

for the n_1 region ($-T \leq x \leq 0$), and

$$\epsilon_z = \left[A \cos (Tq) - B \sin (Tq) \right] e^{p(x+T)}$$

for the n_3 region ($-\infty \leq x \leq -T$).

The solutions must be such that continuity must be maintained at the boundary for ϵ_y and $d\epsilon_y/dx$.

$$\frac{d\epsilon_y}{dx} \begin{cases} -hAe^{-hx} & x \geq 0 \\ q(-A \sin qx + B \cos qx) & 0 \geq x \geq -T \\ p(A \cos Tq - B \sin Tq) e^{p(x+T)} & -T \geq x \end{cases}$$

The continuity condition yields the following equations for $x = 0$:

$$-hA = qB$$

and for $x = -T$:

$$q(A \cos Tq + \sin Tq) = p(A \cos Tq - B \sin Tq)$$

This equation may be solved for the ratio A/B and $\tan Tq$, or $\sin Tq/\cos Tq$, as follows:

$$q\left[\left(-\frac{q}{h}\right) + \tan Tq\right] = p\left[\left(-\frac{q}{h}\right) - \tan Tq\right]$$

Solving for tan Tq, we have the following eigenvalue equation for the TE modes:

$$\tan Tq = \frac{q(p + h)}{q^2 - ph} \tag{7-16}$$

Omitting the time and Z dependent factor, the only nonzero field components for TE modes are E_y, H_x, and H_z. Equations 7-17 and 7-18 express the two H fields in terms of the nonzero E_y field:

$$H_x = -\frac{\beta}{\omega\mu_0} E_y \tag{7-17}$$

$$H_z = -\frac{i}{\omega\mu_0}\left(\frac{\partial E_y}{\partial x}\right) \tag{7-18}$$

Transverse Magnetic (TM) Guided Mode Derivation:

The z and t dependent factors are eliminated, and the nonzero field components are derived in terms of H_y as in the previous section for E_y. The nonzero relationships are expressed in Eqs. 7-19 and 7-20. First, for the nonzero E_x field we have

$$E_x = \frac{\beta}{\omega n_j^2 e_0} H_y \tag{7-19}$$

and for the nonzero E_z field:

$$E_z = \frac{i}{\omega n_j^2 e_0} \frac{\partial H_y}{\partial x} \tag{7-20}$$

As in the previous section, the solutions may be written for H_y as follows:

$$H_y \begin{cases} Ce^{-hx} & x \geq 0 \\ C\cos qx + D\sin qx & 0 \geq x \geq -T \\ (C\cos Tq - D\sin Tq)\, e^{p(x+T)} & -T \geq x \end{cases}$$

The equations for E_z may be written using the relationship of Eq. 7-20. Note that in this case the n_j^2 is present, and the equations become a bit more complex because of these factors:

$$E_Z = \frac{i}{\omega e_0} \begin{cases} \dfrac{-hC}{n_3^2} e^{-hx} & x \geq 0 \\[2ex] \dfrac{q}{n_3^2} (-C \sin qx + D \cos qx) & 0 \geq x \geq -T \\[2ex] \dfrac{p}{n_3^2} (C \cos Tq - D \sin Tq) e^{p(x+T)} & -T \geq x \end{cases}$$

The continuity of E_z at $x = 0$ and $x = -T$ results in the following equations:

$$-\frac{hC}{n_3^2} = \frac{qD}{n_1^2}$$

and

$$\frac{q}{n_1^2} (C \sin Tq + D \cos Tq) = \frac{p}{n_2^2} (C \cos Tq - D \sin Tq)$$

Eliminating the ratio, C/D, and solving the equations for $\tan Tq$ results in the eigenvalue equation for the TM modes, as follows:

$$\tan Tq = \frac{(n_3^2 p + n_2^2 h) n_1^2 q}{n_2^2 n_3^2 q^2 - n_1^4 ph} \tag{7-21}$$

MODE NUMBERS AND CUT-OFF DERIVATIONS

The notation, TE_N and TM_N, used here refers to a mode possessing N nodes in the field distribution. The value for N is obtained by taking the argument of eigenvalue Eq. 7-16 or Eq. 7-21 to be $Tq - N\pi$. The cut-off condition is $\beta = kn_1 = kn_2$; i.e., $q = p = 0$. For this condition, substituting these values into Eq. 7-16, we have

$$\tan Tk(n_1^2 - n_2^2)^{1/2} - N\pi = \left(\frac{n_2^2 - n_3^2}{n_1^2 - n_2^2} \right)$$

For $v = (T/2) k(n_1^2 - n_2^2)^{1/2}$ normalized frequency, we have

$$v_C = \tan^{-1}\left[\left(\frac{n_2^2 - n_3^2}{n_1^2 - n_2^2}\right)^{1/2}\right] + \frac{N\pi}{2} \qquad (7\text{-}22)$$

where $\tan^{-1}$ has the range of $0 - \pi/2$. Then, the integer M_{TE} is the number of guided modes represented by Eq. 7-23:

$$M_{TE} = \left\{\frac{1}{\pi}\left(2v - \tan^{-1}\left[\left(\frac{n_2^2 - n_3^2}{n_1^2 - n_2^2}\right)^{1/2}\right]\right)\right\}_{\text{integer}} \qquad (7\text{-}23)$$

N_{TM} for the TM guided modes is derived from Eq. 7-21 and given by Eq. 7-24:

$$N_{TM} = \left\{\frac{1}{\pi}\left(2v - \tan^{-1}\left[\left(\frac{n_1}{n_3}\right)^2 \left(\frac{n_2^2 - n_3^2}{n_1^2 - n_2^2}\right)^{1/2}\right]\right)\right\}_{\text{integer}} \qquad (7\text{-}24)$$

The proof will not be shown here, but a reference to Hunsperger[1] will reveal that if β is solved graphically for the values in Eq. 7-16, the argument for the tangent will be $\tan 2Tq$ and at cutoff,

$$\tan 2Tq = 0$$

$$2Tq = M\pi$$

$$2Tk(n_1^2 - n_2^2)^{1/4} = M\pi$$

$$k = \frac{\omega}{c} = \frac{2\pi f}{c} = \frac{2\pi}{\lambda_0}$$

Therefore,

$$\frac{4T}{\lambda}(n_1^2 - n_2^2)^{1/2} = M$$

Then,

$$\frac{16T^2}{\lambda_0^2}(\Delta n)\, 2n_1 \approx M^2$$

or

$$\Delta n \geq \frac{M^2 \lambda_0^2}{32 \, n_1 T^2}$$

where $M = 1, 3, 5$.

The equation can be rewritten in terms of $M = 0, 1, 2, 3, 4 \ldots$, which result in Eq. 7-25:

$$\Delta n \geq \frac{(M + 1)^2 \lambda_0^2}{32 n_1 T^2} \tag{7-25}$$

For the condition $M = 0$ for various materials, Δn can be calculated as shown in Table 7-1. Since many waveguides are fabricated at a thickness from 1000 Å to 10,000 Å, $T \approx \lambda$.

The analysis of fiber-optic dielectric strip guides is much more complex than that presented for planar types. No exact analytic solutions are available for the modes of strip guides, but numerical calculations have been made for rectangular cross-section cores embedded in a medium with a lower index of refraction. Schlosser and Unser[2] describe a numerical method for guides with high aspect ratios; i.e., their ratio of width to thickness is large. See also Refs. 3, 4, and 5, the latter for a method commonly called the "effective index method."

An example of the effective index method will be presented here to illustrate its application to a ridge waveguide similar to that in Fig. 7-1(d). The guide has the following parameters for Ti diffused in LiNbO$_3$:

$\lambda = 0.8 \ \mu m$
$n_f = 2.234$
$n_s = 2.214$
$n_c = 1$
$N_h > N_f$

Table 7-1. Materials Commonly Used for Waveguides.
(M = 0 for all calculations)

Material	n_3	$\Delta n \geq$
GaAs	3.6	0.0087
Ta$_2$O$_5$	2.0	0.015
Quartz	1.5	0.021
LiNbO$_3$	2.203	0.014
ZnO	2.0	0.015

where λ is the wavelength of operation, n_f is the waveguide, n_s is the substrate, N_h is the effective index of the ridge, and N_f is the effective index of the surrounding material.

The equations that define the various parameter relationships are as follows:

$$V = kh\sqrt{n_f^2 - n_s^2} \tag{7-26}$$

$$b = (N^2 - n_s^2)/(n_f^2 - n_s^2) \tag{7-27}$$

Equation 7-26 is the normalized frquency and Eq. 7-27, the normalized guide index. Both are general equations not related to the figure. Solving Eq. 7-27 for N, we have

$$N = \sqrt{n_s^2 + b(n_f^2 - n_s^2)}$$

If $\Delta n = n_f - n_s$ is small, b approaches unity at $\lambda \gg \lambda_c$ and $\lambda = \lambda_c$ ($b = 0$). An approximation for finding N is then given by Eq. 7-28:

$$N \approx n_s + b(n_f - n_s) \tag{7-28}$$

An additional expression for asymmetric optical guides is represented by Eq. 7-29:

$$a = (n_s^2 - n_c^2)/(n_f^2 - n_s^2) \tag{7-29}$$

For the ridge waveguide proposed, the specific equations are shown below:

$$V_n = kh\sqrt{n_f^2 - n_s^2}$$

$$V_f = kf\sqrt{n_f^2 - n_s^2}$$

$$b_h = (N_h^2 - n_s^2)/(n_f^2 - n_s^2)$$

$$b_f = (N_f^2 - n_s^2)/(n_f^2 - n_s^2)$$

Using the above equations and Fig. 7-4, the normalized ($\omega - \beta$) diagram for $a \to \infty$, the values for V_h, V_f, b_f, b_h, N_h, and N_f can be calculated. These values are $V_h = 4.2$, $V_f = 2.3$, $b_f = 0.2$, $b_h = 0.65$, $N_h = 2.227$, and $N_f = 2.218$. These values are obtained for the TE mode.

Figure 7-4 may be used to find the TM modes if the asymmetric equation is modified as shown in Eq. 7-30:

$$a = \frac{n_f^4(n_s^2 - n_c^2)}{n_c^4(n_f^2 - n_s^2)} \tag{7-30}$$

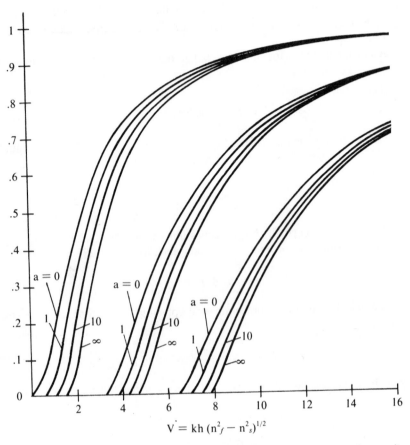

Fig. 7-4. Normalized ω-β for a planar slab waveguide with b as a function of normalized frequency for various values of symmetry, V.

COUPLING TO INTEGRATED OPTICS

The discussion in the previous section has addressed construction techniques and analysis of the waveguides. These devices carry all optical signals to the various signal-processing devices on the integrated optical circuit. A logical next step for discussion prior to analysis of the devices is a review of the techniques for coupling into or out of the integrated optic circuit. In some cases, the source will be located on the IC substrate, and, in others, optical signals brought into the IC will be processed and converted to electrical signals that are then transferred via electrical conductors. We have several combinations of signals, which can be tabulated as follows:

Electrical → optical IC → electrical
Optical → optical IC → electrical
Electrical → optical IC → optical
Optical → optical IC → optical

The first of these signal combinations occurs when optical ICs can be used to process signals for reasons of speed or economy. During surface acoustic wave (SAW) analysis, situations will arise where optical devices are superior for both of these reasons.

The second case occurs when optical signals must be preprocessed prior to detection, e.g., wavelength dimultiplexed, power split, etc. The assumption here is that detection is also performed by the optical IC (OIC).

The third situation occurs when sources are part of the OIC, and optical signals or signal (WDM) are produced.

Last of all is the processing of optical signals in and out of the OIC, which may already have some electrical processing on board. A repeater, for example, would require optical signals in and out, but electrical inputs would also be necessary. When the total OIC is studied, we will return to this subject. It is sufficient here to examine the optical coupling commonly used.

Note that the objectives of optical coupling are to produce low loss across the coupling itself, create as little disturbance as possible so as not to produce extraneous modes, and in some situations, to retain phase continuity. The first two are fairly straightforward requirements. The third is necessary when coupling to a circuit with $\Delta\beta$ switches where phase is very important. Phase must be retained or a correction must be made to the optical signals, which, of course, implies that some amount of insertion loss will be present and signal corruption can be a result.

Coupling of light beams to planar waveguides is the technique of converting laser beams into a surface wave and, the reciprocal operation, i.e., converting surface waves into a beam. In some situations, the laser may be a part of the OIC, and in others, coupling to the OIC is necessary only because detectors are onboard.

As mentioned previously, there are several techniques for fabricating waveguides onto an integrated optic circuit. Strip and planar guides were two alternatives discussed. Planar couplers are classified into two categories: the first, transverse; the second, longitudinal. For the former case, the source beam is focused on the exposed cross section of the guide and, for the latter, the beam is coupled obliquely onto the guide. The first technique involves head-on coupling, and the latter requires grating prisms or tapered couplers.

The first of the transverse couplers is shown in Fig. 7-5(a), which illustrates the conversion of beam energy into a surface wave. The shape of the laser beam is Gaussian, and matching must be adequate to make the input beam width conform to the surface wave field to prevent large energy losses. These losses are attributed to higher surface modes being generated and to radiating modes. Since contour matching is not a difficult task, very low-loss couplers can be constructed. The boundary at $Z = 0$ (end of the guide) must be made flat and free of particulate and dirt. Since the thickness of the waveguide structure is about 1 μm, precise alignment is critical. (At the time this book was being written, this method for coupling laser beams to waveguides was not too promising.) The technique shown in Fig. 7-5(b) requires a little less precision, but index mismatches are nevertheless present, i.e., n_c (guide) > n_s (substrate) and $n_g > n_c$ (cover material).

The prism coupler is more commonly used for coupling beams to planar waveguides. Relatively speaking, this is a rather old technique, which began with theoretical studies in 1962[6] and laboratory experiments in 1964.[7] A prism coupler is shown in Fig. 7-5(c).

The theory of operation for prism couplers is as follows: There is a critical angle at which the ray strikes the prism, θ_c, which has the denser medium, i.e., $n_p > n_a$. Total internal reflection within the prism is governed by Eq. 7-31, as follows:

$$\theta_m > \theta_c = \sin^{-1}\left(\frac{n_a}{n_p}\right) \qquad (7\text{-}31)$$

The refractive index $n_f > n_a$, n_s. To insure complete energy interchange, the two modes must be phase-matched, and the relationship defined by Eq. 7-32 must be observed:

$$k_p \sin \theta_M = \beta_{sw}$$

$$k_p = \frac{2\pi n_p}{\lambda_0} \qquad (7\text{-}32)$$

where k_p is the plane-wave propagation of the uppermost medium and β_{sw} is the surface-wave propagation factor.

The complete interchange of energy occurs along the interaction length, L,

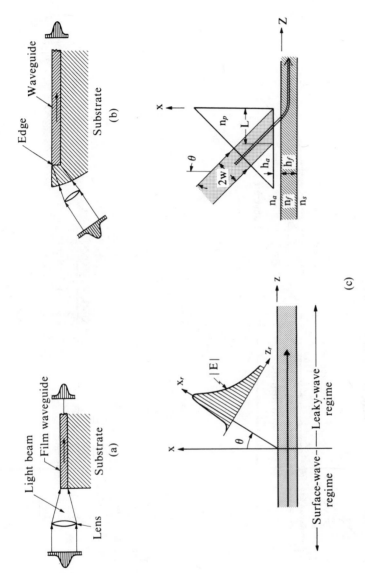

Fig. 7-5. Signal coupling techniques for waveguides: (a) coupling the source with a lens to a waveguide flush with substrate; (b) coupling the source to the waveguide with its edge embedded; (c) prism coupler.

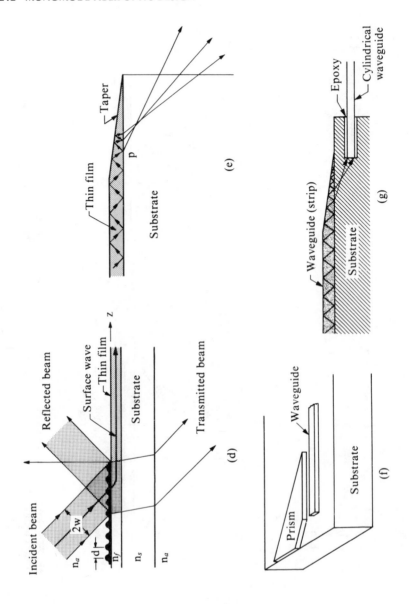

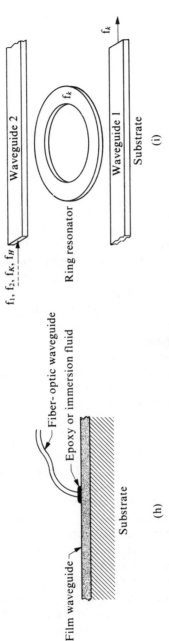

Fig. 7-5. Signal-coupling techniques for waveguides (*continued*): (d) grating coupler; (e) tapered waveguide coupling (waveguide embedded in substrate); (f) prism and gap coupling to waveguide with integrated optics; (g) taper coupling to external cylindrical waveguide; (h) simple pigtail connection to integrated waveguide film; and (i) ring resonator discriminator for wavelength.

along the Z axis in the film waveguide. The equation for this expression is as follows:

$$KL = \pi/2 \qquad (7\text{-}33)$$

where $L = w \sec \theta_M$.

Equation 7-34 defines the coupling coefficient, K, as a function of beam width and angle:

$$K = \frac{\pi \cos \theta_M}{2w} \qquad (7\text{-}34)$$

Some of the basic characteristics of the prism will now be examined. Because of small losses and nonperfect Gaussian beam profiles (effect of beam form), only about 80 percent of the energy is coupled to the planar waveguide. If maximum coupling is to occur, the righthand beam boundary must intersect the prism corner as shown in Fig. 7-5(c). For the condition when the righthand beam boundary falls to the right of the prism corner, this portion of the beam is lost. On the other hand, if the beam falls to the left of the prism corner, a portion of the surface wave is coupled back through the air gap into the radiation mode of the prism, which induces a loss.

Prism couplers must be constructed of dense material compared to n_s, n_a, and n_f. In most situations that arise, $n_s \approx n_f$; then $\beta_{sw} \cong k_f$, $\sin \theta$ is smaller than unity, and $n_p > n_f$, a value that may be difficult to obtain with GaAs. GaAs is a material with a rather high $n(\approx 3.55)$. Because of this problem, transverse end-coupled devices are often used. For couplers with air gaps between the film guide and prism, the presence of dust or grit will have a large impact on the performance. The air gaps are defined by $h_a \leq \lambda_0/2$, where λ_0 is the operating wavelength. The air gap spaces can be filled with a bonding material and adjusted until maximum coupling occurs, or they can be adjusted to make the coupling somewhat similar to a tuning element.

If epoxy is used with bonded couplers, care must be taken to prevent shrinkage, or the coupling may change during curing. Faster drying epoxies usually exhibit this characteristic but slower drying types may be substituted if they are oven cured. One should consult the epoxy manufacturers prior to making a choice because adhesives are a rapidly growing technology.

This discussion has dealt primarily with coupling a beam onto a film waveguide. The reverse is also true; i.e., surface waves may be coupled from the planar waveguide into the prism, thus producing a beam at the output. The couplers are not exactly reciprocal; the beam forms may be different. If exactly the same beam profiles are used, they can be considered reciprocal, but for the practical application this is usually not the case.

Some aspects of prism coupler design must now be considered. If a coupler is poorly designed, incoming to surface wave or outgoing to beam conversions will be very lossy. To be considered viable, therefore, designs must promote high coupling efficiency.

Film waveguides that are thin are able to support only a single mode and may thus be desirable when only single-mode operation is desired. For multimode operation, however, film thickness will become of prime concern. The radiation angle, θ, in an output beam varies with mode number for a given prism configuration and wavelength. Hence, if the incoming surface wave on the film waveguide is multimode, many output beams may appear; these are commonly called "M lines." However, only that single-mode operation is being considered that will help simplify the analysis. Most of the integrated optic devices deal with single-mode operation. When multimode operation is to be considered, each mode may be treated independently. One must be careful because any imperfection in the coupling device can cause mode coupling, and the independent-mode technique may not be used. As mentioned previously, single-mode operation may be insured by choosing a thin film guide since such a guide can support only a single mode.

For readers who wish to delve into the subject in more detail, see Tamir and Bertoni[8] and Tamir.[9] Their analysis involves leaky-wave analysis of the prism. Presented here are the resulting equations and their restrictions, as follows:

$$E = E_0 \exp\left(-jk_a x_r - \alpha Z_r \sec \theta_r\right) \operatorname{erfc}\left(\frac{\alpha LW}{2} - \frac{x_r}{W}\right) \qquad (7\text{-}35)$$

where θ_r, x_r, and Z_r are the beam coordinates and

$E = f(k_a, x_r, Z_r, \theta, \alpha, W)$
$E_0 = $ constant amplitude of the beam
$k_a = $ propagation factor in air $= 2\pi/\lambda_a$
$W^2 = \dfrac{-2jx}{k_a} \cos \theta_r$
$\alpha L = $ leakage factor $\gg 1$

As one may observe, Eq. 7-35 is quite complex, but all is not lost. If the fringing term, $\operatorname{erfc}(\alpha LW/2 - x_r/W)$ is neglected, a reduction in the complexity results, as follows:

$$E = E_0 \exp\left(-jk_a x_r - \alpha Z_r \sec \theta\right)$$

Then, for $Z_r > 0$,

$$E \approx E_0^1 \exp \left(-\alpha Z_r \sec \theta \right) \qquad (7\text{-}36)$$

and for $Z_r < 0$, $E = 0$.

The beam profile is exponential, with width approximately equal to the length along Z_r where the amplitude of E decreases by $1/e$. The relationship between leakage and beam width is

$$\alpha W_b = \cos \theta \qquad (7\text{-}37)$$

where W_b = equivalent width and

$$\sin \theta = \frac{\beta}{k_0 n_p} \cong \frac{\beta_{sw}}{k_0 n_p}$$

Then θ and W_b can be determined by leaky-wave α and β characteristics. Since α is controlled by adjusting the air gap of the prism, h_a—where α gets larger for smaller h_a and smaller for larger h_a—and since it is proportional to e^{-k/h_a} ($\alpha \propto e^{-k/h_a}$), one may readily observe how critical the air gap is for fabrication purposes.

If beam reciprocity is to be maintained, then the incident beam must meet the following criteria:

1. It must be oriented exactly at angle θ.
2. Its width must be as given in Eq. 7-37.
3. Its shape must be identical to that described by Eq. 7-36.

Several aspects of design are expressed in Tamir and Bertoni.[10] They will not be covered here in any detail. Coupling efficiencies, which range from 70 to 80 percent, are affected by a combination of parameters.

Phase matching, as we have discovered with $\Delta\beta$ switches, is a very important parameter. It is affected by the amount of offset the beam has, i.e., the amount of beam overlap at the corner of the prism.

Beam shaping by varying the air gap will also affect the efficiency of the coupler. The air gap can be used for tuning, similar to the way transformer coupling is used in electrical tuning. By sliding, clamping, or changing the coupling instead of the air between the prism and the film waveguide, the shape, leaky-wave characteristics, and phase of the beam may all be varied.

The next type of coupler, shown in Fig. 7-5(d), is the grating coupler. These couplers function in a way similar to prism types except that the prism and air gap are replaced by a grating layer. The grating is usually formed photograph-ically. A photoresist film that has been exposed to the interference patterns produced by two opposite traveling waves is used to produce the grating. The waves are generated by splitting a laser beam into two and recombining the waves. Gratings may have various profiles, e.g., sinusoidal, triangular, trape-zoidal, or other less regular shapes. Profiles of the grating are dependent on the photoresist and developing methods. Research and development are discussed in a number of sources.[11, 12, 13, 14, 15]

Since gratings can be easily fabricated using standard VLSI practices, the technique is viable for coupling laser beam energy to the OIC. The distance between peaks to wavelength, d/λ, must be kept small to prevent loss, which is a result of beam diffraction.

Due to the periodic nature of the grating, space harmonics are produced that have the longitudinal propagation factors expressed by Eq. 7-38:

$$\beta_v = \beta_0 + v \left(\frac{2\pi}{\Lambda}\right) \tag{7-38}$$

where $v = 0, \pm 1, \pm 2, \ldots$

Without the grating, the following relationship holds:

$$\beta_v = K_a n_1 \sin \theta_m$$

The phase match condition can be satisfied when $\beta_v > K_a n_1$ for guided modes ($K_a = 2\pi/\lambda_0$).

Grating equation 7-38 expresses coupling of a beam to the film waveguide in a rather superficial way. Due to lack of space, and in-depth analysis cannot be presented here. Since grating geometry has an impact on the performance of the coupler, various shapes, such as rectangular, triangular, trapezoidal, sinusoidal, etc., result in various desirable and undesirable effects, and these should be studied if a grating is to be fabricated by the engineer. Most gratings will be if they are a part of an integrated optic circuit.

Film waveguide leakage is another issue that should be examined if multimode technology is to be studied, but most of our work focuses on single-mode technology. Couplers other than single-beam couplers are obviously available; in fact, an entire book may be written on them (the references already provided should suffice for the reader who wishes to delve further into the subject.)

Some additional information on couplers will be provided when integrated optic devices are addressed. Among the coupling techniques presented will be film-to-film waveguide coupling, which is important for routing optical connec-tions on an OIC. The reader must be aware that nothing is cast in stone in integrated optics; it is a technology undergoing continuous change. Prior to any design, a literature search should be initiated; as a general rule, timely information is that less than a year old.

The tapered coupler shown in Fig. 7-5(e) is a thin-film tapered waveguide. Readers who wish to pursue the subject are referred to Tien and Martin[16] and Sohler.[17] The film tapers down as it comes closer to the terminating point. The decreasing angle of incidence of the light as it zigzags down the taper will eventually reach or surpass the critical angle. After it does so, the light energy will be refracted as it passes through the substrate and at some point in the substrate will be transformed into a beam.

The light rays emerge at different angles because of the tapered coupler, and the beam is divergent. Beam divergence can be between 1 and 20 percent, with coupling efficiencies up to 70 to 75 percent. Divergence may not be a problem in certain applications—at detectors, say, or in close proximity to guides (guide-to-guide coupling). Lens systems may be necessary, and these increase fabrication costs.

Taper couplers are usually used for coupling light beams out of film waveguides, rather than into them. Efficiency is drastically reduced when coupling light into a tapered guide because it is difficult to align and match because of the beam's divergent shape.

Figure 7-5(f) shows a film prism and gap coupling to a waveguide. This device is considered experimental but has potential. The gap may be controlled geometrically—i.e., by altering its shape or the gap distance itself. The critical angle may be altered through the use of dopants. One must always beware, however, of the following restrictions for prism couplers:

$$n_p > n_2 \qquad n_2 > n_1, n_3$$

To observe these restrictions may not always be feasible or practical. Research is being pursued on this coupler because it can be fabricated as an integral part of an integrated optic circuit and coupling can be closely controlled by dopants. Other possibilities—such as variations of the geometry—will no doubt produce effects that will enhance this coupling technique.

The coupling method shown in Fig. 7-5(g) shows a typical application of a film-tapered coupler used for coupling a beam to a cylindrical fiber-optic waveguide. Note that the light beam is penetrating several materials with different indices of refraction, as follows: the substrate, n_s; the epoxy, n_e; the core, n_c; and the film waveguide itself, n_f. Each will have an effect on the incident angle.

Let us examine some of the values: $n_s \approx 3.50$ (for GaAs), $n_e \approx 1.3$ to 1.5; $n_f \approx 1.5$ (for silica). The coupling technique is primarily used for multimode coupling; with careful selection of materials, good coupling efficiency can be achieved. For example, GaAs is not a good choice of material. Some other materials with an index of refraction close to that of silicon, such as $LiNbO_3$ (lithium niobate) may be better utilized for the substrate. Another possibility is a doped area where the coupling is to take place. There are numerous possibilities for tapered film couplers. The objective here is to show some of them.

Consider now a simple pigtail waveguide connected to a film waveguide using a couplant or epoxy, as shown in Fig. 7-5(h). This is the more direct approach, but the coupling efficiency is rather low (40 to 50 percent). Tien et al.[18] discuss this technique, which is called a "leaky wave coupler" in the literature. It can be used when losses are not important because the integrated optics application is necessary for a particular signal-processing enhancement. A further discussion of these devices will be reserved for the integrated optic circuit section of this chapter.

Figure 7-5(i) illustrates a ring resonator. Note the potential filter applications of this optical component. These devices can be fabricated simply by using VLSI technology, which is ideal for WDM demultiplexing applications or other wavelength discrimination applications. The equations and calculations required for various applications are presented by Chang et al.[19] Further discussions will be presented in the section on integrated optic circuits at the end of this chapter, where all components discussed in Chaps. 6 and 7 will be shown as part of integrated optic circuits.

Figures 7-6(a), 7-6(b), and 7-6(c) show planar-to-planar couplers. These methods of coupling are particularly useful in integrated optics because optical circuits may often be in more than a single layer of the OIC.

The first of these coupling techniques is similar to the $\Delta\beta$ coupler; i.e., a certain coupling length, distance between waveguides (n_1, $n_3 > n_a$, n_s, n_2), and wavelength will produce the desired coupling. This coupling can be used to alter phasing of waveforms or may be designed to be phase-independent. Again, only single-mode technology is considered.

Figure 7-6(b) is an efficient method of coupling two planar waveguides; efficiencies approach 100 percent. This method requires an index of refraction of $n_2 > n_s$, as is the case for each of the other coupling methods. Note that the film mismatch is present and the efficiency is high. For single-mode operation, the mismatch does not present a problem, but for multimode operation, 100-percent efficiency is impossible. The taper in the two waveguides must be gentle in order to prevent excessive radiation into the surrounding medium, i.e., into n_s and n_a (air) surrounding the strip guides.

The third and last technique of planar-to-planar waveguide coupling, with a grating, as shown in Fig. 7-6(c), can yield efficiencies approaching 65 to 70 percent. This coupler will be similar to the $\Delta\beta$ switch; i.e., the overlap must be equal to the coupling length, L, and the height of the grating, the periodicity, n_g, and the material between the grating (air for this case) will all affect performance. Also, one must consider the effects of various grating geometries—

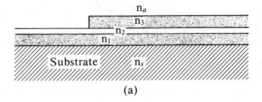

$$(a)$$

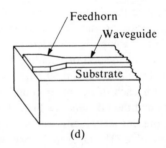

$$(b)$$

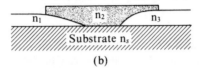

$$(c)$$

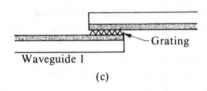

$$(d)$$

Fig. 7-6. (a) Two optical waveguides, n_1 and n_2, sharing same substrate; (b) two tapered waveguides connected to the third layered guide; (c) two waveguides coupled with a grating, (d) feedhorn coupling to waveguide.

such as triangular sinusoidal, trapezoidal, rectangular, etc.—on performance. Work done on these couplers is discussed in Chang et al.[20]

A tapered feedhorn is shown in Fig. 7-6(d), which has a long taper of several wavelengths. Short tapered sections cause excessive losses. A theoretical analysis of the feedhorn is given in Pinnow.[21] The results of the analysis indicated that 90-percent coupling efficiency is possible for 50-μm input width with a 2-mm (2000-μm) taper transition to a 3-μm film waveguide. Since these are large dimensions for integrated-circuit VLSI technology, this feedhorn could not be installed on an OIC easily, but it would be adequate for use on some hybrid devices, i.e., one consisting of a combination of OIC and IC components collocated on a substrate. Photoresist and substrate materials are used to fabricate the interconnects for both optical and electrical circuit wiring. Most integrated-circuit technology for transistor circuits measures 1.5 to 5 μm on a side. With some of the new X-ray fabrication techniques, 0.3-μm devices have been fabricated, with the entire die measuring 1 to 3 mm. Therefore, it would be difficult to impossible to fabricate feedhorns on integrated-optic substrates unless large dies are used (10 to 20 mm on a side).

FIBER-OPTIC MODULATORS

Modulation can be accomplished either directly or indirectly. The direct technique, which is used in the majority of applications, requires the modulation signals to be impressed on the source, which modulates the source intensity. This is a straightforward approach. The indirect technique uses a source that is continuous and either disrupts the beam (intensity modulation) or changes the wavelength (optical version of FM). This section will address both direct and indirect modulation, but more emphasis will be placed on the latter because the other types of modulation have been covered to some extent in the chapter on fiber-optic transmitters.

Here, the discussion of modulation techniques will be limited to optical angular modulation, i.e., phase or frequency. Most of these types of modulation are more difficult to implement than intensity modulation, which is currently by far the most widespread. The advantages of frequency modulation such as conversion gain, however, may make the complexity worthwhile to improve overall performance of the transmission system. Phase modulation has some inherent advantages, but each component and the waveguide itself must pass the optical signals without phase corruption. This can be a rather tall order. Phase corrections are built into devices; predistortion of the phase or postdistortion of the phase are required to make phase appear uncorrupted at the receiver detector.

Optical wavelength modulators (frequency modulation) have not had widespread use. Optical wavelength is shifted in response to an applied signal. For the small shifts in frequency available from optical-frequency modulators, detection requires the use of heterodyne type systems, and this presents implementation difficulties. Bragg modulators are the most suitable for this type of modulation.

As was already discussed in regard to $LiNbO_3$, switch phase correction and modulation can both be accomplished using these types of devices.

Before going into modulation techniques in more detail, a discussion of some of the figures of merit is in order. Performance is the end result of any design; i.e., performance must be maximized for the particular design constraints given in the specifications. Design figures of merit often allow the engineer some type of performance trade-off or combination of economic and performance trade-offs. The latter combination is more prevalent in industry. Four widely used measures of modulators are extinction ratio, bandwidth, drive power required per unit bandwidth, and insertion loss. Some of these terms are almost self-explanatory. Extinction ratio is not. The term refers to the maximum depth of modulation. The drive power per bandwidth is usually given at a particular extinction ratio. Also, a term somewhat neglected here is linearity. Where non-linearity occurs in modulation processes, harmonics will be present. The extent of these harmonics will determine such phenomena as cross-talk and channel noise.

For intensity modulation, the depth of modulation is an important factor. Intensity modulation refers to a device that varies the intensity of a coherent light wave (lasers are implied here) in response to a time-varying signal (modulation). For the condition where no modulation signal is applied to the modulator, the light intensity at the detector is I_0. When the maximum signal is applied, we have I_m. The extinction ratio is defined as follows:

For $I_m \leq I_0$,

$$\eta_m = \left(\frac{|I_m - I_0|}{I_o} \right)$$

For $I_m \geq I_0$,

$$\eta_m = \left(\frac{|I_m - I_0|}{I_0} \right)$$

For signals less than the maximum modulation intensity, the I_m and η_m values are replaced with I and η, respectively, and the extinction ratios are as follows:

For $I_m \leq I_0$,

$$\eta = \left(\frac{|I - I_0|}{I_0} \right)$$

For $I_m \geq I_0$,

$$\eta = \frac{|I - I_o|}{I_m}$$

The modulation depth described previously can also be related to phase changes as in polarization and Bragg grating devices with the following equation:

$$\eta = \sin^2\left(\frac{\Delta\phi}{2}\right) \tag{7-39}$$

Often the modulator is designed with phase bias so that $\Delta\phi = \pi/2$ with no signal and $\Delta\phi$ has a maximum excursion of ± 1 radian. Then Eq. 7-39 has a maximum excursion of 70 percent (intensity change).

Waveguide devices that use phase changes to couple energy from one guide to another or couple energy from one distinct mode to a second distinct mode have a relationship expressed by Eq. 7-40:

$$\eta = 1 - \left[A + (B\Delta\phi)^2\right]^{-1} \tag{7-40}$$

The 2-radian excursion of the phase modulator is equivalent to a 70-percent modulation index for intensity modulators. This excursion often cannot be tolerated in lasers that are directly modulated electrically. For phase modulation using external techniques, the 70-percent excursion presents no immediate problems.

The extinction ratio can be used as a figure of merit because the SNR at the detector is dependent on it. If the extinction ratio is too low and most of the detected light carries no signal, then the background noise of most detectors will be dominated by the noise in the unmodulated received light. This condition is realized when modulation is on the order of 3 to 5 percent. For lasers that are directly modulated, the modulation is usually about 50 percent or less. Although some lasers presently allow for a modulation index of 60 percent, overmodulation—i.e., greater than 60 percent for a few cycles—can result in laser source destruction.

The next figure of merit to consider is the power per unit bandwidth. Some manufacturers prefer the term known as "specific energy" $(P/\Delta f)^2$. Units for this measurement are expressed in milliwatts per megahertz. The bandwidth is defined as the frequency difference between two frequencies for which the intensity modulation drops from 70 to 50 percent. Most FM systems allow a 10-percent modulation bandwidth for systems operating at optical wavelengths (10,000 gHz is common). Desirable bandwidths are 8 to 10 gHz, but present-day modulators cannot achieve these values. For those engineers familiar with FM techniques, the analysis becomes simpler when the deviation is small as compared to the carrier frequency (wavelength in fiber-optic systems). As new modulation methods are discovered, optical type FM communications will undoubtedly become more appealing. Also, wide bandwidths are necessary when performing fast switching, which is one of the most important applications. With the advent of GaAs technology, higher speed serial transmission is necessary for both local and long haul networks. Since this new technology can operate in the 500 MHz to 1 GHz region, time division multiplexing several channels of information in time can require many gigahertz of bandwidth. A great deal of research is being conducted to increase the limits of present methods.

When modulators are inserted into networks, the insertion loss must be known accurately to determine loss budgets. The insertion loss is defined as follows:
For $I_m \geq I_0$,

$$L_i = 1 - \frac{I_m}{I_{in}}$$

For $I_m \leq I_0$,

$$L_i = 1 - \frac{I_0}{I_{in}}$$

or

$$dB = 10 \log \left(1 - \frac{I}{I_{in}} \right)$$

for the losses defined above.

The correct method of defining insertion loss is to include all connectors, filters, or other items that induce an optical attenuation. Some systems cannot tolerate high insertion loss because the length of the link will be reduced, thus causing a need for more repeaters. This, of course, will have a great deal of impact on system economics. The modulator insertion loss must be considered, but one should also be aware that demodulation produces an insertion loss, and this can sometimes be as large as the modulator loss.

The last item is a figure of merit for switches: switching time. It is related to bandwidth as follows:

$$T = \frac{2\pi}{\Delta f}$$

Other miscellaneous items of importance, such as SNR, noise, crosstalk, etc., will be considered when specific devices are discussed. Often, optically induced noise components are small compared to electrical noise such as for wide bandwidth receivers, and thus neglected. On the other hand, SNR variations in indirect modulation schemes are entirely due to optical noise. Therefore, each device must be evaluated depending on the nature of its functionality. When dealing with intensity modulation, crosstalk is similar in many respects to extinction ratio. Crosstalk will be covered later in the chapter.

The Bragg cell modulator will be the first technique to investigate because of its utility as a fiber-optic component. The first approach to finding the relationship between acoustic and optic waves will be intuitive. Let us begin by examining some of the physical quantities in Table 7-2.

For the condition that momentum is conserved during the collisions shown in Fig. 7-7, where K is a wave vector, Eq. 7-41 (for phonon annihilation) and Eq. 7-42 (for phonon creation) are implied:

$$K' = K + K_a \qquad (7\text{-}41)$$

$$K' = K - K_a \qquad (7\text{-}42)$$

The vectors for Eq. 7-41 may be reduced to their component levels as shown in Eqs. 7-43 and 7-44:

$$K' \cos \theta' = K \cos \theta \qquad (7\text{-}43)$$

$$K' \sin \theta' = K_a - K \sin \theta \qquad (7\text{-}44)$$

Then, by dividing Eq. 7-44 by 7-43 and rearranging the terms, we have Eq. 7-45:

$$\tan \theta' = \frac{K_a}{K} \left(\frac{1}{\cos \theta} \right) - \tan \theta$$

$$\theta' = \tan^{-1} \left[\frac{K_a}{K} \sec \theta - \tan \theta \right] \qquad (7\text{-}45)$$

Table 7-2. Comparison of Photon/Phonon.

Photon	Phonon	Physical Value
tr K	tr K_a	Momentum
tr ω_0	tr ω_a	Energy
ω_0	ω_a	Angular frequency

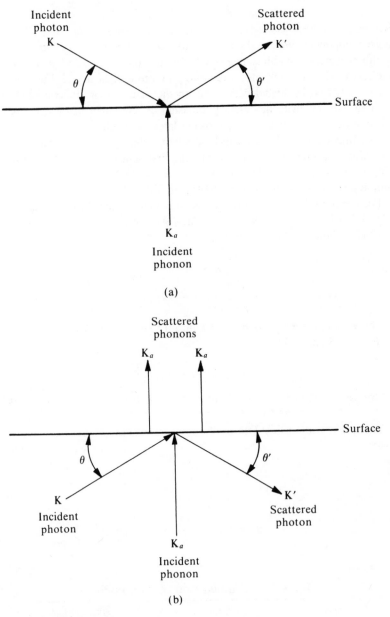

Fig. 7-7. Phonon and photon interaction.

For the restrictions, $\theta \ll 1$ rad and $K_a/K \ll 1$, Eq. 7-45 can be further reduced to the approximation expressed in Eq. 7-46:

$$\theta' = \frac{K_a}{K} - \theta \qquad (7\text{-}46)$$

Optical and acoustic velocities are expressed from wave theory as $c = \omega_0/K$ and $v_a = \omega_a/K_a$, respectively. Then Eq. 7-46 may be rewritten in the alternative form:

$$\theta' = \left(\frac{c}{v_a \omega_0}\right) \omega_a - \theta \qquad (7\text{-}47)$$

The Bragg angle, θ_B, is defined by Eq. 7-48 for $\theta' = \theta = \theta_B$:

$$\sin \theta_B = \frac{K_a}{2K} \qquad (7\text{-}48)$$

The conservation of energy is approximately valid for photon–phonon collision, and the frequency relationship for a scattered photon is

$$\omega = \omega_0 + \omega_a$$

for phonon annihilation and

$$\omega = \omega_0 - \omega_a$$

for phonon creation.

The frequency of the diffracted wave will satisfy the two previous frequency relationships for most cases when practical devices employ Bragg cells.

When the incident and diffracted refractive indices are different, a departure from isotropic diffraction characteristics occurs. The phenomenon is due to birefringent diffraction. In an anisotropic medium, a change in direction and polarization between the incident and refracted wave causes birefringent diffraction.

In most practical situations, $\omega_a \ll \omega_0$ and $\omega \approx \omega_0$, with the relationship between K' and K expressed as follows:

$$K' \cong rK$$

where r is the ratio of refractive index associated with the diffracted wave to the refractive index associated with the incident wave.

The relationship between the parameters, K_a, K, and r as a function of $\sin \theta'$ (diffracted beam) and $\sin \theta$ (incident beam) is derived as follows. Using Eq. 7-44,

$$\sin \theta' = \left(\frac{K_a}{K'}\right) - \left(\frac{K}{K'}\right) \sin \theta$$

Therefore,

$$\sin \theta' = \left(\frac{K_a}{2Kr}\right) - \left(\frac{1}{r}\right) \sin \theta$$

$$= \left(\frac{K_a}{2Kr}\right) - \left(\frac{K}{2K_a r}\right) (1 - r^2)$$

Factoring,

$$\sin \theta' = \left(\frac{K_a}{2Kr}\right) \left[1 - \left(\frac{K}{K_a}\right)^2 (1 - r^2)\right] \tag{7-49}$$

Also,

$$\sin \theta = \left(\frac{K_a}{2K}\right) \left[1 + \left(\frac{K}{K_a}\right)^2 (1 - r^2)\right] \tag{7-50}$$

Equation 7-49 is the desired result for the diffracted beam, and Eq. 7-50 is the desired result for the incident beam, shown without a derivation. Some interesting results can be observed from these equations. The first is that both equations require the magnitude of the righthand side to be less than or equal to 1. The restrictions can be shown as follows:

$$1 - \left(\frac{K_a}{K}\right) \leq r \leq 1 + \left(\frac{K_a}{K}\right)$$

where $K_a < K$.

Let us examine the case when $\theta' = \theta$. Then, $r = 1$, which is associated with the Bragg angle of incidence. Another solution does exist when $f = (K_a/K) - 1$, but the condition, $K_a < K$, cannot be satisfied; therefore, only the condition, $r = 1$, need be considered.

The Bragg cell interactions must now be considered to complete the discussion of these devices. Shown in Fig. 7-8(a) is a schematic diagram of a Bragg cell. An acoustic wave is generated by impressing the electrical input signals on the piezoelectric transducer attached to one end of the medium that has an acoustic termination that will suppress any reflected waves. As a result of acousto-optic interaction, the diffracted waves shown in the figure are modulated by the electrical information impressed on the piezoelectric transducer, which in turn modulates the medium and the diffracted beam. The modulated light has a spatial variation due to the modulation that is equivalent to a time variation of the signal; therefore, parallel processing of the optical signal is possible.

Let us now examine the figure and note that Λ is the acoustic wavelength. The optical frequency is again given by ω_0, and K is the wave vector of the incident light [see Fig. 7-8(b)]. The acoustic wave has similar parameters, where ω_a and K_a are the acoustic-wave angular frequency and wave vector, respectively. Due to acoustic-optic interaction, the angular frequency is characterized by the expression, $\omega_m = \omega_0 \pm m\omega_a$, and $K_m = \pm mK_a$ is the wave vector relationship. The value of m is represented by the integers, $m = 1, 2, 3, \ldots$

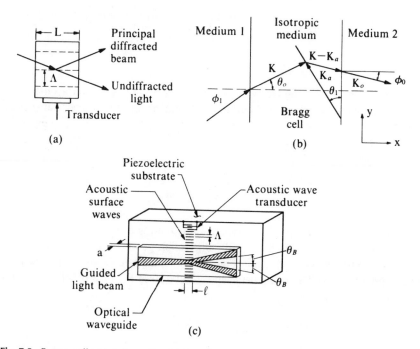

Fig. 7-8. Bragg cell: (a) Schematic, (b) Ray diagram, and (c) fiber-optic film version of Bragg cell that may or may not be operated in the Bragg regime.

n. These representations describe the constituent plane of the optical field in a medium.

A factor defined as Q determines which regime the Bragg cell is operating in. The first is called the Raman-Nath Q, which is < 1; the second is the Bragg Q, which is $\gg 1$; and the last regime is a combination of the two, $1 < Q < 10$. The Raman-Nath operation of the cell is characterized by several beams being diffracted with significant amounts of power. For Bragg operation of the cell, only two beams are emanated: the undiffracted main beam and the principal diffracted beam.

The value Q is defined by the following equation:

$$Q = \left(\frac{K_a^2}{K}\right)L = \left(\frac{\lambda}{2\pi}\right)\left(\frac{w_a}{V}\right)^2 L \qquad (7\text{-}51)$$

Let us evaluate a simple example using the following values: $c = 3 \times 10^{10}$ cm/sec; $v = 3.5 \times 10^5$ cm/sec; $f_a = 20$ MHz; and $\lambda_0 = 0.5$ μm (0.5×10^{-4} cm). Note that if the wavelength of light is 1.3 μm and the cell is made 5 cm long, then the cell will be operating in the Bragg region. Then,

$$Q = \left(\frac{0.5 \times 10^{-4}}{2\pi}\right)\left(\frac{4\pi \times 10^7}{3.5 \times 10^5}\right) \times 1$$

$$= 1$$

The cell must be designed with the operating wavelengths and acoustic operating frequencies determined before the L dimension and the material (which determines the acoustic velocity) are selected. The Bragg angle was previously defined as follows:

$$\sin \theta_B = \frac{K_a}{2K} = \frac{1}{2}\left(\frac{\omega_a c}{n\omega_0 V}\right)$$

where the velocity of light in an isotropic medium is c/n.

A beam-trajectory plot showing the diffraction geometry of a Bragg cell for a tilted acoustic wave is shown in Fig. 7-8(b). Using Snell's law, Eq. 7-52 may be written for the relationship between ϕ_1 and θ_0 [see Fig. 7-8(b)]. For medium 1, that of air, we have Eq. 7-53.

$$n \sin \theta_0 = n_1 \sin \phi_1 \qquad (7\text{-}52)$$

$$n \sin \theta_0 = \sin \phi_1 \qquad (7\text{-}53)$$

The vector equations for K and K_a can now be written as Eqs. 7-54 and 7-55, respectively, where i and j are unit vectors in the x and y directions, respectively:

$$\overline{K} = K \cos \theta_0 i + K \sin \theta_0 j \qquad (7\text{-}54)$$

$$\overline{K}_a = -K_a \sin \theta_1 i + K_a \cos \theta_1 j \qquad (7\text{-}55)$$

where $K = n\omega_0/c$ (optical) and $K_a = \omega_a/V$ (acoustic). Then

$$\overline{K}_0 = K_{0x} i + K_{0y} j$$

$$\overline{K}_0 = \overline{K} - \overline{K}_a = (K \cos \theta_0 - K_a \sin \theta_1) \, i$$

$$+ (K \sin \theta_0 - K_a \cos \theta_1) \, j$$

but

$$\sin \phi_0 = \frac{K_{0y}}{K_0} = \frac{K \sin \theta_0 - K_a \cos \theta_1}{\left[(K \cos \theta_0 - K_a \sin \theta_1)^2 + (K \sin \theta_0 - K_a \cos \theta_1)^2 \right]^{1/2}}$$

Using trigonometric reductions,

$$\sin \phi_0 = \frac{K \sin \theta_0 - K_a \cos \theta_1}{K - K_a}$$

Then,

$$\sin \theta_0 = \frac{\dfrac{1}{n} K \sin \theta_1 - K_a \cos \theta_1}{K - K_a}$$

For $K_0 \approx \dfrac{\omega_0}{c}$, $\omega_a \ll \omega_0$, and $K_0 \approx K$, we have

$$\sin \phi_0 = \sin \theta_1 - \left(\frac{K_a}{K_0} \right) \cos \theta_1 \qquad (7\text{-}56)$$

Equation 7-56 shows an important relationship between the angle at which the optic and acoustic beams are injected into a Bragg cell and the output of the diffracted beam, provided the acoustic and light have the relationship that $\hat{K}_a \times$

$\hat{K} > 0$ (where $\hat{K}_a$ and $\hat{K}$ are unit vectors in the K_a and K direction). If angles ϕ_1 and θ_1 are much less than 1, Eq. 7-56 can be approximated by Eq. 7-57, which holds for $\theta_1 > 0$, $\theta_0 > 0$, $\phi_1 > 0$, and $\phi_0 < 0$:

$$\phi_0 = \phi_1 - \left(\frac{c}{V\omega_0}\right)\omega_a \qquad (7\text{-}57)$$

Another special case to consider is when the acoustic beam angle is twice that of the light beam with respect to the references shown in Fig. 7-8(b), i.e., when $\theta_0 = 2\theta_1$ and the Bragg angle is defined as shown in Eq. 7-58:

$$\omega_B = \left(\frac{2nV\omega_0}{c}\right)\sin\theta_1 \qquad (7\text{-}58)$$

If the θ_0 value is substituted into the K_{0y} component of $\overline{K}_0$, we have Eq. 7-59, as follows:

$$K_{0y} = K\sin 2\theta_1 - K_a\cos\theta_1 \qquad (7\text{-}59)$$

Substituting the values for K, K_a, and ω_B into Eq. 7-59, the result is Eq. 7-60:

$$K_{0y} = \left(\frac{\omega_B - \omega_a}{V}\right)\cos\theta_1 \qquad (7\text{-}60)$$

Note that in Eq. 7-60, if $\omega_a = \omega_B$, then the diffracted beam emerging from the Bragg cell is along the X axis, i.e., it is perpendicular to the cell–air interface. From the equations describing the Bragg cell, it is evident that for $\theta_0 = 2\theta_1 = 2\theta_B$, the angle between the incident light beam and the acoustic wave front is equal to the Bragg angle, θ_B.

In the Bragg regime of operation, the amplitude of the principal diffracted light beam is approximately proportional to the amplitude of the acoustic wave under the following conditions: The acoustic wave amplitude must be fairly small and the bandwidth of the acoustic wave must be narrow compared to its center frequency. The change, Δn, in the optical waveguide refractive index brought about by acoustic modulation is expressed by Eq. 7-61 (see Pinnow[21] for the derivation):

$$\Delta n = \sqrt{n^6 P^2 \left(10^7 P_a\right)/2\rho V_a^3 A} \qquad (7\text{-}61)$$

where P, P_a, ρ, A, and n are photoelectric tensor components similar to r_{13}, r_{33}, etc. P_a is the acoustic power in watts; ρ is the density; A is the cross-sectional area $(a \times \ell)$; and n is the refractive index, respectively.

If we let $M_2 = n^6 P^2 / \rho V_a^3$ Eq. 7-61 can be rewritten in a more compact form with all items due to material constants reflected in M_2 (with a typical value for quartz of 1.5×10^{-18}):

$$\Delta n = \sqrt{10^7 M_2 (P_a / \ell a)} \tag{7-62}$$

Equation 7-62 reflects the relationship, as one may observe, between acoustic amplitude and refractive index variation, which may then be related to phase and finally to the power per megahertz variation, $P_a / \Delta f$, a figure of merit.

The equation relationship between phase and Δn is given by Eq. 7-63:

$$\Delta \phi = \frac{2 \pi \ell \Delta n}{\lambda_0} \sin \left(\frac{2 \pi a}{\Lambda} \right)$$

$$= \frac{2 \pi \ell}{\lambda_0} \sqrt{10^7 M_2 (P_a / \ell a)} \sin \left(\frac{2 \pi a}{\Lambda} \right)$$

or

$$\Delta \phi = \frac{2 \pi}{\lambda_0} \sqrt{10^7 M_2 (P_a \ell / a)} \sin \left(\frac{2 \pi a}{\Lambda} \right) \tag{7-63}$$

Equations 7-61, 7-62, and 7-63 represent values and parameters for Fig. 7-8(c), which is an integrated optic depiction of a Bragg cell.

Next, a relationship expressing $P_a / \Delta f$ will be developed that gives a figure of merit for acoustic-optic modulators in watts/MHz. An assumption is made that the diffraction angles of light and sound are matched, thus giving an upper limit on the acoustic-optic modulator. (See Gordon[22] for an in-depth analysis.) The condition previously stated is that $\Lambda / \ell = \lambda_0 / nb$, where Λ is the sound wavelength and b is the beam width. Then,

$$\Lambda = \frac{V_a}{f_a} \tag{7-64}$$

and

$$\Delta f = \frac{V_a}{b} \tag{7-65}$$

If an equation is written to express $\Delta f(\Delta\phi^2)$, we have as a result Eq. 7-66:

$$\Delta f(\Delta\phi^2) = \frac{V_a 4\pi^2}{\lambda_0^2 b} 10^7 M_2 \left(\frac{P_a \ell}{a}\right) \sin^2\left(\frac{2\pi a}{\Lambda}\right) \tag{7-66}$$

Substituting the value for ℓ into Eq. 7-66—i.e., $\ell = (nb/\lambda_0)(V_a/f_a)$—the result is Eq. 7-67:

$$(\Delta f)(\Delta\phi^2) = \left(\frac{4\pi^2 V_a^2 \, 10^7 M_2 P_a n}{\lambda_0^3 f_a a}\right) \sin^2\left(\frac{2\pi a}{\Lambda}\right)$$

or

$$(\Delta f)(\Delta\phi^2) = \left(\frac{4\pi^2 V_a^2 10^7 M_2 P_a n}{\lambda_0^3 f_a a}\right) \frac{1}{2}\left[1 - \cos\left(\frac{4\pi a}{\Lambda}\right)\right] \tag{7-67}$$

Taking the maximum value, where the cosine term is $\pi/2$ and $\Delta\phi = 2$ radians, we have

$$\Delta f = \frac{\pi^2 V_a^2 10^7 M_2 P_a n}{2\lambda_0^3 f_a a} \tag{7-68}$$

Letting $M_1 = V_a^2 n \, M_2$ and solving for $P_a/\Delta f$, we have

$$\frac{P_a}{\Delta f} = \frac{2\lambda_0^3 f_a a}{\pi^2 M_1 10^7}$$

or

$$\frac{P_a}{\Delta f} = \frac{45 f_a \lambda_0 a}{M_1} \text{ mW/MHz} \tag{7-69}$$

Equation 7-69 is adjusted for cgs units, i.e., λ_0 and a are measured in centimeters. An alternative, more appropriate form expresses a and λ_0 in micrometers (see review problem 8). In Eq. 7-69, f_a is expressed in Hz. For the discrete form of the Bragg cell, the acoustic beam can be cylindrical, but this case will not be considered here (see review problem 9).

Table 7-3 is a representation of various materials used in bulk modulators and Bragg cells and also shows the theoretical limits of modulation, including a safety factor that results in a practical limit.

The next modulator to be considered is the Mach-Zehnder interferometer (see

Table 7-3. Acousto-Optic and Physical Properties of Common Optical Modulator Materials.

Material	M_1 10^{-6} $(cm^2\text{-}sec)/gm$	M_2 10^{-18} sec^3/cm	M_3 10^{-11} $(cm\text{-}sec^2)/gm$	Theoretical limit, $P_a/\Delta f\, mW/MHz$	Practical limit, $P_a/\Delta f\, mW/MHz$
TeO$_2$ (S)	6.85	792.75	11.1	0.12	0.6
PbMoO$_4$ (L)	11.4	34.78	3.13	0.41	2.05
LiNbO$_3$ (L)	6.6	6.95	1.0	1.29	6.45
SiO$_2$ (fused quartz) (L)	0.79	1.51	0.13	9.8	49.0

Material	n	Acoustic Attenuation, dB/cm	V_a, 10^5 cm/sec	Δn @ $P_a/A = 100\ W/cm^3$
TeO$_2$ (S)	2.27	4.9	0.617	1.3×10^{-4}
PbMoO$_4$ (L)	2.39	3.3	3.66	6.2×10^{-4}
LiNbO$_3$ (L)	2.2	0.05	6.57	5.8×10^{-5}
SiO$_2$ (L)	1.46	3.0	5.96	2.7×10^{-5}

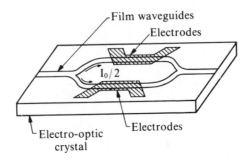

Fig. 7-9. Mach-Zehnder phase modulator or switch.

Fig. 7-9). The transfer function for the light intensity or optical power can be written either way as follows:

$$P = P_0 \cos^2 (\Gamma/2) \tag{7-70}$$

or

$$I = I_0 \cos^2 (\Gamma/2) \tag{7-71}$$

These equations indicate that, for no phase shift between the two legs of the interferometer, the signals are precisely summed with no loss. For the case when Γ is 180°, complete interference results, and the output is zero. The electrodes or length of the arms can be adjusted to produce the phase shift desired. Often, the phase is trimmed using the electrodes to bias the arm.

To insure linear phase modulation, $\pi/2$ bias is added to the phase, thereby producing Eq. 7-72.

$$P = P_0 \cos^2 \left[(\pi/2) + (\Gamma/2)\right]$$
$$= P_0 \sin^2 (\Gamma/2) = (P_0/2) (1 - \cos \Gamma) \tag{7-72}$$

Mach-Zehnder waveguides are fabricated by diffusing photolithographically defined Ti patterns (4 μm wide and 400 Å thick) into a Z-cut LiNbO$_3$ substrate. A 2500-Å SiO$_2$ layer is sputtered on the sample to isolate the waveguides from the electrodes. The electrodes are sputter-deposited or electroplated over the SiO$_2$ layer above the waveguides.

The electrode fabrication is such as to produce a characteristic impedance of approximately 40 Ω. The sputtered electrode dimensions are 18 μm wide, 3 μm thick, and 6 mm long, and the gap between the stripling and ground is 15 μm. The 3-dB bandwidth is approximately 3.5 GHz. Care must be taken during packaging to prevent package-to-devices resonance within the device bandwidth, which, of course, will reduce the usable bandwidth.

The modulator is actually a combination of two devices and may be used as such, i.e., as a beam splitter and as a combiner. It has a great deal of utility when separated into the two devices, each an integrated optic component.

In the next section, other components will be investigated as part of the total circuitry. Filters, for example, have not been covered heretofore because discussing them as part of a total optical circuit will have a somewhat greater impact. Other types of modulators will also be covered. One must always keep in mind that the technology is rapidly changing and that only the rudimentary or basic principles are presented here. But care has been taken to impart sufficient information to the reader to assist him in the analysis of more complex components as they arrive in the marketplace.

VLSI TECHNIQUES APPLIED TO INTEGRATED OPTICS

In Chap. 6 and other sections of the present chapter, the components that make up integrated optics have been introduced. This section will examine three aspects of very large scale integration (VLSI), as follows: The first investigates integrated optic circuitry and some of the associated fabrication techniques. The second addresses GaAs and MOS logic control and drive circuitry for integrated optic circuits. The third and final aspect examines the necessary total integration of the optic, control, and drive circuitry into a viable design solution.

The first approach to integrated optics is to consider some examples of integrated optic circuitry. One must keep in mind that most of the circuits referred to here are not in full-scale production, i.e., some of them have been extracted from R&D papers and articles. The objective is to introduce the reader to the general technology, rather than to actual optical circuit design, because of the rapidly changing nature of fiber optics.

MOS is introduced at this time because this technology is useful for slow-speed switching circuitry and because in many local-area-networking applications it is the most cost-effective. Some of the fabrication techniques involved are compatible with gallium arsenide on a one-for-one basis.

Figures 7-10(a) through 7-10(d) depict some of the circuits that will be examined. The first of these, Fig. 7-10(a), is an electro-optic circuit with a 4:1 multiplexer. A FET driver is provided to drive a laser located in a well in the integrated-circuit substrate. The laser is the only optical segment of the circuit. (Details of the fabrication of the circuit will be explored in the next section.) Circuits similar to this one have operated at 8 G bits/sec, which is fairly impressive.

The circuit depicted in Fig. 7-10(b) is for the receiver and demultiplexer and

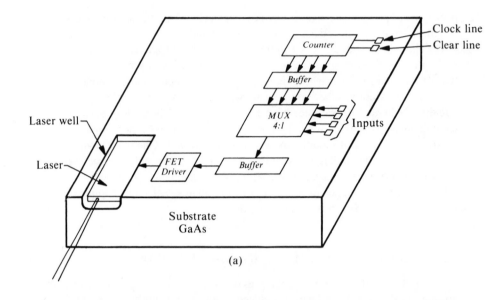

(a)

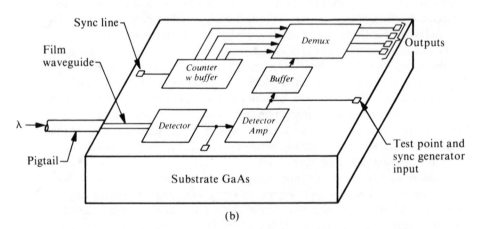

(b)

Fig. 7-10. Integrated optics: (a) Typical integrated optical and electronic circuit depicting a 4:1 multiplex fabricated on the same substrate with laser and its drive circuitry; (b) demultiplexer and receiver integrated optic and electronic circuit; (c) wavelength division multiplexer with two analog and two digital channels, all of which are optical circuits except for the electrodes; and (d) 4×4 optical integrated switch.

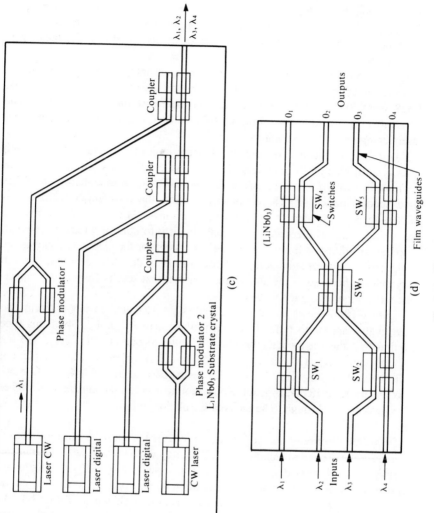

Fig. 7-10. (Continued)

complements that in Fig. 7-10(a). All of the high-speed circuitry is located on the GaAs IC with the detector; this allows lower-speed devices to process the demultiplexer output ports. One of the previously discussed coupling techniques may be used.

These two devices may be used for short-range interconnects or local-area-network transmitter and receiver pair. For the low-cost version of these circuits, MOS technology may be used, and the substrate material will be silicon. The transmitter can be fabricated with a LED source for short-range transmission, lasers being reserved for long distances.

Figure 7-10(c) shows a wavelength division multiplexer. The output of each laser is controlled by the couplers. The two analog channels have the lasers on continuous wave (CW). The Mach-Zendler type phase modulators are either modulated or biased off. The couplers will allow the wavelength of the digital devices to be modulated in an on/off fashion, or the lasers themselves may be modulated. The couplers allow external circuitry to control which wavelength is to be transmitted. All circuits are optical except for the electrical connections to the lasers, phase modulators, and coupler electrodes. The electrodes switch very quickly because of the small time constant that results from small capacitance and large resistance values.

Figure 7-10(d) shows an integrated-optic 4×4 switch. In this diagram, the wavelengths appear to differ from each other; actually, a 4×4 of this type requires that the wavelengths be close in value to one another, e.g., 830 nm (λ_1), 850 nm (λ_2), 870 nm (λ_3), and 890 nm (λ_4). Otherwise, the couplers or switches will become very lossey. The worst-case losses of a 4×4 are approximately 12 to 18 dB, depending on coupling loss and the number of switches the signal passes through. The cross talk isolation is fairly high—from 30 to 50 dB.

The switch electrodes and active region length, L_0, must be carefully chosen to produce the desired switching effect. The $\Delta\beta$ switch with all design parameter variations was discussed in Chap. 6. Let us assume that the wavelengths are similar and crossover (switching) occurs when voltage is applied to the electrode of the switch. A truth table of the switch action is shown in Table 7-4. An examination of this truth table reveals several redundancies; for example, when switches 1, 2, 4, and 5 are on, the outputs will be identical to those when all switches are off. When an even number of switches are on, the switching action can sometimes defeat the purpose of the device; for example, 1 and 4 or 2 and 5 on (in each of these cases, the switching action is negated). The objective is to switch the incoming wavelengths with a minimum number of switches. All of the possible switching combinations are shown in Table 7-4. Note that a number of outputs are produced by redundant inputs.

Table 7-5 reduces the entries of the various inputs; all possible output com-

TABLE 7-4.   4 × 4 Switch Truth Table.

Inputs				Switches (On = 1; Off = 0)					Outputs ($\lambda_n = \eta$)			
λ_1	λ_2	λ_3	λ_4	SW1	SW2	SW3	SW4	SW5	O_1	O_2	O_3	O_4
1	1	1	1	0	0	0	0	0	1	2	3	4
1	1	1	1	1	0	0	0	0	2	1	3	4
1	1	1	1	0	1	0	0	0	1	2	4	3
1	1	1	1	1	1	0	0	0	2	1	4	3
1	1	1	1	0	0	1	0	0	1	3	2	4
1	1	1	1	1	0	1	0	0	2	3	1	4
1	1	1	1	0	1	1	0	0	1	4	2	3
1	1	1	1	1	1	1	0	0	2	4	1	3
1	1	1	1	0	0	0	1	0	2	1	3	4
1	1	1	1	1	0	0	1	0	1	2	3	4
1	1	1	1	0	1	0	1	0	2	1	3	4
1	1	1	1	1	1	0	1	0	1	2	4	3
1	1	1	1	0	0	1	1	0	3	1	2	4
1	1	1	1	1	0	1	1	0	3	2	1	4
1	1	1	1	0	1	1	0	0	4	1	2	3
1	1	1	1	1	1	1	1	0	4	2	1	3
1	1	1	1	0	0	0	0	1	1	2	4	3
1	1	1	1	1	0	0	0	1	2	1	4	3
1	1	1	1	0	1	0	0	1	1	2	3	4
1	1	1	1	1	1	0	0	1	2	1	3	4
1	1	1	1	0	0	1	0	1	1	3	4	2
1	1	1	1	1	0	1	0	1	2	3	4	1
1	1	1	1	0	1	1	0	1	1	4	3	2
1	1	1	1	1	1	1	0	1	2	4	3	1
1	1	1	1	0	0	0	1	1	2	1	4	3
1	1	1	1	1	0	0	1	1	1	2	4	3
1	1	1	1	0	1	0	1	1	1	2	4	3
1	1	1	1	1	1	0	1	1	1	2	3	4
1	1	1	1	0	0	1	1	1	3	1	4	2
1	1	1	1	1	0	1	1	1	3	2	4	1
1	1	1	1	0	1	1	1	1	4	1	3	2
1	1	1	1	1	1	1	1	1	4	2	3	1

binations are shown (as one may observe, four output combinations cannot occur). This circuit has a great deal of utility provided the loss can be eventually reduced; it behaves similarly to common PBX switching.

The overview of the circuits previously discussed addressed their functionality only. Now, the fabrication techniques will be examined for the equivalent electro-optic circuits. Figure 7-11(a) diagrams the laser located in the well shown

TABLE 7-5. Reduction in Switching Operations.

		Switches				Outputs ($\lambda_n = n$)		
SW1	SW2	SW3	SW4	SW5	O_1	O_2	O_3	O_4
0	0	0	0	0	1	2	3	4
0	1	0	0	0	1	2	4	3
0	0	0	1	0	1	3	2	4
0	0	1	0	1	1	3	4	2
0	1	1	0	0	1	4	2	3
0	1	1	0	1	1	4	3	2
0	0	0	1	0	2	1	3	4
1	0	0	0	1	2	1	4	3
1	0	1	0	0	2	3	1	4
1	0	1	0	1	2	3	4	1
1	1	1	0	0	2	4	1	3
1	1	1	0	1	2	4	3	1
0	0	1	1	0	3	1	2	4
0	0	1	1	1	3	1	4	2
1	0	1	1	1	3	2	4	1
1	0	1	1	0	3	2	1	4
.... Cannot occur					3	4	1	2
.... Cannot occur					3	4	2	1
0	1	1	1	0	4	1	2	3
0	1	1	1	1	4	1	3	2
1	1	1	1	1	4	2	3	1
1	1	1	1	0	4	2	1	3
.... Cannot occur					4	3	1	2
.... Cannot occur					4	3	2	1

in Fig. 7-10(a) and shows the various layers of its heterostructure. Figure 7-11(b), on the other hand, shows the diffused strip laser structure within the well. A complete description of the various masking steps, the types of process used, and some of the other finer details can be found in Refs. 23 and 24.

The details of FET fabrication are shown in Fig. 7-11(c). Although the FET shown uses GaAs technology, this technique is rather generic; i.e., it applies to MOS or other technologies where FETS are fabricated on a substrate.

An investigation of some of the design subtleties is warranted at this point to determine how design effects performance. Previously, the transit time of the devices was shown to be related to the operating frequency. The transit time equation is repeated here as Eq. 7-73.

$$\tau = \frac{L^2}{\mu \overline{V}_{DS}} \tag{7-73}$$

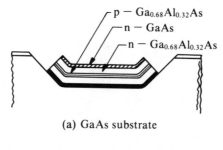

(a) GaAs substrate

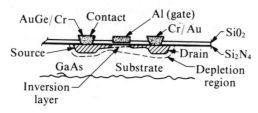

(b) Diffused stripe added

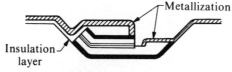

(c) FET detail

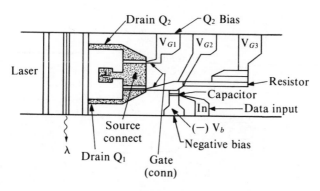

(d) Top view

Fig. 7-11. Integrated optic construction details: (a) Laser fabrication layers, (b) diffused strips added to the laser, (c) FET detail, (d) chip layout of laser and driver, (e) circuit diagram of laser and driver, (f) symbol for FET, (g) layout of a FET, (h) common NOR gate construction and logic symbol, and (i) electrical circuit equivalent of Fig. 7-11(h).

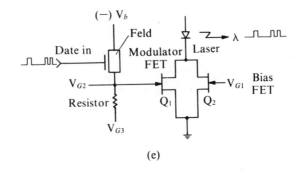

(e)

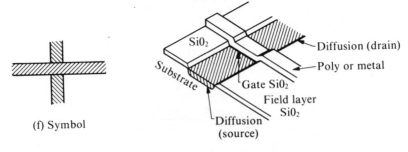

(f) Symbol

(g) FET Layout detail

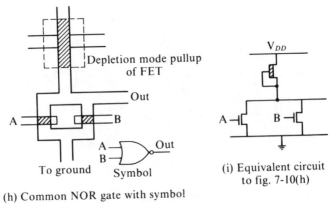

(h) Common NOR gate with symbol

(i) Equivalent circuit to fig. 7-10(h)

Fig. 7-11. (*Continued*)

Eventually, the channel length, L, can be as small as 1 μm because GaAs is a better insulator. Thus the voltage need not be changed, i.e., reduced. If the mobility, μ, for GaAs is 5×10^7 and the V_{DS} is 3 V, then $\tau = 6.67 \times 10^{-20}$ sec; this implies bandwidths in the thousands of GHz. However, there are a number of limiting factors: I/O pin spacing, intrinsic capacitance inherent in the FET, interwiring capacity (wiring on the IC itself), external stray capacitance, and many others. The bandwidth of the IC itself will no longer be limited by material constraints but imparted by external components.

The FET shown in Fig. 7-11(c) has two layers of insulation material. The first is a thin layer of silicon nitride, which will assist in preventing surface impurity contamination of the gallium-arsenide substrate material. A heavier layer of silicon dioxide is then deposited. The wells that form the source and drain structures are a result of doping the substrate using diffusion or ion implant techniques. In the structure with the laser well of Fig. 7-11(d), epitaxial growth is necessary to construct the laser. When used, the ion-implanted areas are annealed during the thermal cycle. A more complete description of the fabrication technique can be found in Ref. 23. The gate connection can be fabricated either of metal or polycrystalline silicon. The latter has fairly high resistance and is commonly used in MOS technology.

Figure 7-11(e) is the circuit diagram of the wafer layout shown in Fig. 7-11(d). The sources of Q_1 and Q_2 are connected as shown in both diagrams and connected to a ground pad. The drains of Q_1 and Q_2 are wired to the cathode of the laser diode. These two FET transistors will handle currents as high as 50 mA and are thus quite large compared to most transistors used for switching. The common switching FET will switch approximately 0.1 mA or less, depending on what it is driving (e.g., fanout). In Fig. 7-11(d), the source and drain areas, which consist of diffusion or ion-implanted areas, are lightly shaded. Note that the gate dimensions are long and narrow: the two unshaded strips with a thin layer of insulation underneath and a substrate below that. The pads for connection to the exterior are metal. Pad V_{Gl} is used for biasing the FET to maintain the laser diode just above threshold. Not fabricated on the chip shown are the bias and temperature compensation circuits that must be used to stabilize the laser for large temperature excursions. Note that a large amount of space is necessary for resistor fabrication. In common silicon or GaAs technology, resistors are not fabricated because they require large amounts of space. They are warranted, however, for high-frequency specialized circuits of this type.

After all circuits are fabricated and the metal connects are deposited, a cap of SiO$_2$ is deposited to seal the circuit from contaminants.

The logic circuitry—not shown in IC layout form in Fig. 7-11(d) but shown in pictorial form in Fig. 7-10(a)—is constructed using common MOS fabrication techniques. Figure 7-11(f) is a symbol used for a FET in a typical chip layout, and Fig. 7-11(g) shows how a FET is actually constructed [the cross-sectional view is given in Fig. 7-11(c)]. Note that the two wells depicting source and drain in Fig. 7-11(g) do not have metal contacts, and the field oxide insulates the polymetal gate contact from the source and drain contacts shown.

Figure 7-11(h) is a layout of a NOR gate, which is the preferred circuit in GaAs or MOS technologies. Its electrical representation is shown in Fig. 7-11(i). Note that the pullup device is a FET rather than a resistor, a FET being much smaller and easier to fabricate than a resistor. The remaining logic is fabricated from a series of these NOR gates, which are the necessary building blocks for all the logic. For the reader interested in delving into fabrication techniques, design rules for layouts, and insight into how performance is affected by design, see Mead and Conway[25] and Allen.[26]

The final topic to be discussed is detector fabrication on an IC substrate. Logic fabrication will be left to the reader using the references just mentioned as a study guide or other appropriate texts.

A detector fabricated in a silicon substrate is shown in Fig. 7-12(a). A hole is etched in the SiO_2 layer, and a P+ well is diffused as shown or ion implant methods are used to form the PN junction. Optical glass waveguide ($n = 1.53$)

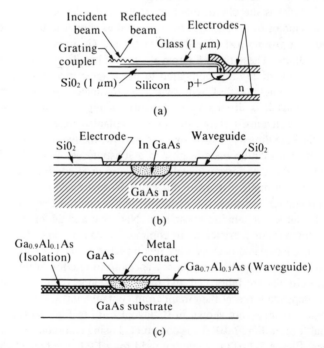

Fig. 7-12. Detector integrated-circuit construction details: (a) Silicon detector photo-diode, (b) GaAs detector, and (c) GaAlAs detector.

is sputtered onto the SiO_2 layer and down into the P+ well to form the optical circuit. The junction is about 1 μm below the surface of the silicon substrate. The metal electrodes are deposited as shown. Another consideration is that such structures must be protected from large FETs by guard rings. Guard rings are established by enclosing or partially enclosing, a circuit with conductors to prevent high concentrations of electrons or holes from wandering through the substrate. Such rings will prevent coupling crosstalk from the FETs.

The detector structure is confined to short wavelength (600 to 900 nm) and bandwidths of under 1 GHz, although larger bandwidths are possible. For wide bandwidths, the structures must be sufficiently small—less than 50 μm in diameter, which is actually large for MOS technology.

The GaAs detector in Fig. 7-12(b) is formed using a combination of epitaxial growth techniques that may be described as follows: The waveguide layer of GaAs (5 μm thick; $n \sim 10^{13}$) is grown on the GaAs substrate material ($n \sim 10^{18}$) using vapor-phase epitaxy. Silicon dioxide (SiO_2) is then deposited pyrolitically and used as a mask. Holes (125 μm in diameter) are formed in the mask using photolithography. Chemical etching is required to remove the GaAs in the detector well (hole), where InGaAs is to be grown. The SiO_2 mask allows growth only in the well area. After the growth process is finished, the mask is removed and the InGaAs is polished to be flush with the waveguide. A contact is deposited and a protective layer of SiO_2 is sometimes placed over the structure.

Such detectors are Schottky barrier photodetectors. They have avalanche gains of up to 250, quantum efficiencies as high as 50 percent, and bandwidths in the neighborhood of 2 GHz. They operate well at wavelengths of 1.06 μm, which makes them ideal for use as detectors for YAG sources.

Development of GaAlAs integrated-optic detectors such as the one diagrammed in Fig. 7-12(c) is similar in many respects to that of the previous detector except that liquid-phase, instead of vapor-phase, epitaxy is used for growth.

These detectors should give the reader an idea of what is available at present. A multitude of other techniques are covered by Tamir.[27]

Usually one will find that the density of fiber-optic circuits is somewhat lower than that of their electronic counterparts by about 60 percent. Some functions, however, are better performed optically than electrically because of the simpler optic circuitry. This is what the engineer should strive for; i.e., to use integrated optics for the advantages it offers, not for its own sake alone.

REVIEW PROBLEMS

1. Derive the following expression, rewritten from Eq. 7-24 using Eq. 7-21:

$$N_{TM} = \left\{ \frac{1}{\pi} \left(2v - \tan^{-1} \left[\left(\frac{n_1}{n_3} \right)^2 \left(\frac{n_2^2 - n_3^2}{n_1^2 - n_2^2} \right) \right] \right) \right\}$$

2. Calculate the thickness of a planar waveguide for $\Delta n = 0.015$, $\lambda_0 = 1 \mu m$, $n_1 = 1.53$, and $M = 0$. What is the T/λ_0 ratio?

3. Which of the materials in Table 7-1 has the largest propagation time?

4. Predict the guiding characteristics of the ridge guide shown in Fig. 7-1(d) in the y direction, where $N_h = n_h$, $N_f = n_f$, and $N^2 = N_f^2 + b(N_h^2 - N_f^2)$. Find V_y, b, and the β, k relationship for $k = kN$. Predict the number of guided modes as $V_y = f(w, \lambda)$.

5. Find the TM modes in the previous example for $\omega = 2 \mu m$.

6. What is the Q value in Eq. 7-51 for a Bragg cell with the following parameters: $\lambda_0 = 1.55 \mu m$; $v = 3.5 \times 10^5$ cm/sec; $f_a = 10$ MHz; and $L = 2$ cm?

7. In which regime is the cell in problem 6 operating? If all parameters but the wavelength are constant, what are the limits for each regime and for the mixed regimes, i.e., when $1 < Q < 10$?

8. Derive an alternative form for Eq. 7-69 when a and λ_0 are given in micrometers.

9. Given the following equation,

$$P_a/\Delta f = 50.8 \, \lambda_0^3 M_3 \text{ mW/MHz}$$

what is the λ_0 value for LiNbO$_3$ at the practical and theoretical limits?

10. Given the following relationship

$$\eta = \frac{(I_0 - I)}{I_0} = \sin^2\left(\frac{\Delta\phi}{2}\right)$$

find the value for η when $\lambda_0 = 1\mu m$ and $a = 1 \mu m$ for the materials listed in Table 7-3.

11. The value for the zeroth-order beam is given by Problem 10; the first-order beam (diffracted) is determined by the Bragg angle, $2\theta_B$, where $\sin \theta_B = \lambda/2\Lambda$. What is the first-order beam value?

12. Why is the Mach-Zehnder phase modulator not used without the $\pi/2$ phase bias? (Hint: Derive with the bias).

13. Draw an optical circuit for a 5×5 switch with a truth table. How many switches are necessary?

14. Draw an optical circuit with switches from a single laser input to four outputs.

15. Derive an equation showing the number of output combinations for the five-switch circuit in Fig. 7-10(b).

16. Draw a diagram for a 4:1 MUX using NOR gates.

17. Prepare a layout using MOS techniques for the 4:1 MUX in problem 16.

REFERENCES

1. Hunsperger, Robert G. Integrated Optics. Study guide, Dept. of Electrical Engineering, University of Deleware, 1978, pp. 3.11–3.14.

2. Schlosser, W. and Unger, H. G. Advances in Microwaves. New York: Academic Press, 1966.

3. Goell, J. E. Bell System Technical Journal **48**:2071 (1969).

4. Marcatili, E. A. *Bell System Technical Journal* **53**:645 (1974).
5. Ramaswamy, V. *Bell System Technical Journal* **53**:697 (1974).
6. Iogansen, L. V. *Sov. Physics-Tech., Phys.* **7**:295 (1962); **8**:985 (1964); and **11**:1529 (1967).
7. Ostenberg, H., and Smith, J. *Journal of Optical Society of America* **54**:1078 (1964).
8. Tamir, T., and Bertoni, H. L. Unified theory of optical beam couplers. *Digest of Technical Papers*, Topical Meeting on Integrated Optics, Optical Society of America, 1972, pp. MB3-1.
9. Tamir, T. *Optik* **37**:269 (1973).
10. Tamir, T., and Bertoni, H. L. *Journal of Optical Society of America* **61**:1397 (1971).
11. Harris, J. H.; Winn, R. K.; and Delgoulle. *Applied Optics* **II**:2234 (1972).
12. Ogawa, K.; Change, S. C.; Gopori, B. L.; and Rosenbaum, F. J. *IEEE Journal of Quantum Electronics* **QE-9**:29 (1973).
13. Peng, S. T.; Tamir, T,; and Bertoni, H. L. *Electron. Letters* **9**:150 (1973).
14. Neviere, M.; Vincent, P.; Petit, R.; and Codilhag, M. *Optical Communication* **8**:113 (1973); **9**:48 (1973); **9**:240 (1973).
15. Peng, S. T., and Tamir, T. Proceedings of Symposium of Optical and Acoustical Micro-Electronics, Polytechnic Press, New York, 1974, p. 377.
16. Tien, P. K., and Martin, R. J. *Applied Physics Letters* **18**:398 (1974).
17. Sohler, W. *Journal of Applied Physics* **44**:2343 (1973).
18. Tien, P. K.; Smolinsky, G.; and Martin, R. J.: Theory and experiment on a new film-fiber coupler. *Digest of Technical Papers*, Topical Meeting on Integrated Optics, Optical Society of America, 1974, pp. WB6-1.
19. Chang, D. C.; Holland, R.; and Kuester, E. F.: Radiation loss from a dielectric channel waveguide bend. *SPIE*, The International Society for Optical Engineering, vol. 317, 1981.
20. Chang, W. S. C.; Muller, M. W.; and Rosenbau, F. J. Integrated optics. In *Laser Applications*, vol. 2. New York: Academic Press, 1974, p. 227.
21. Pinnow, D. *IEEE Journal of Quantum Electronics* **QE-6**:223 (1970).
22. Gordon, E. I. *Proceedings of the IEEE* **54**:1391 (1966).
23. IEEE, GaAs IC Symposium, 1983, pp. 48–53.
24. SPIE Proceedings, Integrated Optics III, vol. 408, April 5–6, 1983, pp. 121–127.
25. Mead, C., and Conway, L. *Introduction to VLSI Systems.* New York: Addison-Wesley, 1980.
26. Allen, J. Introduction to VLSI design (video study guide). MIT Center for Advanced Engineering Study, 1985.
27. Tamir, T. *Topics in Applied Physics Integrated Optics*, vol. 7. Springer-Verlag, pp. 284–302.

8 | LOCAL AREA NETWORKS

INTRODUCTION

Large or long-haul networks, such as telegraph and telephone networks, have been in operation since the turn of the century. Such networks have well established design rules because their technology is older and more mature than that of the local area networks (LANs) of today. Many of the smaller networks such as military command posts and navy shipboard communications can be considered local area networks. Of these, many are little more than scaled-down versions of telephone facilities.

With the advent of microprocessors, local area networks have been designed to handle digital data and video and voice traffic. LANs are networks with installations located less than 10 kilometers but more than 1 meter apart; these installations are connected within the confines of a room, building, campus, etc. There is no hard and fast rule that defines the physical constraints of a LAN, however, because of the rapid changes in technology.

Networks that interconnect locations within a city or that connect cities within a country are considered to be long-haul networks. The distances between their installations are greater than 10 but less than 100 kilometers. Networks connecting countries are thought of as interconnected long-haul networks. As a rule, LANs are small networks under the complete control of a single organization. Some form of interface is used when traffic is leaving or entering a facility (see Fig. 8-1); in this situation, two LANs are connected via a long-haul network. For example, two LANs may be installed in sales offices, but each must use some form of interface to communicate with the standard telephone system as well as to communicate with each other. The point (which will be discussed later on) is that LANs must have some rules (called "protocol") if all terminals need to communicate with computers or each other. The rules are decided upon by the LAN design. The interface (gateway) must abide by both telephone company and LAN regulations to make communication possible. As one can readily observe, the complexity of LANs can increase rapidly if different protocols are involved, which is often the case.

A question one may ask is why use a LAN at all? A few examples of comparing point-to-point wiring without LANs to local area networking will make the answer obvious. Examine the network in Fig. 8-2(a). This is a point-to-point wired network without LAN capability; any communication is accom-

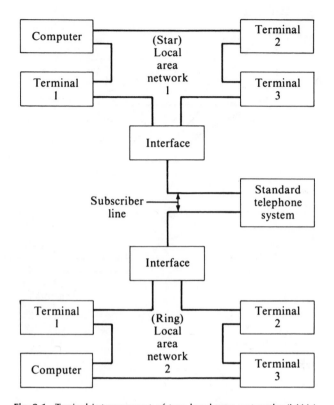

Fig. 8-1. Typical interconnect of two local area networks (LANs).

plished via a wire path between the devices. The number of connections for point-to-point wiring is expressed by Eq. 8.1:

$$C = \frac{N(N-1)}{2} \tag{8-1}$$

where C = the number of connections and N = the number of devices to be connected.

For the LAN shown in Fig. 8-2(b), the number of connections is equal to the number of devices, which is also the case for Fig. 8-1. The two LANs are known as "ring" and "star," respectively and will be discussed in more detail later on.

For connecting small numbers of devices together, point-to-point wiring may suffice, but suppose 100 terminals and a minicomputer are connected much as

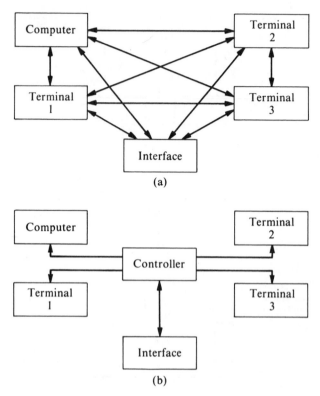

Fig. 8-2. Comparison between a point-to-point wired network (a) and a star network with virtual connections (b).

in Fig. 8-2(a). This will require 4950 connections compared to only 100 for a LAN network. Imagine the task of installing the necessary wiring, which must also be shielded for high-speed data transmission. For all the advantages of a LAN, however, it does not allow one to get something for nothing. The complexity of controlling the network increases with the number of LAN configurations, because as each pair of devices is connected their path is shared by other connected pairs. This is commonly known as a "virtual connection"; i.e., the devices communicate as though individually wired point-to-point but in reality share a common path.

There are almost countless applications for LANs from a simple one or two-room office to a multiple-building complex. Let us consider a simple application such as the resource-sharing network found in most sales offices. In Fig. 8-3, the central sales office has a disk-storage station that contains the sales database. The data is processed by a minicomputer with a production printer for producing

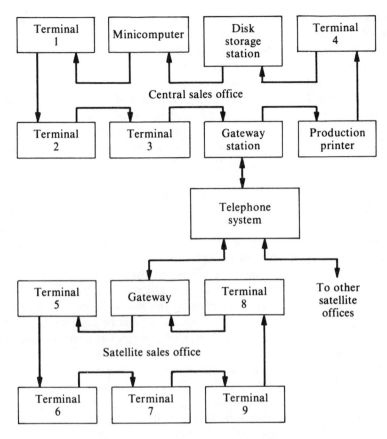

Fig. 8-3. Practical application of a LAN.

hardcopy. This central office services all the satellite sales offices in the region. A system of this type provides all terminals with access to the computer and database. This sharing of the computing resources makes the system highly efficient. Note that gateways are provided to interface the telephone system with the LAN, as was the case in Fig. 8-1.

Very often, a LAN is an internal system without access to the telephone system; i.e., there are no gateway stations. The gateway station is one of the most complex implements of the LAN. If a single manufacturer produces the computer, disk storage station, production printer, and terminals, the interfaces between the LAN and each of these devices are designed as part of the equipment. This is why manufacturers such as Digital Equipment Corporation (DEC), Wang, IBM, etc. make their own LAN. Several other manufacturers may design terminals, computers, and production printers that are DEC, IBM, Wang, etc. compatible. Gateway stations have also been designed to make communication possible between two LANs of different manufacture. As one may observe, the illustrations thus far have not reflected the complexities involved, but as we progress many of these issues will come to light.

If the gateway in Fig. 8-3 were equipped with a microprocessor or a mini-computer, the central sales office and satellites could be connected in a star configuration. Later we will undertake a discussion of topologies such as star, ring, and bus systems. In this chapter the reader will be introduced on a quantitative basis to some of the problems and solutions available in LANs technology.

As one may well imagine, an entire volume could be written on gateways alone. These devices translate the protocol used in one system to that of another. Also, since data transmission speeds may differ, the gateway functions as a transmission speed translator. With each new variation in the LANs, another order of complexity is added to the gateway.

Let us examine what is involved in making two LANs compatible with a gateway. They must be made physically compatible: somehow their electrical or optical connections must be interfaced. The protocols must be translated, and, of course, all of the transmission facilities must be compatible. Finally, the software must be compatible. Indeed, all this seems like a monumental task, and it quite often is, but if a methodical approach is taken to describe networks, some of the complexities can be eliminated. This is one of the reasons for describing LANs using the International Standards Organization ISO model.

Another use for LANs is electronic mail. At times when long distance rates are favorable, such as weekends or after 5:00 PM, message traffic can be conveyed between business establishments automatically. When the business day begins, a local data base can be interrogated for messages. If sales forms are reduced by facsimile machines to digital form, they may be transmitted in a similar fashion with the facsimile as a LAN implement.

A large fast-food chain uses a unique technique for changing the prices of food items. Pricing changes are sent, during cost favorable telephone rate periods, to its restaurants. The pricing in the cash registers is changed during the night. When the restaurant opens the next day, the pricing of all items reflects the changes. Previously, these changes had to be executed by a serviceman. Now they are all accomplished by a computer and LAN in a matter of hours rather than days.

A centrally located computer and telephone-connected LAN can also maintain automatic inventory control, thus relieving managers of this major responsibility.

One of the most challenging applications for LANs is in automated production facilities such as those required by the auto industry. Such LANs must function in harsh environments, e.g., ones with large temperature excursions, large electrical transients, and dust. In applications of this type, the terminal operator will not be a skilled programmer; the application program will thus require menus and prompting. The terminals can also be equipped with speech synthesizers to improve operator interaction with terminals; as speech recognition technology matures, the terminal may cease to require keyboards.

There are a number of goals local area networks will fulfill. The LAN that connects computers at various locations within companies allows for a more efficient use of these resources. Failure of one computer may slow down the facilities, but operation may continue at a reduced rate; system reliability is increased, however, because of redundancy in the system. Organizations that cannot tolerate catastrophic failure must maintain expensive backup computers if they haven't implemented a LAN. Examples of such facilities are the military, industrial process control companies, and banking.

Distributing computing power with LANs is possible because of the availability of relatively inexpensive computers and communication networks, which have been dropping in cost since 1970. Communication facilities now cost more than computers, but prior to 1970 the situation was the reverse.

Most mainframe computers are only a factor of 10 faster than the largest single-chip microprocessor-based computer. But the cost of a mainframe is well over a thousand times greater, which indicates the superior price/performance ratio of microprocessor-based systems.

Several microprocessor manufacturers are making 32-bit data-bus machines. These can either be paralleled for array processing or operated with coprocessors to increase processing power. Many floating-point processors are appearing in printed-circuit-board form; these allow the microprocessor computer user to upgrade systems.

Processing limits LANs operations, not the networks themselves. The cable used for interconnections can be bandwidth-limited, attenuation-limited, or both. Fiber-optic cable plants provide one solution to some of the problems. Fiber optics offers wide bandwidth and low loss per kilometer. The fact that the telephone companies are heavily involved in this technology will insure its steady growth. One advantage of fiber-optic cable is that it is relatively secure from intrusion as compared to copper. Tapping lines with simple clamp-type devices will not suffice.

INTERNATIONAL STANDARDS ORGANIZATION (ISO) NETWORK MODEL

Local area networks are designed in a highly structured manner to reduce their design complexity. Before discussing the various layers in the ISO model, some background in the physical structure of serial data links will be given.

Point-to-point connections are those with copper or fiber-optic cables between the communicating nodes, such as those in Fig. 8-2(a). A virtual connection is when multiple channels of data are sent over the same copper or fiber-optic path. Our main concern will be with this kind of connection rather than with connecting each implement point-to-point as in Fig. 8-2(a).

Two popular methods of forming virtual connections are shown in Fig. 8-4, which, of course, is an oversimplified diagram. That in Fig. 8-4(a) is called "time division multiplexing" (TDM). If the data in terminal 1 is transmitted to terminal 5 one character at a time and the multiplexer and demultiplexer form the path, transmission can occur for one character provided the two terminals are connected during the same time slot, say TS1. Furthermore, suppose that terminals 2 and 4 are also communicating one character at a time, the multiplexer and demultiplexer will connect the two units only during their specific time slot, say TS2. These two connections, as observed by the terminal operators, will be simultaneous. The transmission paths are vitual connections that use TS1 through TS3 as time slot channels.

The other technique in Fig. 8-4(b) is known as "frequency division multiplexing." This diagram again is oversimplified to illustrate a point. When a terminal wishes to communicate, it sends out, on channel f_0, its address and the terminal it wishes to communicate with, using frequency shift keying codes as follows:

$$f_0 - f_M = \text{binary 1}$$

$$f_0 + f_S = \text{binary 0}$$

where f_M is the mark frequency and f_S, the space frequency.

All terminals monitor f_0 when they are not communicating. The answering terminal will tune its receiver to the transmitting terminal, for example, channel f_1 for terminal 1. The transmitting terminal will then tune its receiver to the answering terminals, for example, frequency f_6 for terminal 6. Thus a communication path is established between terminals 1 and 6. Terminal 1 transmits on f_1 and receives on f_6 whereas terminal 6 transmits on f_6 and receives on f_1. Note that this technique allows full duplex communication (bidirectional) and that the transmissions are simultaneous.

This second technique is similar to citizen band radio, i.e., channel 9 is monitored. When a person wishes to call, he/she establishes initial contact through channel 9 and switches to another channel to communicate. Oftern combinations of both techniques are used in LANs. Each technique has tradeoffs to be investigated.

The ISO model is segmented into seven layers, as shown in Fig. 8-5. The layered approach to LANs can best be described by the following principles:

1. Each level should perform a well-defined function.

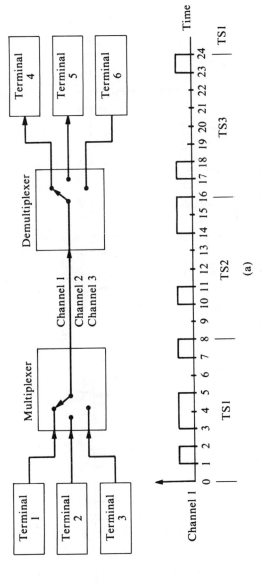

Fig. 8-4. Multiplexing techniques: (a) Time division multiplexing (TDM), and (b) frequency division multiplexing (FDM).

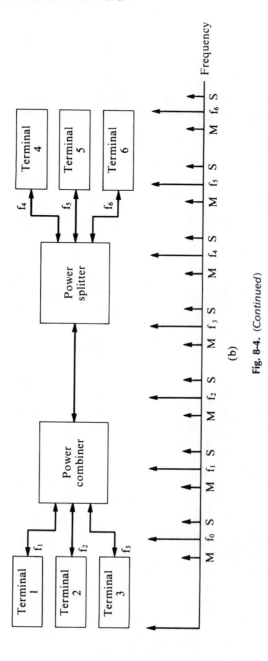

Fig. 8-4. (Continued)

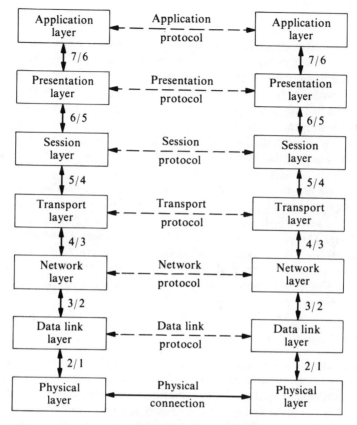

Fig. 8-5. International Standards Organization (ISO) network model.

2. The layers should comply with well-recognized standards if possible.
3. Layer interfaces should be well-defined and should be chosen to minimize information flow across interfaces.
4. The layer size should not be too large because that would make it unmanageable. Also, it should not be too small because that adds unnecessary complexity to the model.

The layers connected with dotted lines in Fig. 8-5 are virtual connections; i.e., each layer communicates as if it were physically connected. Only the bottom layer is actually physically connected, as shown in the figure. Each layer has an interface between the one above and below it. Let us now determine the function of each of the layers.

Physical Layer

The physical layer establishes the physical connection. Design issues include the type of cable plant to be used to implement the design—i.e., fiber optic or copper. Implementation of error-checking hardware is necessary to insure errorfree bit patterns. Note the reference here to bit patterns, not messages, the reason being that message length is of no consequence at this layer. Connectors, types of transmitters and receivers, lightning protection such as avalanche devices, electrical compatibility, noise, and, of course, optical compatibility are all design issues for the physical layer of fiber-optic cable plants. (The design is the province of electrical and mechanical engineers.)

When there are multiple physical connections, this hardware layer may look like a single connection to the layers above. The layers above may also have this effect; i.e., a single connection may be multiplexed to look like several connections. The physical layer will also control whether a channel is full or half duplex—i.e., bidirection or unidirectional—at any one time. This layer is also concerned with mechanical issues, such as the form factor of network interfaces. Collision detection is also one of its tasks, a subject that will be discussed in more detail in Chap. 9.

Data Link Layer

Data is assembled into manageable groups of bytes called *frames*. Note that the beginning and end of each frame must be known. The flags at each end must be distinguishable from data; for long messages, multiple frames must be sent. It is the data link layer that adds the necessary sequence numbers to the frames so that data in multiple frames can be reassembled in its original state at the receiving device.

The flags must be removed when the data is received; as a result, the data link layer becomes transparent to the layer above (network layer). If a frame is damaged during transmission, the data link layer will initiate retransmission of the damaged frame. Also, when frames require an acknowledge (ACK) or not acknowledge (NACK) response, the data link layer will send these messages.

Error-checking algorithms are sometimes employed at this level because burst errors can go undetected for certain types of physical-layer Cyclic Redundancy Checking (CRC). The commonly used CRC will detect burst errors with reduced accuracy when they are greater than 18 bits for a 16-bit CRC register. The data link layer is also responsible for preventing receiver overflows; i.e., the transmitter is kept informed of the receiver buffer status during transmission.

Data management is provided during full-duplex and half-duplex operation. Often ACK/NACK messages are piggybacked on message traffic rather than sending a special ACK/NACK frame. This piggybacking technique will be discussed in more detail in Chap. 9.

Network Layer

The fundamental purposes for the network layer is to establish a virtual circuit and maintain it during transmission. The communication between network nodes takes the form of packets that contain the addressing and supervisory information necessary for network operations. The packet will be discussed in more detail later in this chapter. Another task of the network layer is interfacing with the next layer in the hierarchy—the transport layer. The network layer assembles and disassembles transport messages during transmission and reception, respectively.

Traffic or flow control is another function of this layer. Traffic flow may be such as to cause congestion of the networks, but this layer will regulate and keep it moving in a well ordered manner.

When traffic is monitored for accounting purposes, as it is in larger networks, this layer performs the necessary tasks. It monitors bits, characters, packets, or frequency tune when customers use the facilities—whatever is necessary to produce billing. Also, difference nodes may have different accounting and rates that must be included in billing.

An additional responsibility of this layer is maintaining priorities in a prioritized system, i.e., keeping messages in proper order (for example, making sure that part 1 of a message arrives before part 2 rather than vice versa).

The routes, ordinarily determined through the use of routing tables, are usually hard-wired into the network for small systems; larger networks may establish routing to assist with traffic conditions.

Finally, the network layer may be tasked to keep peer layers informed of traffic status. This may take the form of a readout indicating network status that the operator may monitor. The network layer can also be implemented as part of the input/output operating system drivers for some LANs. As one may easily surmise, this is a very important layer.

Transport Layer

The transport layer takes data from the session layer, the next higher layer in the ISO model, and breaks it up into smaller units. Basically a multiplexing layer, it multiplexes this data and puts it into the network. It can also present the session data to more than one network connection in order to improve through-put.

The transport layer also monitors the quality of service, being responsible for end-to-end error detection and recovery. It addresses the end user without regard to the route or address of machines along the route between end-user machines.

The transport layer multiplexing involves disassembly (transmitting) and assembly (receiving) of session layers. It not only isolates the session layer, which can be part of the host-machine operating system, from changes in hardware, but maintains session-layer segments in their correct order during transmission and reception.

If cost of service is expensive, the transport layer may also multiplex several session layers into a single network connection to relieve congestion. This type of situation may be encountered during heavy traffic flow.

The fast host machine must be prevented from overrunning the slower one. The transport layer is responsible for performing this function at the session level.

Session Layer

The session layer establishes the connections between two users, which is actually the end result of the whole process. Speaking more technically, a connection is established between the presentation layers, which then perform certain transformations on the data before it is transferred to the application layer.

Providing the necessary synchronization between end-user tasks, the session layer sets up communication options, such as whether a connection will be half or full duplex. It provides the necessary addressing to be implemented by a user or a program. It will map addresses to names; i.e., it performs a directory function. As one may readily observe, this is a function of most host operating systems.

The session layer provides smooth operation between users, i.e., the opening and closure of data transfer between users. It also provides file transfer check points and the necessary buffering between two end users. Such buffering may be necessary before the session data is conveyed to the end user or before data can be transmitted.

Presentation Layer

The presentation layer translates or interprets the session layer data for the application layer. When computers are communicating and the application layers must be able to understand the information transferred by two dissimilar machines, a common syntax is needed to represent alphanumerics, file formats, data types, and character codes. The presentation layer will negotiate a syntax, which is often called the "transfer syntax," that will allow the application layers of the two machines to communicate.

As one may surmise, the transfer syntax will make exchange of information possible even when the originating and the destination application processes are of different formats. To negotiate, the application layer must supply the presentation layer with all necessary information about the data to be exchanged and the rules for the procedure. The application layer need not provide certain specific information, such as file formats, but only the type of data to be exchanged: financial, display, etc. The originating presentation layer must provide a transfer syntax that both are capable of using, and both must agree on this syntax or communication cannot proceed. If local mapping between the originating and transfer syntax can be performed, the two application processes can communicate, but this is not a communications issue. The ISO has not standardized all possible syntaxes because, as one might well imagine, this would become a life's work for some committee. The ISO, however, does recommend some of the more popular syntaxes and is standardizing the negotiating protocols. More of these issues are discussed in Chap. 7.

Application Layer

The application layer is the one the end user is acquainted with since it allows the user access to the computer. The application layer is responsible for making all other layers transparent to the operator of the equipment and provides such services as removing line feeds, carriage returns, and control characters from the text; it also inserts appropriate subscripts, superscripts, tabs, etc., into the text.

The application layer may provide such services as log in and password checking that might be needed in banking or time sharing of computer services. One would need this type of service, moreover, to maintain network security, which has become of the utmost importance with the advent of computer "hackers," as individuals involved in breaking into secure networks are called. This, of course, is an even greater problem for the banking industry with the increased use of automatic banking.

The application layer is responsible for setting up negotiations for an agreement on semantics, such as the case of a bank communicating with a loan company on the information to be transferred, e.g., credit references rather than money transactions. During a communication period, the application layer may be capable of switching between semantics, a procedure commonly called "context switching." An example would be two communicating partners first transferring credit checking and then switching to transferring airline reservations. The ISO has agreed on the standardization of service protocols, such as management, file transfer, and virtual terminal types.

The ISO standards are being widely accepted, but by no means are they the only standards. A compilation of general standards is presented in Table 8-1;

Table 8-1. Standards.

Sponsor Organization	Affiliation	Membership	Influence in Marketplace
ISO (International Standards Organization)	Voluntary compliance	Participating nations with U.S. representation	Close relationship with CCITT
CCITT (International Consultative Committee on Telegraph and Telephone)	International Telecommunications Union (a United Nations treaty organization)	Most of Western Europe, with U.S. representation	Enforced by law in countries with nationalized communication
ECMA (European Computer Mfg. Association)	Computer suppliers selling in Western Europe	Trade organization of suppliers	Contributes to ISO and issues ECM standards
ANSI (American National Standards Institute)	Voluntary compliance	Manufacturers, users, communications companies, and other organizations	U.S. has representation in ISO

NBS (National Bureau of Standards)	U.S. Government agency	U.S. Government network users; ARPANET implementers of DoD (Department of Defense)	Federal information processing standards purchased by Government; DoD need not comply
IEEE (Institute of Electrical and Electronic Engineers)	Professional society	Society members	Contributes to ANSI and issues standards: IEEE 802, IEEE 488
EIA (Electronic Industry Association)	U.S. Trade Org.	Manufacturers	Contributes to ANSI: Physical Layer RS422, RS232-C
DoD (Dept. of Defense)	Gov't. agency	Gov't./MIL	Military establishment
Others	Voluntary defacto standards	Various manufacturers	Standards for own equipment

these are the most popular, but others have been generated by manufacturers and various organizations. As one can imagine, standards are compulsory for certain types of communications—for example, throughout a country or for government equipment. In time of war or national emergency, components within systems must be easily replaceable should there be equipment failure.

Most of the standards in Table 8-1 are not similar. Network designers must be careful when designing interfaces for standards, therefore, because a standard may not be well recognized. If this is the case, the product will have a limited lifetime and market.

COMMON NETWORKS

This section will acquaint the reader with a cross section of local area networks. The first is ARPANET, which is the earliest design of a packet switching network. It was pioneered by the Advanced Research Projects Agency (ARPA), which provided a test bed for packet-switched type networks such as LANs. The next three networks to be investigated are baseband LANs that represent a good share of the LAN market. Wang and Sytek networks represent broadband local area networks which are very useful for transmitting diversified data.

Messages may be broken up into manageable segments and transmitted after the total message is complete. The transmission will consist of a burst of data, and network control functions will process packets in an efficient and timely manner. Packet switching is primarily used in data communication networks, but digital voice has made the possibility of packet-switched voice more feasible.

One, of course, may ask, why a packet-switched telephone? The answer is that telephone plant facilities are not used very efficiently. As an example, the average telephone call lasts approximately three minutes, and the circuits used ordinarily operate half duplex; i.e., only one person talks at a time. As a result, only 10 percent of the voice circuit is in use during peak load periods. As one can readily observe, packet voice will promote more efficient use of plant facilities and prevent possible tariff increases.

ARPANET

The ARPANET network provides host-to-host protocols for data exchange between the machines. The specifications of format and procedure for data exchange have been well documented; the user protocol has been only functionally specified. The functions—open, close, receive, send, and interrupt servicing—allow the user to tailor interfaces, which makes possible the use of other computers on the net if they comply with host-to-host protocols. This is an important established principle for protocol specifications.

When ARPANET is compared with the ISO model (see Fig. 8-6), it is seen

Layer	ISO (Model)	ARPANET	DECNET	ETHERNET	SNA	TOP V1.0	MAP V2.1
7	Application	User	Application layer	UNET mail protocol	End user	FTAM (DP) File transfer	FTAM (DP) File transfer
6	Presentation	Telenet interactive terminal interface parameters, FTP (File Transfer Processor)	Application layer	Unet file and terminal protocol	NAU service (network, addressable units)	NULL ASC II and Binary encoding	ASC II and Binary encoding
5	Session	(None)	(None)	(None)	Data flow control	Basic combined subset and session kernel	Basic combined subset and session kernel
4	Transport	Host to host	Network services	UNET transmission control protocol	Transmission control	Transport (IS) Class 4	Transport (IS) Class 4
3	Network	Source to destination IMP (Interface Message Processor)	Transport	UNET INTERNET protocol	Path control	Internet (DIS) Connectionless and for X.25	Connectionless and for X.25
2	Data link	IMP to IMP	Data link control	ETHERNET controller	Data link control	LLC (IEEE 802.2) Type 1, Class 1	LLC (IEEE 802.2) Type 1, Class 1
1	Physical (baseband)	Physical (baseband)	Physical (baseband)	Transceiver and cable (baseband)	Physical (baseband)	CSMA/CD (IEEE 802.3)	Token bus (IEEE 802.4)

Fig. 8-6. Comparisons of ISO model with other common networks.

to have a physical layer, but its second layer is a combination of layer 2 and a part of layer 3 of the ISO model. The ARPANET third layer is composed of parts of the ISO network and transport layers, and its fourth layer is composed of the rest of the ISO transport layer. ARPANET has no session layer, but its sixth and seventh layers have an ISO equivalent.

SNA (IBM's Systems Network Architecture)

SNA is a LAN devised by IBM to allow large customers such as banks, loan companies, manufacturers and other businesses to integrate their data-processing facilities so that they may be used much more efficiently, and, if computer failures occur, some backup computing power will be available. As an example, consider a bank with two mainframes and fifty terminals throughout the main banking facility and several branches. All fifty terminals can access either computer via a LAN. If a computer fails, the bank will have one computer that may be overloaded with the fifty terminals. The terminal operators will note only that terminals are reacting more slowly in processing traffic. Prior to the advent of local area networks most products and interfaces had protocols that were unique to their own product line. IBM alone had hundreds of communication products with a few dozen access methods and a dozen or more protocols. Imagine then, the chaos that existed when a network of a new sort was configured, each being a fresh engineering challenge with no doubt new interfaces and protocol modifications. It must have been rather refreshing to have all components of a network drop-shipped and integrated at the premises, and, lo and behold, with only minor modifications, it worked!

Interfaces and protocols must allow for change if a LAN is to remain viable. Innovations that increase performance and expand facilities are required if a network is to succeed. The SNA has these attributes.

SNA local area networks consist of nodes that are machine types; these are given in Table 8-2. Observe that microcomputer terminals with preprocessing capability incorporate several of these node attributes in a single machine. Many supermicrocomputer have the computing power of a minicomputer.

Figure 8-7 depicts a two-domain SNA network that can be extended to form larger networks. The domains are managed by the Systems Services Control Point (SSCP). There is an SSCP for each type 5 node; it monitors and supervises the domain. At least one or more logical units (LU) is required for user processes. The node is provided with a physical unit (PU) that the network uses to perform administrative functions, such as test, or to put the node on or off line.

Table 8-2. Machine-type Nodes.

Node Type	Description	Comment
1	Terminals	
2	Controllers	Supervisor peripherals
3	Undefined	
4	Front-end Processors	For preprocessing data
5	Host Processor	Mainframe computer usually

The first two SNA layers function similarly to those of the ISO model. Layer 2 of the SNA structure closely resembles the Synchronous Data Link Control (SDLC) protocol, which is produced in integrated-circuit form, such as the WD1933 manufactured by Western Digital.

The path control (PC) layer shown in Fig. 8-6 performs the networking and also part of the transport function. It deals with routing and congestion control issues as does layer 3 of the ISO model, but it also performs data multiplexing/demultiplexing. It delivers messages and performs message multiplexing/demultiplexing from origin to destination—the transmission control (TC) layer. Each link has a queue that contains all the data destined for a particular adjacent

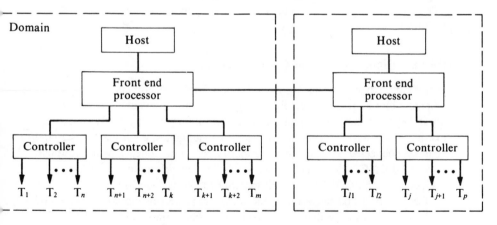

Fig. 8-7. Two-domain SNA networks.

node; this is the primary station when the queue is served by a secondary station, and vice versa. Unrelated packets may be assembled in this queue to enhance transmission efficiency and cope with hierarchical SNA addressing.

The next layer is transmission control, which provides the necessary transport service to sessions. When the transmission is complete, it will delete sessions. This layer in SNA is not the same as the session layer in the ISO model. Once the SNA transmission control service is established, the session (SNA type) manages message priorities, regulates the rate of data flow between processes, provides buffer allocation, affords an interface to upper layers, and performs, when required, encryption and decryption. (Note, one must be careful when reading articles about SNA not to get the term "session" confused with the same term as used in ISO terminology.) The properties of the subnet, which serves as a connection between hosts, are kept transparent to the peer layers by the transmission control function.

The data flow control, as one can observe in Fig. 8-6, supervises the session layer (SNA type); i.e., it determines the session flow direction—origin-to-destination or destination-to-origin. This layer also provides for error recovery. The header information normally included in the ISO session layer is embedded in the SNA transmission control layer.

The sixth SNA layer, or NAV service, provides presentation services similar to those of the corresponding ISO layer and also session services (not the same as these of the ISO) for establishing a connection. It also includes additional network services to control the overall network.

The end-user layer is self-explanatory; it corresponds closely to the ISO application layer.

DECNET (Digital Equipment Corp. Network)

The DECNET network was designed to allow DEC's customers to interconnect computers into a private LAN. Previously, DEC computers could communicate only if they were hardwired point-to-point, i.e., if they are physically connected. Now they may communicate through intermediate nodes; e.g., machines A and B can be connected to machine C so that A can communicate with B through C. With DECNET, nodes need only be on the network to communicate.

A comparison of DECNET with the ISO model is found in Fig. 8-6. Note that the transport and network layers are interchanged in the two models and that DECNET has only five layers. Also note that there is no session layer.

The DECNET physical layer is similar to the other physical layers previously discussed. The data link control layer exhibits some differences in frame construction, being byte oriented. Discussion of it will be deferred until later in the chapter. Whereas some LANs, typically SNA networks, require the same route for all packets of a message, DECNET message packets may be routed independently through the subnet. DECNET does not have a session layer, and transformation of data must be performed indirectly; i.e., encryption, decryption, text compression, and various code conversions must be accomplished with remote files.

ETHERNET (Xerox Corp.)

ETHERNET is another baseband network dedicated to LAN. The physical layer consists of a 50-ohm cable with an adjacent transceiver; the transceiver is connected to a data-link-layer controller (ETHERNET controller) on the backplane of the equipment. This latter connection, made with an interface cable, supplies power and receives, transmits, and monitors collision-detection signals. A simplified block diagram of a typical ETHERNET network is shown in Fig. 8-8. Note that signal isolation is required because the system has a single grounding point; i.e., similar to a CATV application, each station must not be grounded independently. The coaxial cable thus becomes an antenna that is responsible for excessive amounts of electromagnetic radiation. This provides an ideal application for fiber optics and will be discussed more thoroughly during the discussion of the physical layer.

ETHERNET standards and specifications cover the first two layers of the ISO local area network model. In the model detailed by Fig. 8-6, the layers above the ETHERNET controller are implemented in the Unet Protocol (which provides communication between Unix base operating systems). The Internet Protocol (IP) and the Transmission Control Protocol (TCP) are both implemented by using United States Department of Defense standards. Given with fill-in blanks these upper layers are actually defined by the customer or user himself. The upper-layer Unet File Transfer Program is a technique for transferring Unix files between two host machines. This layer is only superficially covered because it too can be defined by other agencies than the Unet File and Terminal Protocol. The upper-layer Unet standards allow for direct access of Unet Internet Protocol, which is not possible with the other models discussed.

ETHERNET has been implemented in several DEC installations by systems integrators. The DEC machines include PDP-11 and some VAX series computers with controllers designed using LSI-11 microcomputers. The Unix operating system has been used extensively for these applications. ETHERNET has an inherent shortcoming, since its maximum end-to-end length is limited to 2.5 km. This particular dilemma can be solved using fiber optics. The increase in length can be rather dramatic, in fact, by the order of a magnitude. Such an increase is highly dependent on the wavelength used: 1300 nm is preferable with LED transmitters in conjunction with PIN FET receivers. For longer dis-

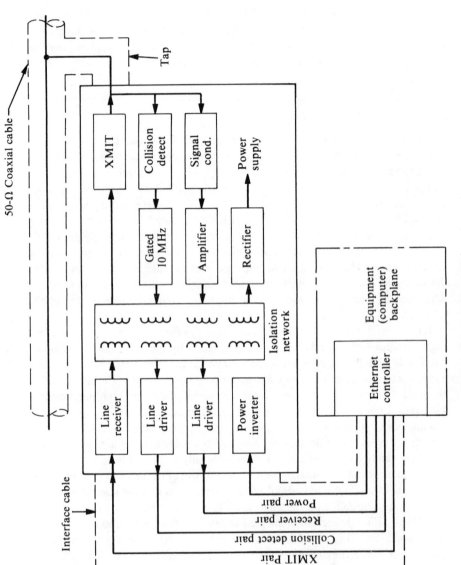

Fig. 8-8. Ethernet-connected equipment.

tances, lasers and monomode transmission techniques are more suitable, but monomode networks are expensive.

A large number of other baseband networks are beginning to saturate the LAN market. Many manufacturers have incorporated features form the networks presented here. By studying them carefully, the reader may come to appreciate their design subtleties.

WANGNET

WANGNET is not included in Fig. 8-6 because it is a broadband network able to manage rather diverse forms of data simultaneously. Since its frequency division multiplexed (FDM) channels can handle video, audio, or digital type data, this LAN obviously has a great deal of utility in office environments. Before implementing a broadband LAN, certain questions must be examined. The basic ones are whether a simple baseband system will suffice when the data being manipulated is all digital and whether low data-transmission rates are adequate.

One of the advantages of WANGNET is its use of a mature technology: CATV technology has been around for 20 years and the RF components are highly reliable. Also, the cost of CATV components is less than that of many baseband LANs. The bandwidth of Wangnet is 340 MHz, with a bandwidth of 42 MHz reserved for user RF devices. This band is located from 174 to 216 MHz (channels 7 to 13 in the TV band).

WANGNET consists of a number of communication services labeled Wangband, interconnect bands, utility band, and peripheral attachment service; Fig. 8-9 shows the band allocations. The 10 to 12-MHz and 12- to 22-MHz bands are point-to-point network components with fixed-frequency modems to provide the virtual connections. The band from 48 to 82 MHz is controlled by a data switch that, in turn, is supervised by a computer; its functions are similar to those of a miniature telephone exchange. A call is established when the data switch selects an unused frequency and initiates the tuning of a frequency-agile modem to establish the channel.

A utility band is provided for user RF devices; for example, TV channels 7 to 13 on the standard broadcast band may be provided on the cable by the user. Other channels may be broadcast via this cable if they are mixed down or up to this band.

Perhaps the most important band is the CSMA/CD (Carrier Sense Multiple Access/Collision Detection) Wangband. This FDM channel has a fairly high transmission speed (12M bits/sec), and it provides LAN service. The WANGNET protocols emulate the lower five layers of the ISO model, i.e., the physical to session layers.

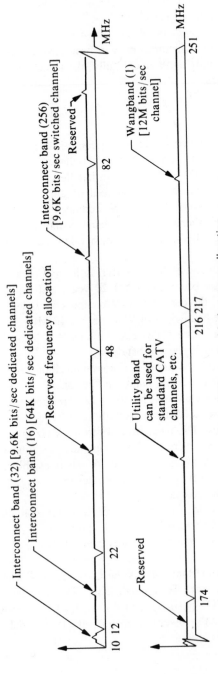

Fig. 8-9. Wangnet cable frequency allocation.

LocalNet

This particular broadband network is manufactured by Sytek Inc. The RF coaxial cable used for implementing LocalNet is a standard CATV type with a 5- to 400-MHz bandwidth. The broadband frequency allocation is a midsplit system. Outward signals from the head end are in the 226 to 262-MHz band and inward signals to the head end are in the 70- to 106-band. When a terminal conveys data to the computer (see Fig. 8-10), the digital data is passed to the PCU (Packet Communication Unit) via an RS232 link. The PCU's modem will impress the data on the RF carrier with frequency shift keying (FSK). The data is now in RF form. It will pass through the 50/50 head end to be up-converted in frequency to the 226- to 262-MHz band and will then be detected by the receiving PCU and presented to the computer via a RS232 link. Note, when the computer answers, that the outward signal must pass through the two amplifiers and to the head end to be up-converted to the receive frequency band. This is a highly simplified version of how the equipment actually functions.

Each of the bands has 120 (300 KHz) channels. The modems are arranged in groups by letters A through F, and each of the letters has 20 channels per group. These lettered groups have a bandwidth of 20 × 300 kHz or 6 MHz. Each of the channels within the group operates at a 128 kbit/sec using the CSMA/CD protocol. With time division multiplexing of each of the 128-kbit/sec channels, 200 or more terminals can be serviced by a single channel. T

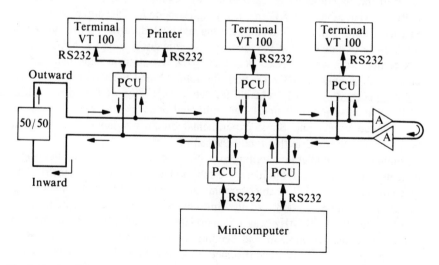

Fig. 8-10. Small local net installation with four network implements connected to four minicomputer RS232 ports.

bridges can be used to implement more than 20 channels. Use of bridges will permit many thousands of terminals to be serviced. As one can readily observe, this network can easily be expanded to meet most user needs.

LocalNet 20, which has a transmission rate on the cable of 1 Mbit/sec, is useful for low-duty cycle users such as terminals connected to minicomputers. The LAN will cover a geographic radius of up to 50 kilometers.

LocalNet 40 is useful for higher duty cyle users because it operates at 2 Mbits/sec. Both LocalNet 20 and LocalNet 40 are unsuited for intercomputer transfers.

Sytek has several other implements that attach to their cable. Their network control center, NCC, serves a multitude of functions. It includes a network monitor that will statistically monitor network channels and record traffic density; it can also be used to forward and store electronic mail—i.e., the station to receive the mail will be notified and can then interrogate the NCC. Ths device can store a directory routine that allows names to be translated into terminal addresses.

The NCC hardware consists of a motorola MC68000-based microcomputer with a streaming tape drive (for the operating system) and a 20-megabyte Winchester disk drive. The connection to the broadband cable is via two LocalNet 200/100 PCUs, which have eight RS232 ports each. This equipment has a Unix-based operating system to increase its utility.

The 20/100 LocalNet PCU follows the ISO model with some exceptions in regard to the functionality of each layer. For example, the physical layer has an RF coaxial cable and RF frequency agile modem. The digital transmitter must have FSK modems and transmit the modulated carrier, while incoming data must be RF-detected at the receiver and converted to digital form.

Note that we have discussed only that Sytek data transmission band on the cable which is only 36 MHz for both the transmit and receive bands. A great deal of spectrum obviously remains for future expansion of facilities, such as CATV channels, audio channels, facsimile and other user services, as well as Sytek-related implement. Some experimental digital telephone facilities have been successfully used in this type of system.

Commercially available remoting has been accomplished using fiber-optic modems that operate to 5 kilometers. Such modems consist of an electrical-to-optic conversion at the RF cable drop and an optical-to-electrical conversion at the far end (remote location). At the time this book was being written, only a transmit remote at 70 MHz was commercially available, but by the time of publication a receive modem had become available. Also, advances in fiber optics can be expected to extend distances to a minimum of 20 to 30 kilometers; in fact, the necessary technology is already available.

Fiber-optic remotes are useful where an excessive amount of electrical noise is present—such as near mill machinery, punch presses, and high-voltage switch yards—and, to some extent, where radiation is present. Since fiber-optic cable may be completely nonconductive, it may be routed in the same cable runs as electrical power cable with no adverse effects.

X.25

The last network to be addressed, the X.25, is by no means the least important. It is more commonly referred to as the CCITT Recommendations X.25 (for public packet switched networks). This is a standard being adopted by many countries, including the United States, for the physical, data link, and network layers of the ISO model of LANs. Although the X.25 standards are similar, some subtle differences will be examined in the appropriate section of this text. X.25 provides a virtual connection between computers; this, of course, is only half the job of communication, software compatibility being the other half. Unix operating systems will help alleviate some of the headaches of communications engineers, but many manufacturers that tout Unix operating systems are actually offering Unix-*like* operating systems. The disparity lies in the fact that the communication between the computers is only almost correct. "Almost," of course, counts only in horseshoes and handgrenades!) The standards are meant eventually to eliminate all loopholes and make the equipment completely compatible. This has both good and bad effects. The good is that equipment may be purchased anywhere and will operate or interface correctly. The negative is that a better interfacing design may be found, but cannot be used because of the standards. One of the most successful organizations in terms of standards is the United States Military with their MIL standards.

The material already presented in this chapter is meant to give the reader the necessary background material for a reasonable understanding of LANs. For want of space, and also because of the software nature of the upper layers, only the first three layers of the ISO model will now be addressed. The physical layer will be emphasized because this is where the differences between fiber optics and more conventional networks lie. The reader interested in an in-depth treatment of local area networks should consult the references.

PHYSICAL LAYER

The physical layer will be approached here from a general point of view, i.e., specific network topologies will be discussed without reference to manufacturers or trade names. These generalities will then be applied to specific LANs that are available off-the-shelf. Single-mode technology as applied to LANs were limited at the time this book was being written. Many of the applications presented here will represent a custom-design point of view; as the technology matures, of course, more off-the-shelf designs will become available.

Since these devices will be designed as custom VLSI, they may be implemented using discrete components. Discrete replacements cannot be made on a one-for-one basis because VLSI designs usually require FETs for load resistance; they take up much less space than monolithic resistors.

SIMPLEX LINKS

Two simplex single-mode fiber-optic links are shown in Figs. 8-11(a) and 8-11(b), which illustrate time-division and frequency-division multiplexing, respectively, at the transmitter, and demultiplexing of the two multiplexing schemes at the receivers. The messages in Fig. 8-11(a) are transmitted on a single baseband channel (no carrier) by assigning time slots to each. Moreover, they are sent sequentially—M_1, M_2, and M_3—and the multiplexer will repeat the sequences. TDM as presented is rather oversimplified in that temporary storage may be necessary while the message is waiting to be transmitted, the incoming messages (M_1, M_2, and M_3) may be at different transmission rates than that of the fiber-optic transmitter, and some sort of signaling between the MUX and message channel is required. As one may observe, the term "simplex channel" does not imply a simple channel; rather, it is a description of the unidirectional nature of the channel. TDM will be covered in more detail for specific LANs in Chap. 9. The reverse process occurs at the receive end.

Figure 8-11(b) is an analog communication link in which the digital signals—f_1, f_2, and f_3—are at different frequencies. The modulated carriers, in turn, intensity modulate the fiber-optic transmitter, which is similar to AM. At the receive end, carriers f_1, f_2, and f_3 are separated by filtering after they have been optically detected (optical-to-electrical conversion). Each carrier is then demodulated to extract the digital data. Although this process also appears to be very simple, it usually isn't.

In both Figs. 8-11(a) and (b) two wavelengths could have been used to produce full duplex transmission. Moreover, a third multiplexing scheme analogous to FDM could have been instituted—i.e., wavelength division multiplexing (WDM). This technique will be examined in Chaps. 9 and 10 when specific networks are discussed.

A block diagram of the proposed transmitter of Fig. 8-11(a) is presented in Fig. 8-12(a). The progression of the design will be from a block diagram to a logic representation of the various parts of the block diagram, and finally to a VLSI implementation that may be fabricated in either NMOS or GaAs technology, depending on the transmission requirements. The timing generation is not included in the analysis and design, but an overview of how it may be accomplished is given.

The theory of operation for the multiplexer, cyclic redundancy check CRC, and the Manchester encoder is as follows. Messages M_1, M_2, and M_3 are im-

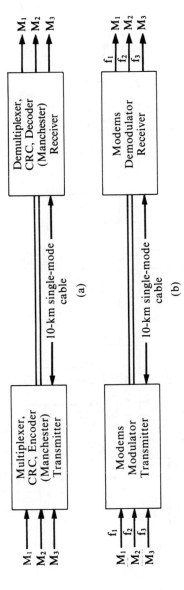

Fig. 8-11. Simplex TDM and FDM links: (a) Digital TDM link implemented with a single-mode fiber-optic cable, and (b) analog link with FDM channels separate in frequency.

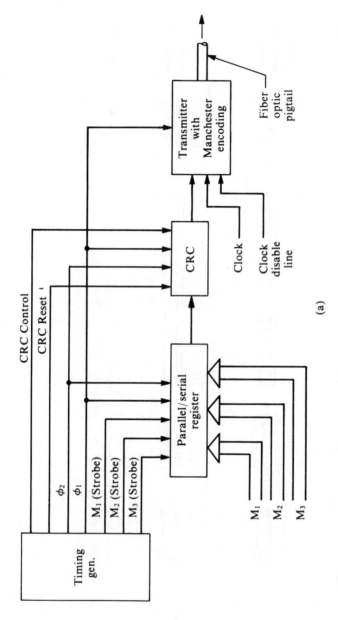

Fig. 8-12. Circuit details and diagrams for a digital link: (a) Transmitter detail with multiplexer, CRC, and transmitter; (b) transmitter, logic diagram, and timing diagram for a Manchester encoder; (c) Manchester encoder and transmitter schematic in NMOS or GaAs VLSI.

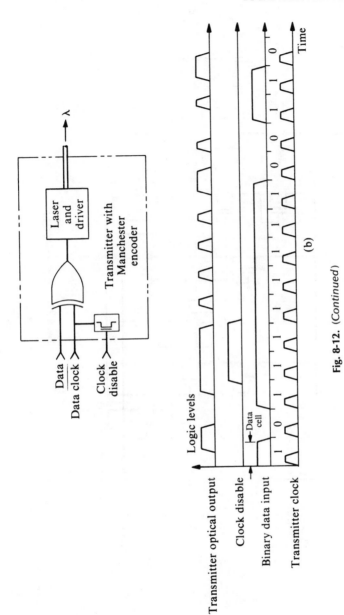

Fig. 8-12. (*Continued*)

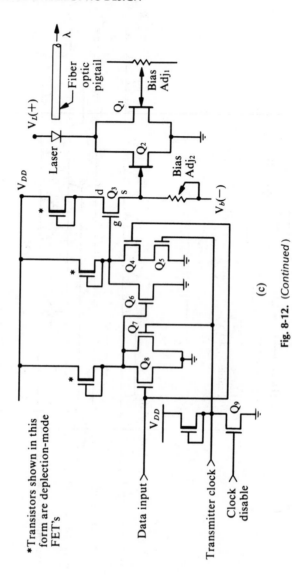

Fig. 8-12. (*Continued*)

pressed on the input lines for parallel-to-serial conversion. To maintain generality, the word length is not specified, but 8-, 16- or 32-bit words may be used. The timing generator will strobe the appropriate M line to load the register, and the other strobes must be blocked for one word length in time (ϕ, clock time multiplied by the word length). After the message word M has been strobed in, it will be shifted by the ϕ_1 and ϕ_2 clocks into the CRC register. The CRC will compute the FCS word, which is a serial-computed checksum.

Prior to transmission of the message, the Manchester coder clock, ϕ, is interrupted for eight consecutive bits, or for a pattern of two, followed by one uninterrupted, two interrupted, and the final three uninterrupted. The interrupted bits are used for signaling the receiver that the beginning or end of a message is being received. The interrupted clock cycles cause a pulse duration increase in the transmitted output that lasts longer than two clock times. Normally, if the clock isn't interrupted, the Manchester coder will produce a transmitter output transition at least once during each bit time. Therefore, by sending various clock-interrupted patterns, the beginning and end of a pocket may be determined, reported, and signaling transmitted with the delimiter, or "flags" as they are sometimes called (the start and stop are eight-bit words). Also, the delimiters need not be only eight bits long, but eight bits are used in the IEEE 802 standard. The data between two delimiters is considered to constitute a "frame." Discussion of frames will be deferred until later in this chapter.

Let us now turn our attention to Fig. 8-12(b), which is the transmitter block of Fig. 8-12(a) composed of a Manchester encoder, clock disable, and laser-driver circuitry. The data input to the transmitter is 101111111100110_B, with the subscript B indicating binary data. Note that the data is exclusive ORed with the clock; each time the clock and data are different, the transmitter output is high (full on), and when they are identical, the transmitter output is low (it can be off or at a very low output state). The only exception occurs when the clock is disabled; i.e., the clock is always low. For this later condition, the output will follow the data input; i.e., the output will be an exact replica of the data waveform. These clock-interrupt signals will be used for transmitter and receiver signaling. Note that the output pulses have longer time durations when these clock interrupts occur. The condition is easily detected at the receiver with a monostable multivibrator (one-shot) or a simple counter circuit.

The transmitter circuitry in Fig. 8-12(c) shows an overutilization of FET transistor; in MOS and GaAs technology, however, transistors are used in place of resistors whenever possible because the former take up much less space on the IC die. FETs Q_1, Q_2, and Q_3 form the laser driver, as previously discussed. The two external bias adjustments allow the drive voltage excursion and threshold current to be adjustable. These potentiometers may be replaced by fixed resistors and fabricated as part of the IC. The FET in the drain circuit is a depletion-mode device that is easily fabricated in monolithic circuitry. The remaining circuitry in Fig. 8-12(c)—i.e., Q_4 through Q_9—forms the exclusive OR and clock disable logic.

The exclusive OR will be examined more closely because it appears on several other occasions in this chapter. If the clock and data-rate input lines are both high, the gate of Q_3 is held low, thus switching off Q_2 and the laser. When both data inputs are low, Q_7 and Q_8 are off which turns on Q_6 thus holding the gate of Q_3 low and keeping Q_2 and the laser off. For the situation when the clock and data are different, Q_6 is held off and either Q_4 or Q_5 is off; this allows Q_3 and Q_2 to turn on, which will turn on the laser.

The preceding is a description of the operation of the exclusive OR function that ignores the clock disable line, which must be low. A ground on the Q_9 gate will keep it off. If the clock disable line is held high, Q_9 will turn on, switching both Q_7 and Q_4 off; as a result, a logic 1 on the data line will turn on Q_8, thus switching off Q_6 and turning on the laser. Therefore, since only binary data 1s will turn on the laser, it operates as a binary transmitter instead of a Manchester-coded device. As the reader may surmise, the transmitter can be used to transmit uncoded data as well.

The circuitry shown in Fig. 8-12(c) is by no means the only method of monolithic implementation; the technology is rapidly changing. The amount of space circuitry will occupy on a chip can be estimated. The laser and drive circuitry will take up approximately 0.6 mm^2; Q_1 through Q_9, 2.5 μm^2 per device for a total of 22.5 μm^2; and the depletion-mode devices, which are approximately twice as large as the enhancement-mode devices, 5 μm^2 each, or 20 μm^2 total. The total area required for the transmitter, encoder, and clock disable circuitry on the die, therefore, is approximately the same as that for the laser and drive circuitry because the exclusive OR and clock disable circuitry take up such a small amount of space.

The CRC register, briefly discussed before, can now be examined in more detail. Its place in the transmitter is depicted in Fig. 8-12(a); the logic involved is shown in Fig. 8-12(d).

Cyclic Redundancy Checking (CRC)

Cyclic Redundancy checking (CRC) is a technique of detecting errors. It is presented here because integrated-circuit manufacturers have implemented them in monolithic circuits. Depending on the polynomial generator, the common 16-bit integrated-circuit version of CRC circuits will detect 100 percent of burst errors 18 bits in length and 99 percent of burst errors over 18 bits.

Two methods of CRC checking of message traffic will be presented. The first is a rather simplified approach that will show the arithmetic required. The second is more representative of how the hardware actually performs the task.

CRC Method One: This method of checking results in a remainder of zero if no errors are present. The necessary arithmetic is as follows:

$G(x) = x^5 + x^4 + x + 1$ (Message to be sent)
$G(x) = 110011$ (Binary form)
$G(x) = 33_{\text{Hex}}$ (Hexadecimal form)
$P(x) = x^4 + x^3 + 1$ (Polynomial generated for CRC checking)
$P(x) = 11001_B = 19_H$

$$\frac{x^4 \, G(x)}{P(x)} = Q(x) + \text{FCS}$$

This process shifts the message, $G(x)$, four bits to the left prior to division and allows a four bit FCS, as follows:

$$
\begin{array}{r}
21_H \\
19_H \overline{)330_H} \quad \text{-->} \ G(x) \text{ shifted four bits} \\
\underline{32_H} \\
10_H \\
\underline{19_H} \\
-9_H
\end{array}
$$

$F(x) = x^4 G(x) + \text{FCS}$
$\text{FCS} = |\text{Remainder}| = 9_H$
$F(x) = 339_H$

The incoming message, $F(x)$, is then checked for errors:

$$\frac{F(x)}{P(x)} = \frac{G(x) + \text{FCS}}{P(x)}$$

$$
\begin{array}{r}
21_H \\
19_H \overline{)339_H} \\
\underline{32_H} \\
19_H \\
\underline{19_H} \\
00
\end{array}
$$
$\leftarrow$ -- Zero remainder indicates no errors present

The transmission integrity of a received message is determined by using a *frame check sequence* (FCS). The FCS is generated by the transmitter, inspected by the receiver, and positioned within a frame in accordance with the diagram in Fig. 8-13. The procedure for using the FCS assumes the following:

1. The k bits of data being checked by the FCS can be represented by a polynomial, $G(x)$.

   Examples:

   (a) $G(x) = 10100100 = x^7 + x^5 + x^2 = x^2 (x^5 + x^3 + 1)$
   (b) $G(x) = 00...010100100 = x^7 + x^5 + x^2 = x^2 (x^5 + x^3 + 1)$
   (c) $G(x) = 101001 = x^5 + x^3 + 1$

   In general, leading zeros don't change $G(x)$, and trailing zeros add a factor of x^n, where n is the number of trailing zeros.
2. The address, control, and information field (if it exists in the message) are represented by the polynomial $G(x)$.
3. For the purpose of generating the FCS, the first bit following the opening flag is the coefficient of the highest degree term of $G(x)$ regardless of the actual representation of the address, control, and information fields.
4. There exists a generator polynomial, $P(x)$, of degree 16 for a 16-bit CRC that has the form, $P(x) = x^{16} + x^{12} + x^5 + 1$, or some other commonly used form.

Generation and Use of FCS. The FCS is defined as a 1s complement of the remainder, $R(x)$, obtained from the modulo two division of $F(x)$ by the generator polynomial, $P(x)$. That is, when

$$F(x) = x^{16} G(x) + x^k (x^{15} + x^{14} + x^{13} + x^{12} + x^{11} + x^{10} + x^9$$

$$+ x^8 + x^7 + x^6 + x^5 + x^4 + x^3 + x^2 + x^1 + 1)$$

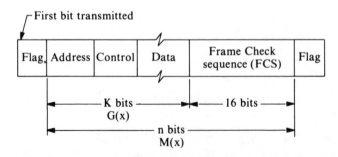

Fig. 8-13. Typical frame encapsulated in flags (delimiters).

then

$$\frac{x^{16} G(x) + x^k(x^{15} + x^{14} + \ldots x^1 + 1)}{P(x)} = Q(x) + \frac{R(x)}{P(x)} \overline{\text{FCS}}$$

The multiplication of $G(x)$ by x^{16} corresponds to shifting the message, $G(x)$ by 16 places and thus providing the space of 16 bits for the FCS.

The addition of x^k $(x^{15} + x^{14} \ldots + x + 1)$ to $x^{16} G(x)$ is equivalent to inverting the first 16 bits of $G(x)$. It can also be accomplished in a shift-register implementation by presetting the register to all 1s initially. This term is present to detect erroneous addition or deletion of zero bits at the leading end of $M(x)$ as a result of erroneous flag shifts.

The complementing of $R(x)$ by the transmitter at the completion of the division insures that the transmitted sequence, $M(x)$, has a property that permits the receiver to detect the addition or deletion of trailing zeros that may appear as a result of errors.

At the transmitter, the FCS is added to the $x^{16} g(x)$ term and results in a total message, $M(x)$, of length $K + 16$, where $M(x) = x^{16} G(x) + \text{FCS}$.

CRC Method Two: In the second method, the incoming $M(x)$ [assuming no errors; i.e., $M^*(x) = M(x)$] is multiplied by x^{16}, added to x^{k+16} $(x^{15} + x^{14} \ldots + x + 1)$, and divided by $P(x)$. We then have

$$Q_r(x) + \frac{R_r(x)}{P(x)} = \frac{x^{16} \left[x^{16} G(x) + \text{FCS} \right] + x^{k+16}(x^{15} + x^{14} \ldots x^1 + 1)}{P(x)}$$

Since the transmission is errorfree, the remainder, $R_r(x)$, will be "0001110100001111" (x^{15} through x^0).

$R_r(x)$ is the remainder of the division,

$$\frac{x^{16} L(x)}{P(x)}$$

where $L(x) = x^{15} + x^{14} \ldots + x + 1$. This can be shown by establishing that all other terms of the numerator of the receiver division are divisible by $P(x)$. Note that FCS = $R(x) = L(x) + R(x)$. [Adding $L(x)$ to a polynomial of the same length is equivalent to a bit-by-bit inversion of the polynomial.]

The division numerator of the receiver can be rearranged to

$$x^{16} \left[x^{16} G(x) + x^k L(x) + R(x) \right] + x^{16} L(x)$$

It can be seen by inspecting the transmitter generation equation that the first term is divisible by $P(x)$; thus, the $x^{16}L(x)$ term is the only contributor to $R_r(x)$.

The second detection process differs from the first in that another term, $(x^{16}L(x))$, is added to the numerator of the generation equation, thus causing a remainder of zero to be generated if $M^*(x)$ is received errorfree.

Implementation

A shift-register FCS implementation will be described in detail here. It utilizes "1s presetting" at both the sender and the receiver. Since the receiver does not invert the FCS, it thus checks for the nonzero residual, $R_r(x)$, to verify an error-free transmission. Figure 8-14(a) is an illustration of the implementation. It shows a configuration of storage elements and gates. The addition of $x^k(x^{15} + x^{14} \ldots + x + 1)$ to the $x^{16}G(x)$ term can be accomplished by presetting all storage elements to a binary 1. The 1s complement of $R(x)$ is obtained by the logical bit-by-bit inversion of the transmitter's $R(x)$.

Although Fig. 8-14(a) shows the implementation of the FCS generation for transmission, the same hardware can be used for verification of data integrity upon data reception.

Before transmitting data, the storage elements, $x_0 \ldots x_{15}$, are initialized to "1". The accumulation of the remainder, $R(x)$, is begun by enabling control "C" and thereby enabling gates G2 and G3. The data to be transmitted goes out to the receiver via G2, and at the same time the remainder is calculated with the use of the feedback path via G3. Upon completion of transmitting the k bits of data, the "C" is disabled and the stored $R(x)$ is transmitted via G1 while G2 and G3 are disabled. G1 provides the necessary inversion of $R(x)$ (assuming that "C" and output x_{15} are ANDed, which is logically equivalent to the two inputs being inverted).

At the receiver, before data reception, storage elements $x_0 \ldots x_{15}$ are initialized to "1s." The incoming message is then continuously divided by $P(x)$ via G3 ("A" enabled). If the message contains no errors, the storage elements will contain "0001110100001111" ($x_{15} \ldots x_0$) at the end of $M^*(x)$.

Table 8-3 is an example of the receiver and transmitter states during a transmission of a 19-bit $G(x)$ and 16-bit FCS.

The implementation of the FCS generation and the division by $P(x)$ as described here is offered as an example only. (Other implementations are possible and may be utilized with different polynomials.) This technique requires only that the FCS be generated in accordance with Rules 3.5 and 12.1 of the American Bureau of Standards and that the checking process involve division by the polynomial $P(x)$. Furthermore, the order of transmission of $M(x)$ is the coefficient of the highest-degree term first and thereafter the powers of x in decreasing order regardless of the actual representation of fields internal to $M(x)$.

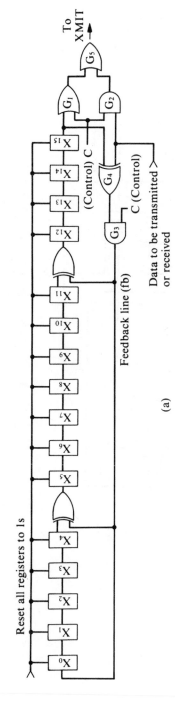

(a)

Fig. 8-14. (a) Diagram of CRC logic: The ϕ_1 and ϕ_2 clock lines that go to each flip-flop, X_0–X_{15}, are not shown (the polynomial generated by this CRC is $X_{15} + X_{12} + X_5 + 1$); (b) VLSI circuitry of this CRC at exclusive OR input X_{11} and feedback with output of the exclusive OR connected to storage cell X_{12} input (output of cell X_{12} input is then routed to input of X_{13}, and all register cells X_0 through X_{15}, are identical); (c) VLSI circuitry showing control portion of the CRC with control signals, X_{15}, and data as inputs and the feedback and transmitter as outputs; (d) logic diagram for the parallel-to-serial register; (e) VLSI circuitry for the parallel-to-serial register; and (f) timing diagram depicting the shifting operation, with MSB of M equal to a logic 1 and all other bits at logic 0.

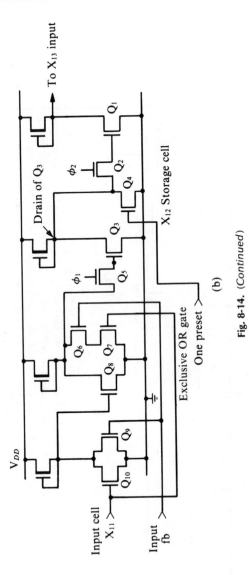

Fig. 8-14. (*Continued*)

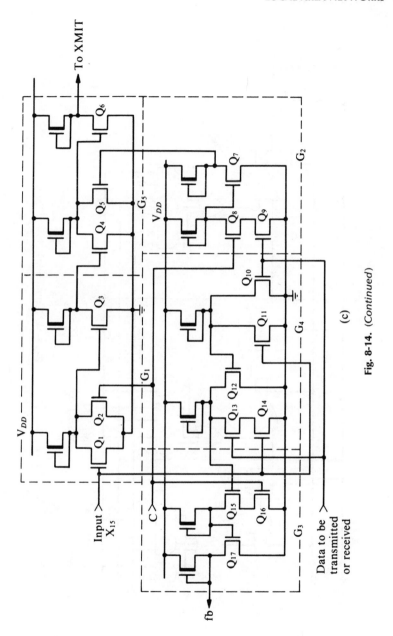

Fig. 8-14. (*Continued*)

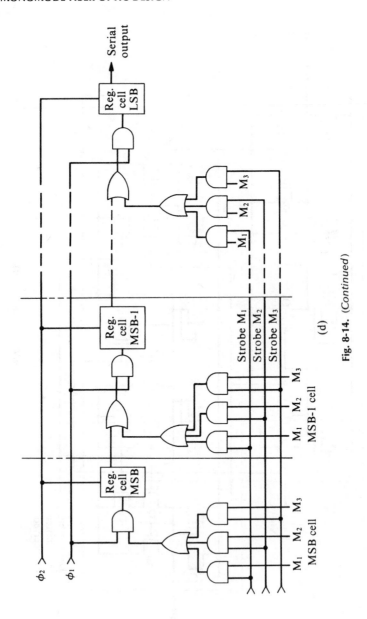

Fig. 8-14. (Continued)

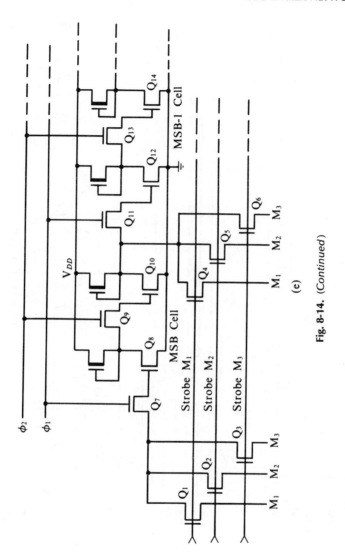

Fig. 8-14. (Continued)

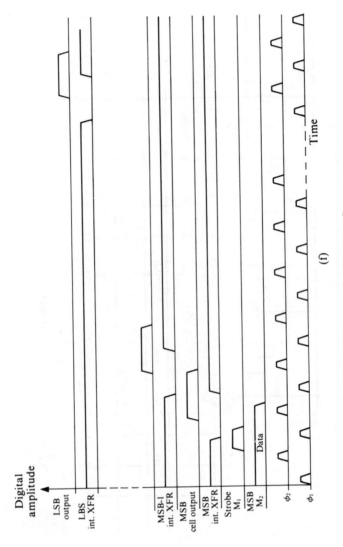

Fig. 8-14. (*Continued*)

TABLE 8-3. FCS Example.

Input To TX	TX CRC	Input To RX	RX CRC
(MSB)	1111111111111111		1111111111111111
0	1111101111110111	0	1111101111110111
1	0111110111111011	1	0111110111111011
1	0011111011111101	1	0011111011111101
1	0001111101111110	1	0001111101111110
1	1000101110110111	1	1000101110110111
0	1100000111010011	0	1100000111010011
0	1110010011100001	0	1110010011100001
1	0111001001110000	1	0111001001110000
1 (G(x))	1011110100110000	1	1011110100110000
0	0101111010011000	0	0101111010011000
0	0010111101001100	0	0010111101001100
1	1001001110101110	1	1001001110101110
1	1100110111011111	1	1100110111011111
0	1110001011100111	0	1110001011100111
0	1111010101111011	0	1111010101111011
0	1111111010110101	0	1111111010110101
1	0111111101011010	1	0111111101011010
1	1011101110100101	1	1011101110100101
1	0101110111010010	1	0101110111010010
	0010110111101001	1	1010101011100001
	0001011101110100	0	1101000101111000
	0000101110111010	1	1110110010110100
	0000010111011101	1	1111001001010010
	0000001011101110 (F)	0	0111100100101001
	0000000101110111	1	0011100010010100
	0000000010111011 (C)	0	0001111001001010
	0000000001011101	0	1000011110011010
	0000000000101110 (S)	0	1100001011000101
	0000000000010111	1	1110011011101010
	0000000000001011	0	0111001101110101
	0000000000000101	0	1011110110110010
	0000000000000010	0	1101101011010001
	0000000000000001	1	1110100101100000
	0000000000000000	0	1111000010111000
	0000000000000000	1	

X^0 .... X^{15} (TX CRC) F C S

x^0 .... x^{15} (RX CRC)

Since the implementation in VLSI of the CRC register has several circuits that are identical, the total circuitry need not be shown, only the typical circuits. Figure 8-14(b) is a cutaway circuit showing one of the modulo 2 addition nodes. This circuitry typifies the nodes depicted by the two exclusive OR functions between x_4, x_5 and x_{11}, x_{12}. The exclusive OR output performs a modulo 2 addition of the feedback and storage cell output, i.e., x_4 or x_{11} as shown in the logic of Fig. 8-14(a). The output of the exclusive OR gate shown in Fig. 8-14(b) will be shifted to the gate of Q_3 when clock ϕ_1 is high. The capacitance of the gate of Q_3 is either charged or discharged depending on the value of the exclusive OR gate output, i.e., whether it is a logic 1 or zero. After clock ϕ_1 goes low, the charge on the gate of Q_3 is isolated and the output of Q_3 will be an inverted version of the logic level outputted at the exclusive OR gate. If Q_4 is inactive, the gate is grounded, ϕ_2 will transfer the inverted exclusive OR output at the drain of Q_3 to the gate of Q_1, and ϕ_2 will then go low. The output shown going to x_{13} will be a replica of the exclusive OR output. Note, it is of paramount importance that ϕ, and ϕ_2 be high at different times and that no overlap occur. The clocks are used to isolate and transfer charge to the gates of the FETs.

When a preset is required, the preset line at the gate of Q_4 is held high for at least two ϕ_1 and ϕ_2 cycles; this will assure that all x_0 through x_{15} outputs are a logical 1 (high).

Let us next consider the control circuitry for the CRC as shown in Fig. 8-14(c). The control circuitry consists of G_1 through G_5. OR gate G_1 consists of Q_1 and Q_2, which form a NOR function of the control input, C, and the x_{15} register element output and Q_3 perform an inversion on the NOR gate to form the OR function. FETs Q_4, Q_5, and Q_6 perform an OR function on the outputs of gates Q_1 and Q_2. Data and the C input are ANDed together by Q_8, Q_9 and Q_7, where Q_8 and Q_9 perform a NAND, and Q_7 the inversion, function. The exclusive OR gate G_4 consists of Q_{10} through Q_{14}; it exclusive ORs data and the x_{15} register cell output. Gate G_3 is formed by Q_{15}, Q_{16} and Q_{17}, which ANDs the output of exclusive OR gate Q_4 and control line C.

The description of the control circuitry seems to be straightforward here, but one should always be aware that logic reduction techniques may be used to reduce the circuitry further. This was not done here so that the reader might become more familiar with the monolithic circuitry.

A diagram of the multiplexer register stages, all of which are similar, is shown in Fig. 8-14(d) in logic form; the timing diagram is shown in Fig. 8-14(f). Data that appears on inputs M_1, M_2, or M_3 is strobed into the cell by clock ϕ_2. This can be observed in Fig. 8-14(f), which shows how a binary word equivalent to 100. . .0 is propagated through the multiplexer after being strobed in the M_1 input. Not revealed by the timing diagram is the fact that strobes for M_1, M_2, or M_3 must not occur before the previous data is shifted out of the register serially. Therefore, the strobe must occur every n bits (word length) $\times$ T_{clock} (clock cycles ϕ_1 and ϕ_2). The strobe pulse width must occur during ϕ_1 clock positive pulses. The ϕ_1 clock performs the internal transfer of data, and ϕ_2 transfers the data to the output of the cell uninverted. For continuous dynamic refresh of the data, a third clock, ϕ_3, may be used to refresh the cell. A pass FET can be added between the output and input of the memory cell, and, after clock ϕ_2, clock ϕ_3 can recirculate the output to the input of the cell.

Figure 8-14(e) is a schematic of a VLSI implementation of the first two stage of the multiplexer. Q_1, Q_2, and Q_3, which are wire OR'd, serve to transfer M_1, M_2, and M_3 data to Q_7 when strobed with the appropriate line. Each of the strobe FETs forms an AND function with the ϕ_1 clock. The gate of Q_8 will charge or discharge to the appropriate logic level of the data. An inverted version of the data appears at the Q_8 drain/depletion transistor node, ϕ_2 will transfer the inverted data during the positive part of the clock cycle via the Q_9 pass transistor, and Q_{10} will invert the data again to the correct value for the data.

The VLSI design solution can be replaced with standard logic family devices, but the VLSI circuits are custom-made and operate at higher clock rates than most MOS family devices. The CRC register is a standard logic part produced by Signetics and Harris Semiconductor. Transmitters and receivers may be purchased off-the-shelf. The objective in this text is to examine circuitry that may be fabricated as part of a VLSI circuit. One should always be aware that this is an engineering enterprise and that each design problem may have a set of solutions that can be totally VLSI, totally discrete logic functions, or a combination of VLSI and discrete logic. The latter case is the usual situation.

Manchester Decoder and Clock Extraction Circuitry

Thus far, only the transmitter portion of Fig. 8-12(a) has been examined; now the receiver must be considered. The first item, of course, is the receiver itself, with its associated gain and detector, but this has already been considered in Chaps. 6 and 7 and the information will not be repeated here. The Manchester decoder and clock extraction circuitry, however, will be examined. Figure 8-15(a) shows a Manchester decoder circuit. Other types are available off-the-shelf in integrated-circuit form. For a number of different designs using counters, see Baker.[5]

This particular decoder is rather simply constructed. Binary data and clock information are Manchester-encoded as shown in the wave and timing diagram of Fig. 8-15(b). The Manchester waveform is injected at the D input of the flip-flop. The leading edge of the clock will transfer the data from the D input to output W, where it is ANDed with a positive clock pulse. The AND gate output

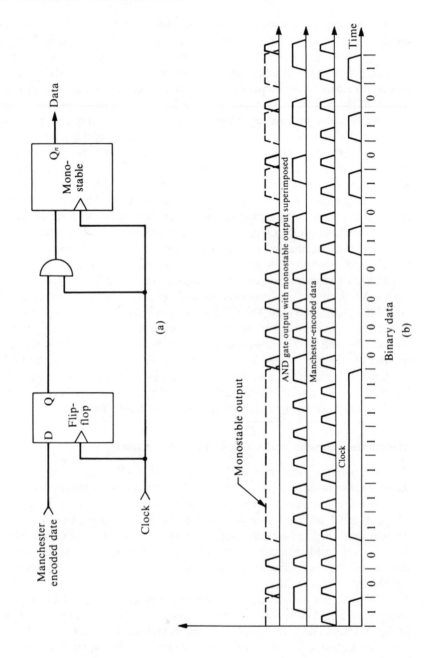

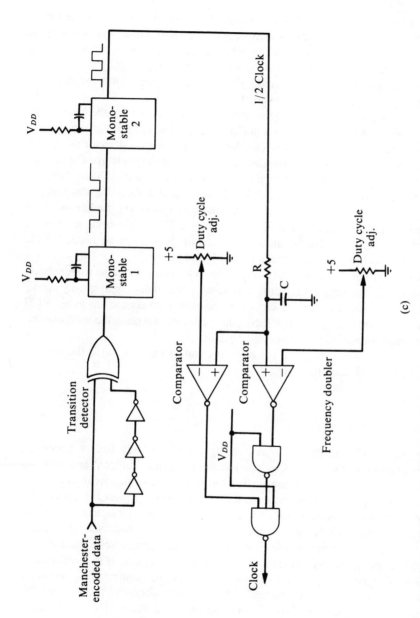

Fig. 8-15. Manchester decoder: (a) Block diagram, (b) wave and timing diagram, and (c) clock extraction circuit [alternative to that in Fig. 8-14(b)].

is represented by a series of pulses (solid-line waveform) shown at the top of Fig. 8-15(b). The monostable multivibrator (one-shot) output will normally be high until the positive-going leading edge of the AND gate triggers it; at that time, the Q_n output will go low. The multivibrator output will go high in one clock time unless it is retriggered. The broken-line waveform indicates the multivibrator operation that recovers the data.

The clock for the Manchester decoder can be extracted from the Manchester waveform. This circuitry is shown in Fig. 8-15(c). The edge of either a positive or negative transition can be detected by using a transition detector as shown. Larger pulse widths may be provided at the output of the exclusive OR gate by adding more inverters to the detector. These pulses from the exclusive OR gate will be stretched using the monostables shown. These devices are nonretriggerable; i.e., once triggered, they will time out and ignore any further pulses at the input. The second monostable will produce a 50/50 duty cycle at one-half the clock frequency. This is an optional circuit; it will not be needed if the frequency-doubling circuit RC has been selected correctly.

There are some problems with monostable multivibrators that the reader should be aware of. The timing may drift with temperature unless high-quality capacitors and resistors that do not change value with temperature are used. VLSI implementation can be accomplished using counters to replace the monostables. Other implementations are discussed by Baker.[5]

The CRC register shown at the receiving end in Fig. 8-12(a) is the same as the one at the transmitting end.

Typical Local Area Networks

The ring network in Fig. 8-16(a) is being considered only from a fiber-optic cable-plant point of view. Other considerations, such as protocol and networking, will be investigated in the section following. This ring circuit consists of several interfaces connected so as to form a ring. For our present purposes, the cable plant will be constructed of single-mode fiber, and the interfaces may use delta-beta ($\Delta\beta$) switches as bypasses to circumvent the node should a failure occur or they may use mechanical switches. Before the cable plant can be implemented, a loss budget must be calculated to determine if the performance will be adequate to meet a set of system performance requirements, and bandwidth calculations must also be made. As we shall observe in Chap. 9, the cable plant may be a ring, star, bus, or combinations of all three.

Star connections using single-mode technology do not have higher order modes sufficient for coupling. At the time this text was written, only four-port stars

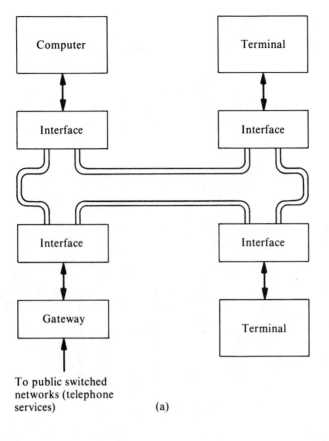

To public switched
networks (telephone
services)   (a)

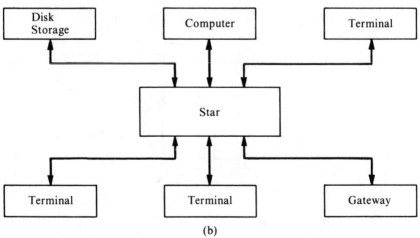

(b)

Fig. 8-16. Typical local area networks: (a) ring network topology using fiber-optic cable for cable-plant implementation, (b) star network, and (c) bus network.

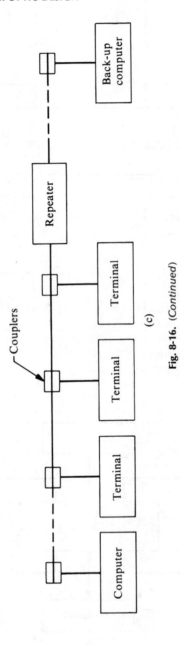

Fig. 8-16. (Continued)

were available. The equation that governs the star output power, P_o, is as follows:

$$P_o = \frac{P_{in}\, m\, K_c}{n} - P_{ex} \qquad (8\text{-}2)$$

where:

    n = number of output ports
    P_{ex} = excess loss
    m = modulation of the transmitter
    K_c = coupling coefficient

Star networks can be simulated by the use of several couplers, but the power output of each coupler must be adjusted to a different tap ratio to produce a true star. Excess loss for these simulated stars will be very high.

If a computer or microprocessor is fitted with several optical channels, it can be considered a hybrid star, i.e., a combination of optic and electric. Most of the stars with single-mode fiber optics will be of the latter design.

The configuration in Fig. 8-15(c) is a bus network that consists of several implements connected by single-mode couplers. Repeaters are often required to increase the power level for long lengths of fiber-optic cable to a usable value.

Let us now consider using a loss budget to determine whether excessive losses will limit a particular cable plant attenuation, i.e., whether performance limitations are dependent on attenuation. Table 8-4 is a loss budget guide drawn up so that no loss will be forgotten or neglected. Obviously, all entries in the table will not be present in every network. For example, star coupler insertion loss will not be found in ring or bus networks.

As an aid to the reader an explanation of each variable in Table 8-4 is given below with typical values:

X_1:   Transmitter output power is self-explanatory; typical range of LED is -20 to 0 dBm and as high as 10 dBm in solid-state lasers.

X_2:   This is the coupling loss; only 0.1 dB to 1.5 dB is typical.

X_3:   Typically, -1 to -2 dB/km, 10 μm at $\lambda = 820$ nm.

X_4:   Range of LAN application is $1/2$ to 20 km.

X_5:   Calculated $X_5 = X_3 \times X_4$.

X_6:   This is the number of switches allowed in the bypass condition between transmitter and receiver.

X_7:   Insertion loss per switch is self-explanatory (typically 1 to 2 dB)

X_8:   Total switch insertion loss $X_8 = X_6 \times X_7$.

TABLE 8-4. Loss Budget Guide.

X_1	TRANSMITTER OUTPUT POWER _____ dBm
X_2	TRANSMITTER TERMINATION LOSS, L_C _____ dB
X_3	CABLE ATTENUATION _____ dB/km
X_4	CABLE LENGTH MAXIMUM _____ km
X_5	CABLE ATTENUATION _____ dB
X_6	NUMBER OF FIBER-OPTIC SWITCHES, N_{sw} _____
X_7	INSERTION LOSS PER SWITCH, L_{IN} _____ dB
X_8	TOTAL SWITCH INSERTION LOSS, $N_{sw} + L_{in}$ _____ dB
X_9	SWITCH TERMINATION LOSS, $2N_{sw} \times L_C$ _____ dB
X_{10}	NUMBER OF SPLICES _____ dB
X_{11}	SPLICE LOSS, L_{sp} _____ dB
X_{12}	TOTAL SPLICE LOSS, $N_{sp} \times L_{sp}$ _____ dB
X_{13}	STAR COUPLER TERMINATION LOSS, L_C _____ dB
X_{14}	STAR COUPLER INSERTION LOSS _____ dB
X_{15}	STAR COUPLER OUTPUT POWER _____ dBm
X_{16}	RECEIVER TERMINATION LOSS _____ dB
X_{17}	RECEIVER MINIMUM SENSITIVITY _____ dBm
X_{18}	TEMPERATURE VARIATION ALLOWANCE _____ dB
X_{19}	CODE FORMAT NRZ − 3 dBm; FORMAT RZ −6 dBm
X_{20}	AGING LOSS _____ dB
X_{21}	TRANSMITTER POWER AT END OF LIFE _____ dBm
X_{22}	REQUIRED SNR _____ dB; BER _____
X_{23}	SOURCE OUTPUT HALVED (TO INCREASE LIFETIME) −3 dBm
X_{24}	MISCELLANEOUS LOSSES _____ dB
X_{25}	TOTAL LOSSES—COUPLING INSERTION, ETC. _____ dB
X_{26}	ADJUSTED POWER OUTPUT _____ dBm
X_{27}	TOTAL AVAILABLE POWER MARGIN _____ dB
X_{28}	POWER MARGIN _____ dB

X_9:   Switch termination loss $X_9 = 2X_6 \times X_2$.

X_{10}:   Self-explanatory.

X_{11}:   Self-explanatory (typically, 0.05 to 0.1 dB).

X_{12}:   Total splice loss $X_{12} = X_{10} \times X_{11}$.

X_{13}:   Input and output termination loss for the star coupler same as that for other terminations (transmitter, receiver, etc.)

X_{14}:   Given by the manufacturer or calculated for the typical range of 1 to 3 dB.

X_{15}:   Calculated or given by the manufacturer; application-dependent.

X_{16}:   Self-explanatory range of values (typically, 0.5 to 1.5 dB).

X_{17}:   Given by the manufacturer or calculated during receiver design (-30 dBm to -50 dBm typically)

X_{18}:   This parameter deals with all temperature-related losses following the transmitter and preceding receiver (such as switch, splice, cable, couplers, connectors, etc.). The transmitter and receiver variations are taken into account when specifying transmitter output and receiver sensitivity. Variations may occur in the alignment of the components of the link that will cause losses. The cable usually has a temperature excursion of $+80°$ to $-55°C$ with large losses at the lower extreme, depending on cable construction.

X_{19}:   Choice dependent on signal format.

X_{20}:   Aging loss also considers only those fiber-optic components that follow the transmitter and precede the receiver. This term deals with cable deterioration, splices, switches, etc.

X_{21}:   Transmitter power will diminish with age; therefore, when the power output has dropped to a specific percentage of the original minimum value (usually 80 percent of the original), end of life is considered to have been reached, and the transmitter should be replaced.

X_{22}:   The SNR or BER must be given to determine if performance can be maintained over the lifetime of the network. If BER is given, the SNR required can be determined from Table 8-5 or calculated.

X_{23}:   Operating LEDs or lasers (in particular) at half their rating will increase device life, but many of the new devices have sufficiently long mean time between failures so that this is no longer required. For installations meant to last longer than five or six years, such as in the telephone industry, however, this may need to be considered.

X_{24}:   Miscellaneous items deal with all items not accounted for in other terms in the budget, such as receiver pigtail and cable waveguide or mismatches. These mismatches can occur on switches, star couplers, Tee couplers, etc.

X_{25}: Total loss must be summed as shown by Eq. 8-3:

$$X_{25} = X_2 + X_5 + X_8 + X_9 + X_{12} + X_{13}$$
$$+ X_{14} + X_{16} + X_{18} + X_{20} + X_{24} \qquad (8\text{-}3)$$

X_{26}: This is the power output available after all adjustments and deratings calculated with Eq. 8-4 below have been made:

$$X_{26} = X_{19} + X_{21} + X_{23} + X_1 \qquad (8\text{-}4)$$

X_{27}: This is the total amount of power margin available before any link losses are considered; it is calculated by using Eq. 8-5:

$$X_{27} = X_{26} - X_{17} \qquad (8\text{-}5)$$

X_{28}: The excess power margin is calculated with Eq. 8-6. If the value is negative, the link will only marginally operate, if at all.

$$X_{28} = X_{27} - X_{25} \qquad (8\text{-}6)$$

A comparison must be made of the excess power margin with the required SNR given the restrictions shown in the calculation of Eq. 8-7. For adequate performance, it is necessary that

$$X_{28} - X_{22} \geq 0 \qquad (8\text{-}7)$$

An example of a fiber-optic ring will now be given to illustrate the use of Table 8-4.

Example:

A LAN is to be designed with the following specifications:

Maximum terminal distance 6 km
Cable attenuation 1 dB/km
Connector loss 1 db/termination
Switch insertion loss 1.5 dB/switch
Maximum number of switch bypasses between terminals 2

Maximum number of splices 3

Loss/splice 0.1 dB

Launch power, 1 mW -10 dBm

Code format NRZ

Receiver worst-case sensitivity -28 dBm

Worst-case BER at 10^{-9} 21.6 dB

A table of values, Table 8-5, has been computed for these specifications. In this loss budget, the label N/A means that the value in question "does not apply." Values are calculated to indicate power margins; if the BER/SNR subtraction is negative, an adjustment must be made to the link. General trade-offs can be made as follows:

1. The transmitter power can be increased.
2. A more sensitive receiver can be substituted.
3. Lower loss cable can be used.
4. Lower loss connectors can be used.
5. Fewer bypass switches can be allowed.
6. Lower loss switches can be used.
7. All switches can be purchased with pigtails and spliced into the fiber-optic circuit, but this technique will make the terminals less portable.

This power margin is fairly small since it should be at least 3 dB. Therefore, some of the loss-reducing techniques should be considered unless the BER requirement can be relaxed. A table of values for converting SNR to BER is provided in Table 8-6 for the reader's convenience.

When configuring a terminal, remember to keep the fiber-optic terminations at a minimum. As an example, connections from the transmitter and receiver to the fiber-optic switch should be made with splices. The switch should be provide with sufficiently long pigtails to allow connections to be made with a bulkhead connector through the terminal enclosure wall. These precautions will keep losses within the machine enclosure minimized. The specification did not include all parameters; those given will allow the reader to focus on one calculation at a time.

The next parameter to consider is rise time. A table of calculations similar to those for the loss budget can be generated, as shown in Table 8-7.

In comparison to system rise time, T_l, in Table 8-7, T_{total} indicates whether the calculated rise time is adequate to meet the system requirements. The transmitter and receiver rise times shown in the table are end results after taking all amplifiers and drive circuits into account. The receiver and transmitter rise times

Table 8-5. Loss Budget for the Example Ring.

TRANSMITTER OUTPUT POWER _−10 dBm_

TRANSMITTER TERMINATION LOSS _1 dB_

CABLE ATTENUATION _1 dB/km_
CABLE LENGTH MAXIMUM _6 km_

CABLE ATTENUATION _6 dB_

NUMBER OF FIBER-OPTIC SWITCHES _2_

INSERTION LOSS PER SWITCH _1.5 dB_

TOTAL SWITCH INSERTION LOSS _3 dB_

SWITCH TERMINATION LOSS $2N_{sw} \times L_C$ _4 dB_

MAXIMUM NUMBER OF SPLICES _3_

SPLICE LOSS _0.1 dB_

TOTAL SPLICE LOSS _0.3 dB_

STAR COUPLER TERMINATION LOSS _N/A_

STAR COUPLER INSERTION LOSS _N/A_

STAR COUPLER OUTPUT POWER _N/A_

RECEIVER TERMINATION LOSS _1 dB_
RECEIVER MINIMUM SENSITIVITY _−28 dBm_

TEMPERATURE VARIATION ALLOWANCE _1 dB_

CODE FORMAT NRZ _−3 dBm_

AGING LOS _1 dB_

TRANSMITTER POWER AT END OF LIFE _−1 dB_ (assuming 80 percent of worst case)

SOURCE OUTPUT HALVED-INCREASE LIFETIME _Not halved_

MISCELLANEOUS LOSSES _N/A_

EQ. 8-3: $\underline{X}_{25} = 1 + 6 + 3 + 4 + 0.3 + 1 + 1 = 16.3$ dB

TOTAL LOSS _16.3 dB_

EQ. 8-4: $\underline{X}_{26} = (-3 \text{ dBm}) + (-1 \text{ dBm}) + (0) + (0 \text{ dBm}) = -4$ dBm

ADJUSTED POWER OUPUT _−4 dBm_
EQ. 8-5: $\underline{X}_{27} = (-14 \text{ dB}) - (-32 \text{ dB})$
          $= 18$ dB

TOTAL AVAILABLE POWER MARGIN _18 dB_

EQ. 8-6: $\underline{X}_{28} = 18 \text{ dB} - 16.3 \text{ dB}$
          $= 1.7$ dB

POWER MARGIN _1.7 dB_ (Actual margin if BER is included in sensitivity specification, which
                    is the usual case for digital receivers)

Table 8-6. Bit Error Rate Versus Signal-to-Noise Ratio.

BER	SNR	BER	SNR	BER	SNR
10^{-2}	13.5	10^{-5}	18.7	10^{-9}	21.6
10^{-3}	16.0	10^{-6}	19.6	10^{-10}	22.0
10^{-4}	17.5	10^{-7}	20.3	10^{-11}	22.2
		10^{-8}	21.0		

can be measured or calculated values. If these devices are purchased as modules, the specification will provide the rise time. If they are to be designed, then the transmitter and receiver values must be calculated or measured.

Bandwidth calculations must be made to determine if a system is bandwidth-limited. If rise time is adequate, the bandwidth will allow correct system performance. Occasionally, bandwidth figures are known only for the transmitter and receiver. Because of bandwidth dependence, the equations for calculating the rise time can be employed.

Cable is specified in MHz-km but is usually given only at a particular wavelength. If a wavelength other than that specified is used, the system rise time can be calculated by using Table 8-7, and the bandwidth can be calculated using either the PPM or PCM equation, depending on the modulation technique.

Table 8-7. Rise Time Calculations.

T_t TOTAL SYSTEM RISE TIME _____ nsec

      DIGITAL RZ, 0.35/bit rate

      DIGITAL NRZ, 0.7/bit rate

      INTENSITY MODULATION ANALOG, 0.35/bandwidth

(PPM) PULSE POSITION MODULATION = [SAMPLING RATE × (PULSE SEPARATION/ PULSE WIDTH) × BANDWIDTH^{-1}

(PCM) PULSE CODE MODULATION = [SAMPLING RATE × (BITS/SAMPLE) × BANDWIDTH]$^{-1}$

T_{2a} TRANSMITTER RISE TIME (analog) _____ nsec

T_{2b} TRANSMITTER RISE TIME (digital) _____ nsec

T_3 RECEIVER RISE TIME _____ nsec

T_4 CHROMATIC DISPERSION (evaluate Eqs. 2-34, 2-35, or 2-36) _____ nsec

$T_{total} = 1.11 \sqrt{T_2 \ (T_{2a} \ \text{or} \ T_{2b}) + T_3^2 + T_4^2} = $ _____ nsec

Data Link Layer

The data link layer provides data flow control across the physical layer. It decodes address information within the data stream and provides address encoding; moreover, it often provides error detection and corrections. The data link layer provides the rules needed to allow the physical layer to communicate efficiently.

There are roughly three types of protocol: no cooperation, limited cooperation, and total cooperation. Examples of each are the following: No cooperation can be best represented by the ALOHA scheme, a technique that allows a node to transmit without regard to the other nodes in the system. A collision when two or more nodes transmit is detected only after the transmitting nodes fail to receive an acknowledgement of the message from the receiver (after a prescribed delay period). Another technique can be employed if the transmitting node does not receive the transmitted data at its own receiver. Repeated transmitting-node collisions are avoided because the delay before retransmission is controlled by random selection. These delays are due to random-number generators in the controllers. Slotted ALOHA is a form that has packets of uniform length. Once a node has seized the channel, it may transmit free of collision for the duration of the slot. This type of ALOHA is an example of a limited-cooperation protocol since a collision may occur during the channel seizure.

Total-cooperation protocols are characterized by no collisions during transmission. Two examples are the token-passing ring and the slotted-line. The token-passing ring allows only the node that possesses the token to transmit. After transmission, the sending node relinquishes the token, and the next node desiring to transmit must capture it. When no nodes with to transmit, the token is passed around the ring. The technique known as the slotted line dedicates a time slot for each node. As the time slot passes the node, the packet is inserted with the destination address, and the receiving node can determine the source of the data by the time slot. It should be obvious that total cooperation between nodes is required to prevent any collisions.

The performance of noncooperative protocols is the poorest, and that of totally cooperative protocols naturally the best. The complexity of the circuits required also follow this same relationship, i.e., noncooperative protocol is less complex than fully cooperative types.

Elementary Protocol

Several different types of protocols will now be examined and their performance compared. Some popular ones have been implemented in integrated-circuit form. Every LAN has certain attributes that must be understood before an analysis can be undertaken; these include the following:

1. Each network has N nodes.
2. The message traffic is of a bursty nature.
3. Message traffic is generated by all nodes with equal probability.
4. All message packets are of equal length.
5. The worst-case delay must be considered, i.e., the delay between the nodes separated by the longest distance.
6. All transmissions are broadcast throughout the network.

Previous definition of the packet has been rather superficial. The packet is an organized assemblage of bits in which data-link layers communicate. (As previously mentioned, the physical layer converses with other nodes in bits.) The packet is composed of address information (i.e., source and destination), control bytes, data, and FCS word, and sequencing numbers. A flag or delimiter is inserted at the head and end of each packet to form a frame.

Figure 8-17 depicts the contents of various types of packets, which will not be discussed in detail until Chap. 9 because they are related to specific LANs.

The address information—destination and source—is self-explanatory; subtleties in design will be covered in the next chapter. The control portion of the packet is used for node housekeeping activities such as ACK/NACK, address extension, type of frame identification (i.e., supervisory or data), and other similar information.

The next item in a packet is data. Packets may be received out of order, and some may require retransmission; these mishaps are characteristic of virtual circuit service, which is representative of most modern LANs. The other technique of message transmission is the *datagram,* in which the data is sent as a message, and no attempts are made to retransmit erroneous messages. An example of a datagram is mail delivery. The letter is received as a total message, and letters may be delivered to the receiving end out of order; i.e., letters one and two may be sent on two consecutive days but letter two may arrive before letter one. This type of service is also characteristic of telegraph service. In this text, only virtual circuit service will be considered.

After the data field, the next item in the packet to consider is the checksum or FCS word. The FCS word can be appended on the checked or the incoming frame FCS can be compared directly with an FCS word generated at the receiver. The technique described previously for the CRC uses the same checksum of each frame, whereas the latter technique uses a different one. The CRC will check the data stream for burst errors, depending on the number of bits in the checking polynomial.

The last items in the frame are the flags, which signify the beginning and the end of the frame. Frames such as Ethernet with no flag at the end sense inactivity on the channel for a timeout period; i.e., no transitions occur.

The previous discussion has centered on the mechanism of data-link-level

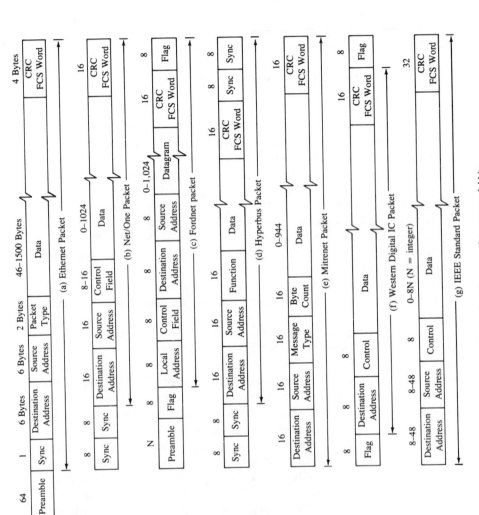

Fig. 8-17. Typical frames of common LANs.

communication between nodes. The issue of how channels are initiated must be considered for protocols that are noncooperative, limited cooperative, or fully cooperative.

The first type to be considered is the ALOHA type (noncooperative). The ALOHA protocol was originally used on radio networks and later adapted for LAN use. These LANs allow nodes to transmit at any time (no cooperation between nodes). The circuitry is simpler to implement, but the performance is poor. When two nodes transmit at the same time, the receiver will note the error but not send an acknowledgement. The transmitting nodes will not receive acknowledgements but after a timeout period must retransmit the data. As one can imagine, during heavy traffic periods, a great deal of collisions occur, thus causing large delays before messages can reach their destination. These long delays will cause losses in the throughput.

If the network access is such that the transmitting node can receive its own transmissions, then collisions that occur after the network delays can be detected by the transmitter. The transmitting nodes will stop transmitting for a random period of time if a collision occurs and begin again if no traffic is sensed on the channel. This type of protocol is commonly called CSMA (Carrier Sensed Multiple Access). There is more than one classification for CSMA protocols: nonpersistent CSMA, which is the type just described, and persistent CSMA. The persistent CSMA is characterized by nodes that begin trying to transmit as soon as no transmission is sensed. Note that this technique senses collisions at the receiver of the transmitting node; because of propagation, the latter requires long delays before the collisions can be sensed.

Suppose that collisions can be sensed at the transmitting node during transmission, i.e., that they can be detected immediately. This technique is called, appropriately, CSMA/CD, where CD stands for collision detect. This is the technique used for implementing the Ethernet protocol.

A technique used by IBM, which is an IEEE 802 standard, is the token ring. A node that wishes to transmit must have the token. When no stations are transmitting, the token is passed from node to node. This technique requires total cooperation between nodes. A variation on token passing is using the CSMA/CD with the channel quiet (no transmission is sensed). When a station finally seizes the channel, it transmits the data frame and appends a token after the flag on the last frame sent. Any node desiring to seize the channel need wait only until the token appears and seize it. The station that has the token can then transmit.

The slotted ring requires each station on a ring topology network to have a unique time slot. The station may transmit only during its time slot. This type of network needs a master station to manage and assign time slots. Throughput will suffer with this type of system but its reliability is very high and therefore a trade-off can be made.

A discussion of network topology will be reserved for Chap. 9, but some performance issues will be addressed here. For example, ring topology may be implemented with token passing, CSMA/CD, CSMA, slotted ring, or register insertion. Bus topology may be implemented with CSMA, CSMA/CD, token passing, or broadband bus with CSMA/CD. A performance comparison of some of these techniques will be presented.

Protocol Analysis

Equation 8-8 defines a parameter, a, which has a large impact on channel utilization. It expresses the ratio of the worst-case length of the data path to packet length, as follows:

$$a = \frac{R \times D_p}{L_p} \tag{8-8}$$

where:

R = data rate, in bits/sec
D_p = worst-case propagation delay, in sec
L_p = maximum packet length, in bits

An example of this calculation follows:

Example 8-1: Given a fiber-optic link with a worst-case distance between nodes of 1 km, a packet length of 128 bytes, and a transmission rate of 10 Mbits/sec, calculate the value of a.

$$\text{Fiber-optic propagation delay} = \tau_d = \frac{n}{c} \approx 5 \text{ nSec/m}$$

$$D_p = \tau_d l = 5 \ \mu\text{sec @ 1 km}$$

Using Eq. 8-8,

$$a = \frac{10\text{M} \times (5 \times 10^{-6})}{128 \times 8}$$

$$= 0.049$$

Another important feature is channel utilization, which is given by Eq. 8-9. The value of a is used in the calculation, as shown below:

$$U = \text{channel utilization} = \text{throughput } (\tau_r) / \text{ data rate } (R)$$

$$U = \frac{L_p}{L_p + D_p R}$$

$$U = \frac{1}{1 + \dfrac{D_p R}{L_p}} = \frac{1}{1 + a} \tag{8-9}$$

where $a = D_p R / L_p$.

As a decreases, the value of U approaches 1, which implies that utilization is 100 percent. This occurs whenever the packet length approaches infinity or the worst-case delay approaches zero—situations, of course, that never arise in practical systems.

The utilization analysis given here takes an intuitive approach. Several features are not considered because they tend to make the model much more cumbersome to derive. Neglected are time delays due to collisions, acknowledgement packets, waiting time for tokens, slotted ring waiting time for slots, register insertion delays. These features are characteristic of CSMA/CD, token bus and ring, slotted ring, and register insertion LANs.

Before investigating more complex calculations, the token passing bus and ring will be examined and, of course, and CSMA/CD protocol. Let us first examine the calculations required to predict the throughput performance of a token passing network, whether bus or ring. The parameters are as follows:

τ_D = time delay due to propagation time
τ_c = average time for one cycle
τ_p = average transit time for a data packet
τ_r = average token passing time
τ_r = throughput
N = number of stations

$$\tau_r = \frac{\tau_p}{\tau_c} = \frac{\tau_p}{\tau_p + \tau_t}$$

For networks with $a < 1$, as previously defined, i.e., with $\dfrac{\tau_D}{\tau_p} < 1$,

$$\tau_c = \tau_p + \frac{\tau_D}{N}$$

Then

$$\tau_r = \frac{\tau_p}{\tau_p + \dfrac{\tau_D}{N}} = \frac{\tau_p / \tau_p}{1 + \dfrac{\tau_D / \tau_p}{N}}$$

$$\tau_r = \frac{1}{1 + \dfrac{a}{N}} \qquad (8\text{-}10)$$

If the networks have values of $a > 1$, then Eq. 8-11 predicts throughput τ_r as follows:

$$\tau_c = \tau_D + \frac{\tau_D}{N}$$

$$\tau_r = \frac{\tau_p}{\tau_D + \tau_D / N} = \frac{1}{a\left(1 + \dfrac{1}{N}\right)} \qquad (8\text{-}11)$$

As would be expected, throughput improves when propagation delays are short and packets are long. Let us examine Eq. 8-11 for a moment. As the number of stations on the link are increased, the throughput will become dependent on the ratio of τ_p / τ_D, or $1/a$. The equation will continue to function in a correct manner; i.e., decreasing delay or increasing packet size will improve throughput.

Next to be considered is the CSMA/CD throughput. These calculations are a bit more involved than those for token passing, and therefore a few additional definitions are in order, as follows:

τ_s = time slots on the medium = $2\tau_d$

P_s = probability of a station transmitting during an available time slot

The time intervals within the medium are of two types: the transmission interval, $a/2$, and the collision interval. The probability, P_A, that only one station will acquire the medium must be calculated. The binomial probability function of the form shown in Eq. 8-12 will provide the expression for P_A shown in Eq. 8-13:

$$f(x) = \binom{N}{1} P^x q^{N-x} \tag{8-12}$$

$$P_A = NP' (1 - P)^{N-1} \tag{8-13}$$

By differentiating Eq. 8-13 and setting the derivative equal to zero, the maximum value of P can be calculated, as follows:

$$\frac{dP_A}{dP} = N(1 - P)^{N-1} - NP(N - 1) (1 - P)^{N-2} = 0 \tag{8-14}$$

Solving the equation for P,

$$P = 1/N$$

Substituting this value of P into the original expression for P_A, the maximum probability of P_A is given by

$$P_A = \left(1 - \frac{1}{N}\right)^{N-1}$$

Next, the mean length of the contention interval must be estimated. The assumption is that the time slots on the link either have alternating intervals with no transmission or have a collision at transmission intervals. The equation describing this function is Eq. 8-15:

$$\tau_r = \frac{a/2}{(a/2) + P(w)} = \frac{1}{1 + 2a \left(\dfrac{1 - P_A}{P_A}\right)} \tag{8-15}$$

where the contention interval $P(w)$ at w time slots is

$$P(w) = P_A (1 - P_A) + 2P_A (1 - P_A)^2 + \ldots kP_A (1 - P_n)^k$$

The value of $P(w)$ at series convergence is

$$P(w) = \frac{1 - P_A}{P_A}$$

The equations for the token bus and ring are evaluated for $N \to \infty$ as shown below for $a < 1$

$$\tau_r \lim_{N \to \infty} = 1$$

and for $a > 1$

$$\tau_r \lim_{N \to \infty} = \frac{1}{a}$$

For the CSMA equation, P_A must be evaluated for the maximum value, $P_A = (1 - 1/N)^{N-1}$:

$$P_A = \frac{\left(1 - \frac{1}{N}\right)^N}{\left(1 - \frac{1}{N}\right)} = \frac{1}{2!} - \frac{1}{3!}\left(1 - \frac{2}{N}\right) + \frac{1}{4!}\left(1 - \frac{2}{N}\right)\left(1 - \frac{3}{N}\right) + \cdots$$

$$P_A \lim_{N \to \infty} = \frac{1}{2!} - \frac{1}{3!} + \frac{1}{4!} - \frac{1}{5!} + \cdots$$

Then P_A can be written as follows:

$$P_A = 1 + (-1) + \frac{(-1)^2}{2!} + \frac{(-1)^3}{3!} + \frac{(-1)^4}{4!} + \cdots$$

or

$$P_A = \frac{1}{e}\Big|_{N \to \infty}$$

Then,

$$\tau_r = \frac{1}{1 + 2a(e - 1)} = \frac{1}{1 + 3.436a} \tag{8-16}$$

For token-passing networks, throughput performance increases as the number of nodes increases, and for $a > 1$, the network performance becomes dependent on a. A good rule to follow is to try to make the packet size large as compared to the delay, which may not always be possible. This same rule applies to CSMA/CD networks. As more stations appear on the network, the likelihood of a collision increases, and performance will suffer.

Various curves showing performance are presented in Figs. 8-18(a) and (b). The first figure shows throughput as a function of a. Note from the curves that if $a = 0.1$, increasing the number of stations on a link improves throughput for token passing and decreases throughput for CSMA/CD links. This was stated mathematically by Eqs. 8-10 and 8-11, but here it is shown graphically. Figure 8-18(b) shows throughput as a function of the number of nodes on the network for token passing and CSMA/CD for two values of a.

A study by the IEEE 802 Local Network Standards Committee produced the graphic results shown in Figs. 8-19(a), (b), (c), and (d). From (a) and (b), it can be seen that the actual data rate is almost at the same level as the channel data rate (channel transmission rate) or both token-passing protocols. The reductions in transmission rate is fairly drastic for the CSMA/CD bus protocol because a great deal of time is lost servicing collisions.

Lightly loaded networks, as shown in Fig. 8-19(c), indicate that token rings and CSMA bus exhibit almost identical performance for larger packets but that the token bus has relatively poor data rates. For smaller packets, the data rate is reduced even further for the token bus.

These curves are useful when making a decision on which type of protocol is best for a particular application because they show how throughput, traffic, protocols, packet size, and data rate impact on networks.

To make a comparison of ring networks would be a rather monumental task. Because of the large number of variables, it is well beyond the scope of this book. To give just one simple example, using small values of a might yield high throughput for one protocol and low throughput for another. A change in this one variable could then have a large impact on other parameters of other protocols. To make a fair comparison of networks, the designer must fix some of the parameters, e.g., minimum acceptable transmission rate, number of stations, packet length, etc. After some of the parameters have been fixed, certain protocols and topologies will drop out as undesirable in terms of performance.

The throughput and channel efficiency will be examined to determine the impact of the header, retransmission, number of ACK bits, etc., on these performance criteria. One obvious question is how many bytes should be in the data field for optimum efficiency? The following calculations will provide the reader with the equation needed to answer this question.

The first protocol to consider is a stop-and-wait type. No piggybacking is

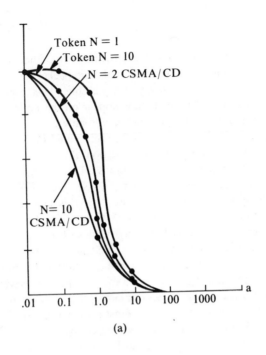

(a)

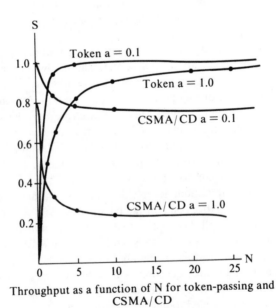

Throughput as a function of N for token-passing and
CSMA/CD

(b)

Fig. 8-18. (a) Curves showing efficiency of CSMA and token-passing networks, and (b) throughput as a function of N for token-passing and CSMA/CD.

allowed; i.e., the control field will not have an ACK type word. The variables are as follows:

A = acknowledge frame, in bits
C = channel capacity, in bits/sec
D = data bits per frame
P_E = probability of a bit error
$F = D + H$ frame
H = number of bits in the header and flags
I = interrupt and service time plus propagation delay
P_o = probability a data frame is lost or damaged
P_A = probability an ACK frame is lost or damaged
R = number of retransmissions
T = time-out interval
U = channel utilization, or efficiency
W = window size
$I = \tau_i + \tau_d$

The first analysis is of the transmission of a frame without any errors. Intuitively, let us consider the time required to send a frame, service an interrupt, and prepare to send an ACK frame. This time is calculated using Eq. 8-17, which is derived as follows:

$$\tau_{TF} = \frac{F}{C} + I = \tau_F + \tau_i + \tau_d = \text{total service time}$$

$$\tau_{TA} = \frac{A}{C} + I = \tau_A + \tau_i + \tau_d = \text{total ACK service time}$$

$$\tau_T = \tau_{TF} + \tau_{TA} = \tau_F + \tau_A + 2\tau_i + 2\tau_d$$

$$= \text{total frame and ACK service time}$$

$$\tau_D = \frac{D}{C} = \text{data transmission time}$$

$$U = \frac{\tau_D}{\tau_T} = \frac{\tau_D}{\tau_E + \tau_A + 2\tau_i + 2\tau_d} \tag{8-17}$$

Equation 8-17 can be expressed as a ratio of bits, as shown in Eq. 8-18. An alternative form is given by Eq. 8-19.

$$U = \frac{D/C}{F/C + A/C + 2I} = \frac{D}{F + A + 2IC} \qquad (8\text{-}18)$$

$$U = \frac{D}{D + H + A + 2IC} \qquad (8\text{-}19)$$

When each transmission of a frame occurs with errors, the total number of retransmission bits is $R(F + TC)$. The value, $F + TC$, will continue until an ACK frame is sent or until the node stops trying to transmit. A successful transmission will require the number of bits and the utilization given by Eqs. 8-20 and 8-21.

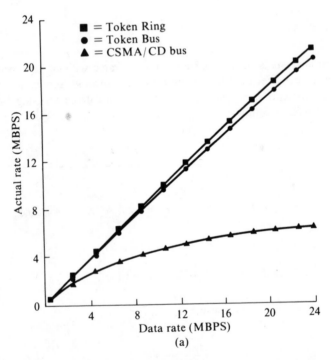

Fig. 8-19. (a) 2000 bits per packet, 100 stations out of 100 active; (b) 500 bits per packet, 100 stations out of 100 active; (c) 2000 bits per packet, one station out of 100 active; (d) 500 bits per packet, one station out of 100 active.

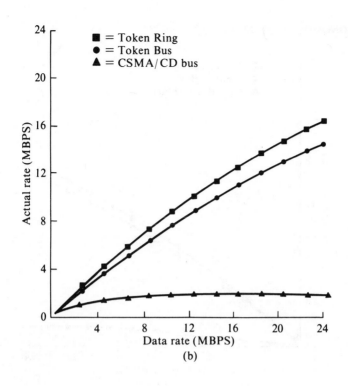

(b)

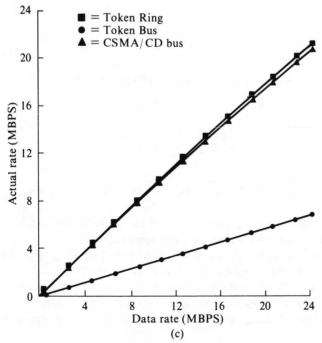

(c)

Fig. 8-19. (*Continued*)

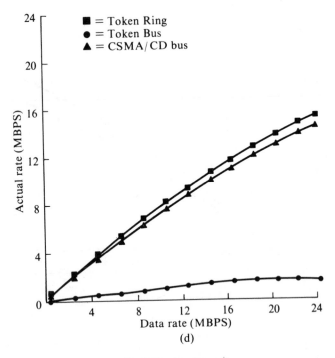

Fig. 8-19. (Continued)

$$B_t = \text{unsuccessful tries} + \text{one successful try}$$

$$= R(F + TC) + (D + H + A + 2IC) \qquad (8\text{-}20)$$

$$U = \frac{D}{R(F + TC) + (F + A + 2IC)}$$

$$U = \frac{D}{R(D + H + TC) + (D + H + A + 2IC)} \qquad (8\text{-}21)$$

The latter form of the equation will allow the reader to differentiate it and find relationships between data capacity and other variables.

Before the equation can be completed, however, the R value must be further examined. The mean number of retransmissions per frame must be estimated based on the probability of the successful transmission of data and its acknowledgement. We will now develop these equations, but first note that to complete a transmission of data, the ACK must also be considered. Parameters are as follows:

$P_s = (1 - P_D)(1 - P_A) =$ probability of successful transmissions
$P_f = (1 - P_s) =$ probability of failure
$n =$ number of transmission attempts needed

Then, $P^{n-1}(1 - P_f)$, where $n - 1 =$ the number of retransmissions, expresses the number of transmissions needed to transmit the frame successfully. The number of transmissions per frame is

$$T_R = \frac{1}{P_s}$$

The number of retransmissions is dependent on the probability of failure as governed by the expression

$$R = T_R P_f = \frac{P_f}{P_s}$$

Substituting this value of R into Eq. 8-21, we have

$$U = \frac{D}{\left(\dfrac{P_f}{P_s}\right)(D + H + TC) + (D + H + A + 2IC)} \qquad (8\text{-}22)$$

For the condition when the timeout, T, is $T = A/C + 2I$, Eq. 8-22 will reduce to the following:

$$U = \frac{D}{\left(\dfrac{P_f}{P_s} + 1\right)(D + H + A + 2IC)}$$

$$= \left(\frac{D}{D + H}\right)\left(\frac{P_s}{P_f + P_s}\right)\left(\frac{1}{1 + \dfrac{A + 2IC}{D + H}}\right)$$

Then,

$$U = \left(\frac{D}{D + H}\right)(P_s)\left(\frac{1}{1 + \left(\dfrac{CT}{D + H}\right)}\right) \qquad (8\text{-}23)$$

For an error occurring in both the ACK and data frames, consider the probability of an error in each bit P_E. Then an error in the ACK and data frame is represented by $(1 - P_E)^{H+D}$ and $(1 - P_E)^A$, respectively. Then, substituting $P_s = (1 - P_E)^{H+D} \times (1 - P_E)^A$ into Eq. 8-23 and letting $A = H$ (i.e., letting the ACK frame be the length of the header), we have

$$U = \left(\frac{D}{D + H}\right) (1 - P_E)^{H+D} (1 - P_E)^H \left(\frac{D + H}{D + H + CT}\right) \quad (8\text{-}24)$$

Taking the derivative of U with respect to D and setting it equal to zero will result in an optimum value for D, or D_{opt}, if the roots are found. The optimum D_{opt} is derived as follows:

$$\frac{\partial U}{\partial D} = \left(\frac{1}{D + H + CT}\right) (1 - P_E)^{2H+D} - \left[\frac{D}{(D + H + CT)^2}\right] (1 - P_E)^{2H+D}$$

$$+ \left(\frac{D}{D + H + CT}\right) \frac{\partial (1 - P_E)^{2H}}{\partial D}$$

$$0 = (D + H + CT) (1 - P_E)^D - D (1 - P_E)^D$$

$$+ D (D + H + CT) \frac{\partial (1 - P_E)^D}{\partial D}$$

$$0 \cong (H + CT) (1 - P_E)^D + [D^2 + D (H + CT)]$$

$$\cdot (1 - P_E)^D \ln (1 - P_E)$$

$$0 = [D^2 + D (H + CT)] \ln (1 - P_E) + (H + CT)$$

Then,

$$D_{opt} = -\frac{H + CT}{2} \pm \sqrt{\frac{(H + CT)^2}{4} - \frac{(H + CT)}{\ln (1 - P_E)}}$$

or,

$$D_{opt} = \frac{H + CT}{2} \left[\sqrt{1 - \frac{4}{(H + CT) \ln (1 - P_E)}} - 1\right] \quad (8\text{-}25)$$

The expression in Eq. 8-25 is used to optimize D (note that the negative radical is ignored when finding the roots of D_{opt} because D can have only positive values. If the probability of a bit error is small—which is the usual case (common values are $P_E = 10^{-8}$ or 10^{-9})—then the value of $\ln (1 - P_E)$ in the equation will reduce the latter to the form of Eq. 8-26:

$$D_{opt} = \frac{H + CT}{2} \left[\sqrt{1 + \frac{4}{P_E (H + CT)}} - 1 \right] \qquad (8\text{-}26)$$

where the term, $4/P_E (H + CT) \gg 1$. Therefore, the value for D_{opt} can be approximated by the following:

$$D_{opt} \approx \sqrt{\frac{H + CT}{P_E}} \qquad (8\text{-}27)$$

An examination of this equation reveals some important observations. As the error rate gets smaller ($P_E \rightarrow 0$), the optimum data field becomes infinitely large. As headers and time-outs become large, the frame size must increase to improve U, which is realistic.

REVIEW PROBLEMS

1. Draw a logic diagram for the demultiplexer shown in Fig. 8-12(a).
2. Draw the VLSI equivalent circuit for this demultiplexer, using a gating scheme at the output of the storage cell; show strobe lines.
3. Check the circuit operation of problem 8-2, using a wave and timing diagram.
4. Do a loss budget calculation for a ring with the following specifications: transmitter output, 1 mw; Connector loss, 1 dB; connectors between transmitter and receiver, 6 max; coupler loss, 1.5 dB; maximum number of couplers between transmitter and receiver, 2; end-of-life power output, 80%; cable loss, 1 dB/km; receiver sensitivity, −38 dBm.
5. Is the ring network in problem 8-4 amplitude- or bandwidth-limited?
6. Make the rise-time calculations for the following cable plant: $\lambda = 1550$ nm; $\lambda_c = 1550$ nm; $\Delta = 0.01$; cable length 10 km; spectral width of the source 2 nm; transmitter operates at 200 Mb/s; receiver rise time 1 nsec; PCM modulation = 8 bits per sample.
7. A network has the following characteristics. It has a single-mode cable plant designed with waveguide material with a refractive index of 1.497. The cable plant has a maximum length of 10 km and a data rate of 10 Mb/s. What is the optimum packet length for the optimum throughput?
8. Which protocol is more efficient for use with the cable plant of problem 7—CSMA, CD, or token passing?
9. If the cable plant in problem 7 has 1000 stations and 10 percent are in use at any one time, which protocol is more efficient for a 400-bit-packet token-passing ring—CSMA/CD ring or token-passing bus?

10. Derive an equation for the optimum header size for the optimum data bits per frame.
11. Do a complete cable plant design for the following specifications:

Laser transmitter:
 wavelength of operation, 1300 nm
 spectral width, 0.1 nm
 power output, 10 dBm
 rise time, 0.1 nsec

Modulation NRZ, PCM

Receiver:
 sensitivity, −48 dBm
 rise time, 1.0 nsec

Data rate, 200 Mb/s

Cable:
 attenuation, 0.8 dB/km
 bandwidth, 1 Ghz-km
 maximum length, 28 km
 coupler delta beta, 3 dB/coupler
 terminations, 1 dB each
 splices, 0.2 dB each
 plant type, ring

Optimize the frame length if there are 200 stations on the ring. Make assumptions for any parameters not given—often the case for industrial designs.

REFERENCES

1. Tanenbaum, A. S. *Computer Networks*. Englewood Cliffs, NJ: Prentice-Hall Inc., 1981.
2. Flint, D. C. *The Data Ring Main: An Introduction to Local Area Networks*. New York: Wiley Heyden, 1983.
3. Baker, D. G. *Local Area Networks with Fiber Optic Applications*. Reston, VA: Reston, 1986.
4. Trapper, C. *Local Computer Network Technologies*. New York: Academic Press, Inc., 1985.
5. Baker, D. G. *Fiber Optic Design and Application*. Reston, VA: Reston, 1985.

9 | LOCAL AREA NETWORKS APPLIED TO SINGLE-MODE FIBER-OPTIC TECHNOLOGY

NETWORK TOPOLOGY

In this chapter, we will investigate techniques for the analysis and design of network topology. Much of the information presented will be used in Chap. 10 for design and analysis. Before taking up topology, however, it would be prudent to determine when the use of broadband or baseband is the more advantageous since the decision will impact on topology and may simplify the analysis to some extent. This chapter will also provide local area network analysis and their applications to common networks.

Let us first examine broadband networks. One of the important advantages of broadband FDM techniques is the ability of these networks to handle analog and digital information and process it simultaneously. The coaxial cable, RF amplifiers, cable drops, modems, and other RF components have been used in the CATV industry with a high level of reliability. In a broadband network, a subset of many TDM channels equivalent to several baseband cables may be implemented. Broadband facilities allow transmission rates of the TDM channels to be upgraded with little or no modifications of the RF components. Future upgrades in facilities may require modem modifications and protocol changes but no changes in the cable plant; the latter, of course, could be quite costly. These cable plants can be easily expanded.

Broadband networks also have a negative side. They have to be implemented with RF modems, and these are more complex and expensive than baseband transmitters and receivers.

Fiber-optic cable plants are now suitable for certain types of broadband systems, but this situation may change as the technology becomes more mature. The problem with present-day fiber optics is that the total broadband spectrum is needed to AM-modulate a single carrier wavelength, and doing so requires at least two levels of demodulations. Also, the AM modulation index cannot be much greater than 50 percent or distortion will result; moreover, if laser diode transmitters are being used, they can be damaged easily. Many transmitter manufacturers incorporate limiters to protect laser sources from being overdriven.

The broadband cable bus uses some sort of head-end unit that can cause a catastrophic failure if it malfunctions. Sytek offers a dual unit that will switch in a backup unit and sound an alarm when a head-end failure occurs. Needless to say, this protection increases the cost of plant facilities.

Baseband LANs are less complex than broadband types. Tramsmitters and receivers lend themselves well to monolithic or hybrid circuit techniques (see Chap. 7), both of which will keep costs low. Some of the baseband transceivers are actually repeaters. This implies that all of the machines must remain powered up unless a network outlet similar to an electrical outlet is equipped with a fiber-optic transceiver. The author has worked on such a system, where fiber optics were used to implement the cable plant and each repeater station was equipped with a relay to allow the outlet to function as a repeater when the equipment was disconnected from the outlet. The outlet cost in small quantities (fewer than 100) is less than $100/outlet.

Other baseband networks use restrictive taps on the cable that are much simpler to implement, but repeaters can be required periodically as needed.

When making a decision on whether to implement broadband or baseband technology, one must first take into account the problem at hand. For example, if standard TV broadcasts are to be distributed by the LAN, the decision is already made (broadband is the only choice). If, however, teleconferencing is to be implemented using slow-scan TV, it is possible for baseband techniques to be used successfully because 9600 bit/sec slow-scan TV is available. If terminals are going to be connected to a minicomputer and used only for interrogating and upgrading the data base, a baseband LAN would be adequate and also the most cost effective.

Types of Topology

We will first examine some of the more common topologies implemented in LANs such as the bus, ring, and star. In some situations, combinations of all three are encountered.

The bus configuration, shown in Fig. 9-1(a), makes use of a passive cable such as an Ethernet bus. The drops are connected to network implements such as terminals, minicomputers, disk storage stations, gateways, etc. If a bus is broken or disconnected between network implements, such as between 3 and 4, the network will function as two busses. Many installations require a backup cable to prevent system failure if a cable is damaged. Another technique is to install accessible connectors at strategic locations and provide long jumpers so that damage can be circumvented. This can be done automatically during emergency situations such as natural disasters or tactical conditions where high reliability is required. The backup switch service is complex and costly, however.

The ring topology, shown in Fig. 9-1(b), will operate under several protocols, but the only types to be considered here are the CSMA/CD and token-passing

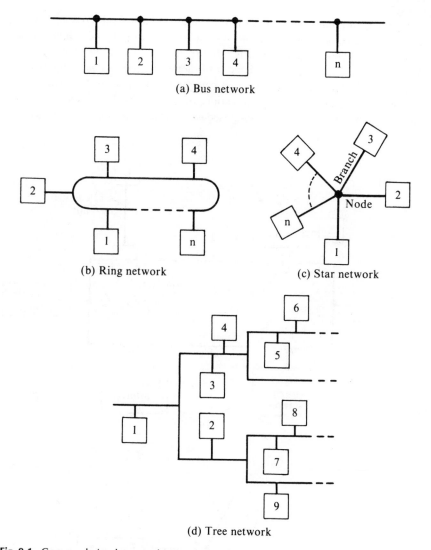

(a) Bus network

(b) Ring network

(c) Star network

(d) Tree network

Fig. 9-1. Commonly implemented LAN topologies: (a) Bus network, (b) ring network, (c) star network, and (2) three network.

types. Ring topologies lend themselves well to baseband TDM transmission techniques. The nodes on the ring serve as repeater stations, which also have error-checking circuitry. The delay at these stations can be as little as one bit. The upper limit is dependent on the buffers allocated at each node.

The third topology is the star, shown in Fig. 9-1(c), which has been used by many computer manufacturers for many years. When several terminals are connected to a computer, a star has been the easiest conformation to implement.

The tree topology, shown in Fig. 9-1(d), can be composed of bus systems similar to Wangnet for the broadband, or Ethernet for the baseband, application.

One should always beware of strict categorization of networks because they can be designed in countless combinations of rings, stars, buses, and trees. One example of a network actually composed of two rings but which may be viewed

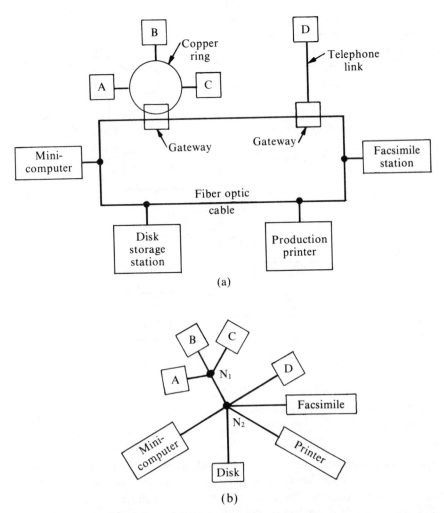

Fig. 9-2. Cable-plant reductions from ring networks to star-connected LANs: (a) Two-ring cable plant, and (b) reduction to star connections.

as a star was constructed by the author; it is shown in Fig. 9-2. The star equivalent is shown in Fig. 9-2(b).

When the transmission delay between nodes is small compared to bit time, the implements will appear to be connected. For optical waveguide, the propagation delay is about 5 nsec/m. If a 1-M bit/sec transmission rate is used on a fiber-optic ring, then 10- to 20-m distances will have 5 to 10 percent of a bit-time delay. See Eq. 9-1 for propagation delay calculations in glass.

$$\tau = \frac{N_{core}}{c} = \frac{1.50}{3 \times 10^8} = 5 \text{ nsec}/\text{m} \tag{9-1}$$

where τ = propagation delay; N_{core} = index of refraction; c = speed of light in a vacuum.

If the copper ring in Fig. 9-2(a) is kept small, the network can be assumed to be the node of a star. The result of the reduction from cable plant diagram to topographical representation is shown in Fig. 9-2(b). The branch connecting N_1 to N_2 is part of the gateway station. The objective of this discussion is to make the reader aware that network architecture allows for variations in topology; i.e., more than one may be suitable for analysis.

GRAPH THEORY

This section is intended to give the reader an appreciation of what can be accomplished on a small-scale network. Larger networks require a computer to perform an analysis, but a study of smaller networks will offer the reader at least some intuitive idea of the larger ones. As one might surmise, the analysis will examine flow of data. Flow analysis is also found in other disciplines— electrical engineering (network synthesis and analysis), fluid mechanics, servo systems, and mathematics, to name just a few.

Let us first define the terms to be used. A *node* is the termination of a branch (such as the work station that terminates a branch radiating from the center of a star) or an intersection of multiple branches [such as the intersection in the star of Fig. 9-1(c)]. In ring topologies, nodes are work stations located around the ring, whereas in bus networks they are taps along the cable. Branches are the line segments connecting nodes; they may be unidirectional or bidirectional, the former being considered a "directed branch" and the latter an "undirected branch." Some networks may be composed exclusively of directed branches, such as the simplex ring, in which the flow of data is in one direction; in others, such as a star, there is communication in both directions. An example of a mixture of both types of branches would be a computer connected in a star with terminals and R.O. (receive only) printer stations.

Let us examine a simple network and determine the shortest path between

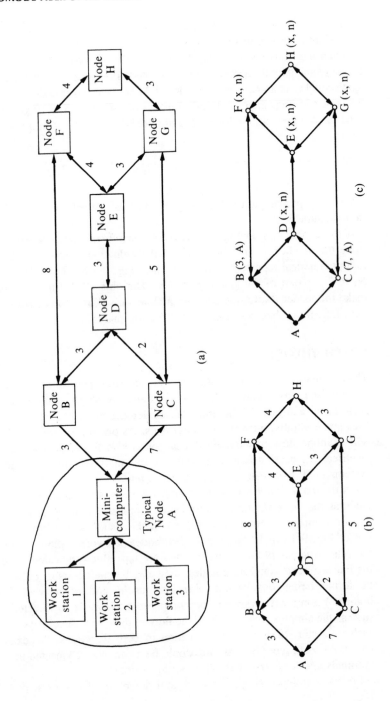

(a)

(b)

(c)

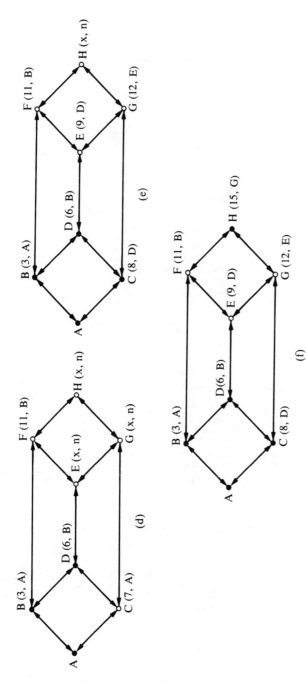

Fig. 9-3. Graphic technique for finding the shortest distance between source and sink: (a) Network topology with distances and nodes labeled; (b) graphic representation of network in Fig. (a); (c) second permanent node selection; (d) third permanent node selection; (e) fourth permanent node selection; and (f) final graph showing all nodes labeled with shortest paths.

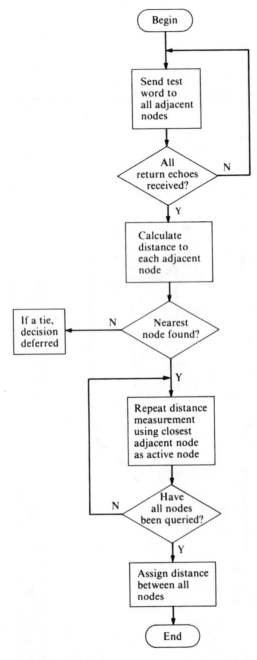

Fig. 9-4. Flow chart to compute shortest path through a network.

two nodes within it. Figure 9-3(a) serves as the physical representation of such a network and Fig. 9-3(b) as its graphical equivalent. The latter will be used to find the shortest path from A to H. The node used as the starting point is filled in, as shown in A of Fig. 9-3(b). The closest adjacent node (B) is next filled in and labeled with its distance from working node (A), as shown by the notation (3, A) in Fig. 9-3(c). Node B is then considered permanent (node C is not because it is further away from A) and all other nodes are considered unknown (x, n). All adjacent nodes are then searched for the one with the shortest distance from the working node. As shown by Fig. 9-3(d), the third permanent node is then seen to be D because the distance from A to D(6) is shorter than from A to C(7) or A to F(11). If the graph is examined for the next permanent node, the distance A to C via node D proves to be the shortest distance from A compared to all other possibilities and is therefore filled in and labeled (8, D), as shown in Fig. 9-3(e). The shortest distance from A to H, however, proves to be through nodes E and G, as shown in Fig. 9-3(f).

Some observations should be made. Occasionally, there are multiple shortest paths, such as between nodes A and F. Paths ABF and ABDEF are both nine units long. Since the former passes through only three nodes, however, it should be the choice because extra nodes often present additional delays in a network and any delay will make a transmission path appear to be longer. (This simplified analysis assumes that all transmission rates are identical, which may not always be the case.) An algorithm in flow-chart form for computing the shortest path is shown in Fig. 9-4. For further study of the subject matter, see Tanenbaum.[1]

As one can imagine, if all transmissions were to take the shortest path, particular branches might become overloaded and thus cause excessive delays. The next topic to be addressed is data flow. This is a useful performance parameter for network segments.

NETWORK FLOW

To compute the information-carrying capacity of a network, the "cut" will be used. This is a technique of removing ("cutting") branches between two nodes until they are disconnected. Let us examine the cuts shown in Fig. 9-5(a), where f.u. indicates flow units:

Cut 1 (AB, AE): 13 f.u.
Cut 2 (AB, ED, JF, JK): 23 f.u.
Cut 3 (BC, FG, KL): 10 f.u.
Cut 4 (CH, LH): 14 f.u.

These are not the only cuts that could have been made; for example, AB, ED, EI or BC, GC, LH would be possible, among others. The maximum

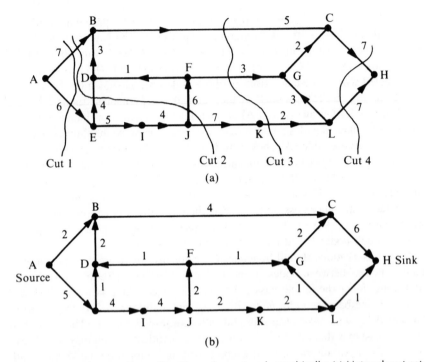

Fig. 9-5. Computing maximum flow through a network graphically: (a) Network cut set, and (b) feasible network from cut set (a).

possible flow across the network is 10 f.u. because the flow units of each branch are considered to be the maximum capacity of that branch.

Before we proceed any further, a glaring problem should be immediately apparent. The maximum flow into B is 9 f.u. and the flow out is only 5 f.u., which implies that the node is a sink; this will not be the assumed case for our analysis, however. Also note that node E shows 5 f.u. in and 9 f.u. out, which implies that it is a source; neither will this be the case for the analysis given here. It will therefore be necessary to state some general rules to aid in the analysis:

1. The source has no inward branches.
2. The sink has no outward branches.
3. No branch has more flow than its capacity but may have less.
4. The inflow is equal to the outflow except for sources and sinks.
5. The outflow of a source must equal sink inflow.

The conditions given above are necessary for the network in Fig. 9-5(a) to be feasible. Let us now examine Fig. 9-5(b) to see how the network can be made feasible. Note that all the preceding conditions are met in Fig. 9-5(b). If the cuts shown in Fig. 9-5(a) are made in Fig. 9-5(b), all have the same flow— 7 f.u.

An enhancement to rule 3 above is that the maximum flow between any two nodes in a graph cannot exceed the capacity of the minimum cut separating these two nodes. [A minimum cut is a cut that has minimum flow across it, such as cut 3 in Fig. 9-5(a).]

A more methodical approach is needed for finding the maximum flow and minimum cut. A maximum flow algorithm, written in Pascal, was published in 1978 by Mathotra, et al.; a flow chart for it is shown in Fig. 9-6. A simplified analysis that assumes only a single source–sink pair appears in Ref. 1. This model is adequate for explaining some of the algorithm's subtleties, but in the real world one must deal with large numbers of source–sink pairs.

Node failure analysis is a rather important topic. The method is to remove paths from the network and examine the realiability issues. Network failures can occur for any number of reasons: normal electronic failures, software glitches in programs, destruction of a facility in a natural disaster or in military situations, and so forth. Degrees of reliability are determined by the end user.

In some instances—for example, in a sales office—node failure presents no immediately pressing problem. For tactical situations, however, such as hospital patient or banking monitors, extremely high reliability is required. In military situations, sufficient redundancy is often employed to prevent catastrophic failure, but such networks operate at reduced performance. Reliability can be increased manyfold by providing 100-percent redundancy, but as one may surmise, the cost increases manyfold also. Most installations rely on the reduced performance technique. Networks that fail gracefully are commonly called networks with "soft failures."

If two nodes are to be disconnected, the minimum number of branches that has to be removed to disconnect them is called their "minimal cut-set." Two paths are considered to be "disjoint"—i.e., "branch-disjoint"—if they *do not* share a common branch. Branches may share a common node and still be considered branch-disjoint. If the network in Fig. 9-7(a) is examined carefully, one will note that paths ADF and ABDCEF are branch-disjoint though they share the common node D, but paths ABDCEF and ADCGF are not because they share the DC branch.

The graph represented by Fig. 9-7(b) is a four-path branch-disjoint network connecting node X to node Y. If at least one branch is removed from each path, X and Y will be disconnected; we would then intuitively expect there to be K branch-disjoint paths, and at least K branches must be removed to disconnect X and Y. Identifying the number of branch-disjoint paths gives a limit to the minimum cut.

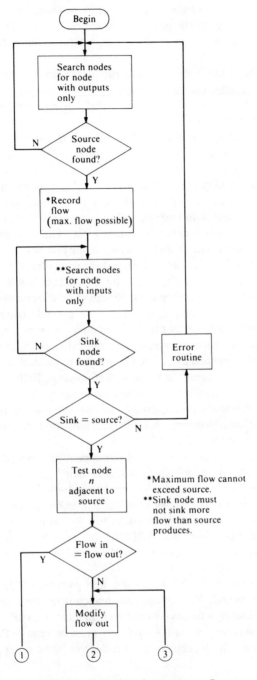

Fig. 9-6. Flow chart for maximum flow.

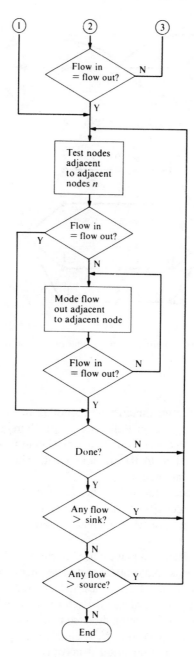

Fig. 9-6. (*Continued*)

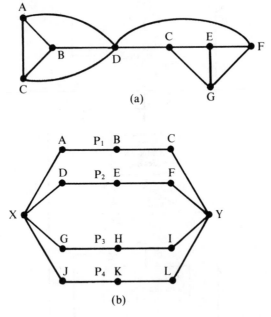

Fig. 9-7. Branch and node disjointed networks: (a) node disjointed, and (b) branch disjointed.

A method of using the maximum-flow algorithm to calculate the number of branch-disjoint paths is given in Reference 1. (In short, if the weight of each branch is replaced with the weight of 1, the maximum flow between the source and sink will be equivalent to the minimum cut, which is also equal to the number of branch-disjoint paths.)

The discussion has been directed to the removal of branches as a means of determining graph connectivity. If nodes are removed instead, the source and sink will eventually be disconnected and thereby determine the node connectivity of the graph.

The maximum-flow algorithm may be used to calculate the node-disjoint paths between a source and sink. An original undirected network of N nodes is modified into a directed network of $2N$ nodes and $2B + N$ branches (where $B =$ the number of original graph branches). Each original node is replaced by two nodes labeled N and $\bar{N}$. All incoming directed branches are connected to N, and all outgoing branches are attached to $\bar{N}$. A simple example of a node-disjoint is provided by Fig. 9-8, where an original network with bidirectional branches connecting all nodes—Fig. 9-8(a)— is expanded into the network shown in Fig. 9-8(b). It can be quickly noted that the node count is 8 and 16 for (a) and (b),

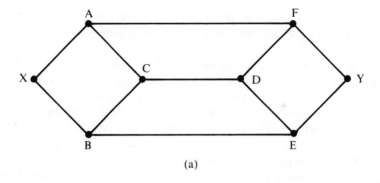

(a)

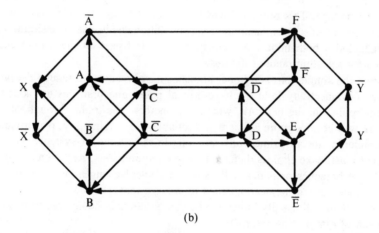

(b)

Fig. 9-8. Expansion of a bidirectional flow network to unidirectional flow: (a) Original bidirectional network, and (b) unidirectional network.

respectively. There are 11 branches in Fig. 9-8(a), moreover, and 30 in Fig. 9-8(b), which is correct since $2B + N = 2(11) + 8 = 30$.

The maximum-flow program will find the node-disjoint paths between the source and sink (X and Y in Fig. 9-8, respectively). The node-disjoint paths in Fig. 9-8(b) are $X\overline{X}B\overline{B}E\overline{E}Y$ and $X\overline{X}A\overline{A}F\overline{F}Y$. Each branch will have a weight of 1 in the analysis. The key to the analysis is that each unit branch can have only one path through it; therefore, no node may be located on two or more paths. The maximum-flow algorithm can determine all the paths through the network that pass through sets of nodes not used by other paths; i.e., each path has a unique set of nodes. To determine the node-disjoint paths for each set of nodes in a large network would require an excessive number of computer runs of this algorithm. [For example, a network of 20 nodes would require 190 runs since for N nodes, $N(N - 1)/2$ runs are needed.] Therefore, the node-disjoint technique is useful for finding node-disjoint paths when there are smaller numbers of nodes.

In a real situation such as those encountered in military tactics, nodes and branches can be removed or destroyed, and the network connectivity must therefore be known. In fact, it must be known at the moment the damage occurs. In other words, the calculation must have been made well in advance, because computers may not be available in an emergency situation. The advent of low-cost microprocessors and electrically erasable, programmable read-only memories (E^2PROMs) makes it possible to store connectivity tables at the nodes.

A particularly useful Pascal program is given in Ref. 1. A flow chart for it is shown in Fig. 9-9 in the event the reader wishes to program the simulation in another language. This program, which was selected because of its popularity, gives a comprehensive explanation of the simulation and its use. The simulation, which includes both node and branch failures, is based on the Monte Carlo method for selecting random failures.[2]

During the simulation, additional branches may be added to reduce the probability of connectivity failures. If additional nodes and branches are added to strategic points in the network, which will increase the traffic throughput, the network will become more efficient and reliable. Cost, reliability, performance, and efficiency are all prime considerations in the design of systems, and the designer must make choices dictated by these requirements. Any increase in the number of branches or nodes will cause only small changes in reliability and performance but will be accompanied by a large increase in costs. A cost performance ratio will usually indicate when an optimum is being approached (the approach of diminishing returns).

A variation of the simulation method can be stored in E^2PROM with a connectivity table. When new branches or nodes are added to the network, the simulation can be executed and the connectivity table updated. The program may actually be stored on disk with only the table stored in E^2PROM. In this case, each node can upgrade its own connectivity graph.

Another variation for upgrading connectivity tables can be accomplished by connecting telephone modems to the node terminals during low traffic hours, e.g., on weekends. The conductivity simulation can be run automatically and the upgrades programmed into the nodes. One fast food chain uses a similar technique to program their cash registers during the night; it is transparent to the node terminal operators unless an actual emergency arises.

Traffic flow is related to bandwidth and message or data transmission time. An important performance criterion is therefore time delay. A study of traffic density is also necessary to show how performance is affected by traffic conditions. Many nodes may use the same, that is, the optimum, path, and the traffic density may be so great that the network may not function properly, if at

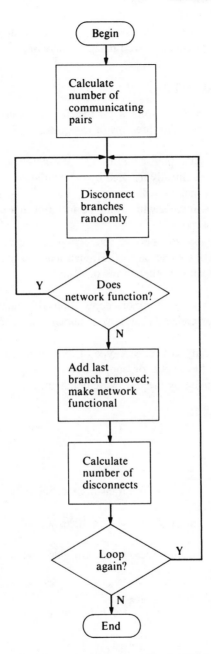

Fig. 9-9. Flow chart for computing network reliability by simulation.

all. Therefore, these factors must be assessed prior to any physical design if one is to produce a functionally well-designed network.

TIME DELAY ANALYSIS

The network must now be examined for the time-delay impact. Delay in networks can be subdivided into three major categories:

1. Delay caused by the physical attributes of a network such as propagation delay or delay specifically induced by logic circuits (digital delay) or delay lines and filters (analog delay).
2. Delays due to software, which may be inherent in a program or designed for a specific purpose.
3. Delays attributable to operator interaction, for example, due to typing answers to queries while on line. Human reaction time is usually considerably longer than the other forms of delay.

Human engineering issues are more applications dependent than the first two types of dealy. For extended discussions of human engineering, see Refs. 3, 4, and 5.

Let us begin our analysis with that type of time delay caused by propagation of a signal down a fiber-optic or copper-transmission line. Propagation in a fiber-optic cable is given by the following equation:

$$T = \frac{n}{c} \tag{9-1}$$

where:

T = propagation time due to the transmission line only, in sec/m
n = refractive index
c = speed of light in a vacuum = 3×10^8 m/sec

If the transmission medium were a uniform glass rod with a refractive index of 1.5, we would have

$$T = \frac{1.5}{3} \times 10^{-8} \text{ sec/m}$$

$$= 5 \text{ nsec/m}$$

$$= 0.5 \text{ } \mu\text{sec/km}$$

In copper lines, the velocity constraint must be considered when calculating transmission line delays. When electromagnetic waves travel down a transmission line that uses spacing insulators or solid dielectrics, velocity decreases occur since the waves are affected by the inductance and capacitance of the transmission line itself. Some typical velocity constants (where c is the speed of light in a vacuum) are as follows:

1. Parallel lines with an air dielectric between them: 0.95–$0.97c$
2. Parallel lines with a plastic dielectric between them: 0.80–$0.95c$
3. Shielded pair with rubber insulation: 0.56–$0.65c$
4. Coaxial line with air dielectric: $0.85c$
5. Coaxial line with plastic dielectric similar to CATV: $0.77c$
6. Twisted pair with rubber insulation: 0.56–$0.65c$

For a coaxial line with plastic dielectric, the velocity of the electromagnetic wave is similar to that of the glass waveguide:

$$v = 0.77 \times 3 \times 10^8 \text{ m/sec}$$
$$= 2.21 \times 10^8 \text{ m/sec}$$
$$T = 4.1 \text{ nsec/m}$$

An interesting note: Water has an index of refraction of 1.33 or a propagation time, T, of 4.4. nsec/m, a value very close to that of coaxial cable of good quality.

As one might imagine, this is a delay parameter that cannot be easily controlled, especially as compared to the other two types. Once a cable plant has been installed, the cost of replacing it with one having a shorter propagation time would be cost prohibitive. The delay, as demonstrated in Chap. 8, will have an impact on frame size and LAN efficiency.

As a simple example of propagation effect, let us consider a LAN with a transmission rate of 10 Mbit/sec and further assume that the data is sent in 500-bit packets. The packet represents a line length of 10 kilometers for the fiber-optic case and 12 kilometers for coaxial cable. This implies that two stations communicating in half duplex must listen for at least 50 μsec before transmitting if the distance between them is 10 to 12 kilometers and no other delays are present. If an acknowledge is required prior to transmission of another packet, the transmitting station will require a 100-μsec delay before transmitting. Since this acknowledge delay is not efficient, there are methods of circumventing it. As packet length increases, the propagation delay begins to become less significant; one might intuitively conclude, therefore, that as transmission distances become greater, packet size should also be made larger. When designing a LAN, the engineer must allow for physical expansion or he may soon have a very inefficient network. Enlarging packets after a network is complete can be almost cost prohibitive and cause a multitude of other problems.

The next type of delay to be considered is delay due to hardware (digital logic) as data passes through a node. There are several reasons for this physical type of delay, as follows:

1. The removal of extra bits in the data stream because of bit stuffing (bit stuffing is a method of assuring data transitions when long strings of zeros or ones occur; it will be discussed later in this chapter)
2. Delays caused by cyclic redundancy checking (CRC)
3. Delays caused by clocking of incoming data to the transmitter in the case of a repeater.
4. Decoding and encoding delays.

As an example of the effect of these delays, let us consider a situation in which each station on a LAN has a five-bit delay as a result of bit stuffing, i.e., no more than five consecutive one's can occur in the data. The bit stuffer will insert a zero to force a transition in the data stream. If the data passes through 20 nodes and each experiences this five-bit delay and if the data transmission rate is 10 Mbits/sec, the delay due to the nodes will be 10 nsec. Note that the logic delay is insignificant compared to the propagation delay, but if the transmission rate is reduced, the reverse is true; i.e., the propagation time is insignificant and the node delay becomes appreciable—100 nsec. This part of the physical delay can be controlled by the designer more easily than cable delays.

The next type of delay is that attributed to software. An example of such delay is when messages are stored in a buffer and assembled for transmission. The packets may be prepared and held in a buffer until the total message is ready for transmittal, or if the node is acting as a repeater, the entire message may be received before being retransmitted. The following example shows the type of calculations needed when a node acts as a repeater and the entire message must be received before it can be forwarded:

Example 9-1: Let the message = 1 page of text.

1 page of text = 78 characters × 48 lines

$$= 4,000 \text{ characters (with overhead)}$$

Characters in ASCII = 4,000 bytes

Number of bits = 8 × 4000

$$= 32,000 \text{ bits}$$

For a 10-Mbit/sec transmission rate, the delay is thus 3.2 msec. From this example, it is obvious that this type of software delay makes all physical delays insignificant.

The third type of delay, those due to human engineering factors, are even larger. Let us examine a typical example.

Example 9-2: Say that a typist types at 60 words/minute. If this speed is encoded in ASCII, the typing rate, TY_R, is calculated as follows:

$$TY_R = 6 \text{ characters/word} \times 60 \text{ words/min} \times \frac{1 \text{ min}}{60 \text{ sec}} \times \frac{8 \text{ bits}}{\text{character}}$$

$$= 48 \text{ bits/sec}$$

If an acknowledgment were to be typed after each message that required only six characters, the 1-sec response would be much larger than any of the previous delays.

Another type of network delay that may be occasioned by hardware, software, or a combination of both is called *queuing*. The queue, which is transparent to the human operator at the node, is a buffer such as that shown in Figs. 9-10(a) and 9-10(b). It allows the node to use the transmission facilities more efficiently. As an example, if only one message were handled at the node at a time, the transmission would have bursts of high-speed data and long idle-channel periods. In Fig. 9-10(a), messages may arrive randomly and be stored at the input queue; they are sorted by the controller for transmission according to time of arrival or priority. Figure 9-10(b) is a queue where the arrival data may be sent to the output queue immediately, or it may be held until the node finishes transmission before being repeated.

Messages (M_i) are composed of packets (P_n), which can be described by the following equation:

$$M_i = \sum_{n=1}^{k} P_n \qquad (9\text{-}2)$$

where M_i may consist of a single packet.

Let us now generate the equation that governs queues. The message arrival rate, M, in messages/second, is the sum of the individual arrival rates, as shown in Eq. 9-3:

$$M = \sum_{i=1}^{k} M_i \qquad (9\text{-}3)$$

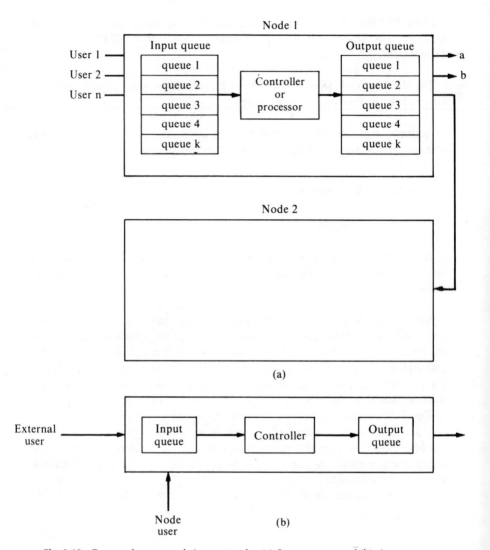

Fig. 9-10. Queues for star and ring networks: (a) Star queues, and (b) ring queues.

If ℓ = average number of bits per message, then the transmission rate of the traffic, τ_r, in bits/sec, is given by

$$\tau_r = M\ell \tag{9-4}$$

If the transmission rate of the channel is operating at full capacity, C_m, then the utilization of the channel, U, is given by

$$U = \frac{T_r}{C_m} \times 100 = \frac{M\ell}{C_m} \times 100 \qquad (9\text{-}5)$$

The average service time, T_s, in seconds, required for the node server (controller) to process a message is given by

$$T_s = \frac{\ell}{C_m} \qquad (9\text{-}6)$$

Thus, from Eq. 9-5, we have

$$U = MT_s$$

To determine the average waiting time, T_w, of a message in the queue, if N = the average number of messages in the queue, then

$$T_w = NT_s \qquad (9\text{-}7)$$

Equation 9-7, of course, is intuitive. Since it takes T_s to service a single message, then if N messages are in the queue, it will take NT_s time to service them all.

If the message rate entering the input queue is equal to the message rate leaving, the length of the queue will remain the same. If, however, the message rate at the input should vary and the output rate remain the same, the probability of new message arrivals must be known if a realistic queue waiting time is to be calculated.

If the messages arrive at k intervals, a Poisson distribution can be used to provide the probability, $P(k)$, of the number of new arrivals in T_s sec, as expressed by Eq. 9-8:

$$P(k) = \frac{\lambda^k e^{-\lambda T_s}}{k!} \qquad (9\text{-}8)$$

where k = number of arrivals in T_s sec and λ = messages/sec. Moreover,

$$\lambda T_s = T_s M = U'$$

and

$$U' = T_s M = \frac{M\ell}{C_m}$$

where U', the channel utilization, unlike U, is not given as a percentage.

Then,

$$P(k) = \frac{(U')^k e^{-U'}}{k!} \tag{9-9}$$

The distribution depicted in Eq. 9-9 is dependent on channel utilization, U', which is a more meaningful concept provided the service time remains fixed, and it usually does in LANs. For networks such as telephone systems, the service time may vary because multiple servers are present at each node (telephone exchanges).

Let us now examine the queue as it exists in a steady-state operation; i.e., it is neither growing nor shrinking in size. The number of messages in the queue, N, can be estimated with the following equation:

$$N = \frac{U'}{\Delta U'} \tag{9-10}$$

where $\Delta U' = 1 - U'$

The waiting time, T_w, is then

$$T_w = NT_s$$

$$= \left(\frac{U'}{\Delta U'}\right) T_s$$

or

$$T_w = \left(\frac{\ell}{C_m}\right)\left(\frac{U'}{1 - U'}\right) \tag{9-11}$$

Equation 9-11 is valid so long as the value of N does not become so large that the node ceases to have adequate buffer space. It is interesting to note that as U' approaches 1, which implies 100-percent utilization of channel capacity, N would approach infinity; this is obviously not the case. Therefore, 100-percent utilization of the channel in Eq. 9-11 is not possible. When various types of protocols are discussed, however, we will discover means for circumventing this problem; e.g., if the queue is full, the network will "busy out" the new arrivals.

The total average delay time attributed to the queue, T_t, is given by the following:

$$T_t = T_s + T_w$$

$$T_t = T_s(1 + N) = T_s\left(1 + \frac{U'}{1 - U'}\right) \qquad (9\text{-}12)$$

Since

$$\frac{U'}{1 - U'} = \frac{(U' - 1) + 1}{1 - U'} = \frac{1}{1 - U'} - 1$$

then,

$$T_t = \frac{T_s}{1 - U'}$$

or

$$T_t = \frac{\ell}{C_m}\left(\frac{1}{1 - U'}\right) \qquad (9\text{-}13)$$

$$T_t = \frac{\ell}{C_m}\left[\frac{1}{1 - (M\ell/C_m)}\right] \qquad (9\text{-}14)$$

or

$$T_t = \left(\frac{\ell}{C_m - M\ell}\right) \qquad (9\text{-}15)$$

Alternative Eq. 9-15 is given to make the reader aware of some of the trade-offs, for example, message length and total average queue delay time. If the derivative of Eq. 9-15 is taken with respect to ℓ and the minimum is calculated to optimize M, we then have the following equations:

$$\frac{dT_t}{d\ell} = \frac{(C_m - M\ell) + M}{C_m - M\ell} = 0$$

and

$$M_{opt} = \frac{C_m}{\ell - 1} \qquad (9\text{-}16)$$

CHANNEL CAPACITY

Channel capacity is another parameter requiring concern since the designer must allow sufficient capacity to expand the facilities, on one hand, and guard against overdesign on the other, since the latter often increases the cost of plant facilities and results in inefficient use. The maximum capacity, C_m, of a transmission system can be calculated using the following equation, as formulated by Shannon[3]:

$$C_m = \text{BW} \log_2 (1 + \text{SNR}) \qquad (9\text{-}17)$$

where BW = bandwidth and SNR = signal-to-noise ratio.

Shannon stated that, for a band-limited channel with $C_i \leq C_m$, a code exists for which the error rate approaches zero as the message length approaches infinity. Also, conversely, if $C_i > C_m$, the error rate cannot be reduced below some positive numerical limit.

Applying Eq. 9-17 to a typical voice channel with an SNR of 1023 and a bandwidth of 3 KHz, we have

$$C_m = 3000 \log_2 (1 + 1023)$$

$$= 30,000 \text{ bits/sec}$$

This value is the absolute maximum, but it cannot be achieved because of intersymbol interference. Nevertheless, Eq. 9-17 does put a upper bound on the system that serves a purpose by ruling out impossible values.

An improved equation for calculating C_m that takes intersymbol interference into account is shown in Eq. 9-18:

$$C_m = 2 \text{ BW} \log_2 M \qquad (9\text{-}18)$$

where M = the number of levels

This equation, which is attributed to Nyquist, does not consider bit error rates (BER). As the number of levels, M, increases, a larger SNR is needed to resolve the signals. As an example, using Eq. 9-18, if $M = 16$ and BW = 3 kHz, then $C_m = 24$ kbits/sec.

When a 16-level system using phase-shift keying is used and the transmission rate is 2400 bits/sec, it is possible to convey information at a rate of 9600 bit/sec. This is accomplished in many present-day audio modems. However, phase coding does not come without a penalty. Each time the phase information is doubled results in a 3dB-increase in noise penalty.

Some of the more sophisticated modems are available with 19.2-kbits/sec data-transmission rates. Such modems do not operate over standard switched networks but require conditioned lines that are leased from telephone companies at a premium cost. The conditioning, which guarantees minimum attenuation and phase distortion, is accomplished through a predistortion of the signals commonly known as "equalization." Some of the newer modems incorporate equalizers that test the line prior to transmission and add equalization as it proves necessary. This is known as "adaptive equalization."

Many of these newer modems will test the line after it has been equalized. If the BER falls below a threshold value, the modem will shift down in transmission rate until the BER is adequate. Although the procedure will take several seconds, it allows the modem user to take full advantage of exceptionally good lines. Quite often business calls may be over short distances. Under such conditions, communications are extremely good as compared to long-distance connections. The cost of these adaptive modems is generally 15 to 20 times greater than that of modems with manually selected transmission rates and no equalization.

Adaptive modems are designed with microprocessors and hosts of interface circuits. The author expects their cost to drop significantly as gate arrays and analog-circuit arrays augment microprocessor interface circuitry. A design is presented in Chap. 5.

The previous discussion dealt mainly with telephone lines, but many LANS do not use these facilities. Since most have a gateway station that is useful for communicating with other LANs via telephone lines, however, the modem arguments given here still hold. The majority of LANs require gateway stations that will allow communication between two different types of LANs, e.g., ETHERNET and ARPANET. Equations 9-17 and 9-18 still hold, but modems are quite different from the telephone type. For example, they require no line equalization and the specifications of cable plant facilities are much more stringent; i.e., delay and amplitude distortion have much closer tolerances.

NOISE CONSIDERATIONS

Noise contributions from various devices in a network will degrade transmission facilities; these devices include amplifiers, receivers, filters, multiplexers, the cable plants themselves, and a host of others. This discussion will address the normal types of noise found in copper-cable as well as optical systems.

Thermal noise occurs in all transmission equipment and limits its performance. This type of noise has a uniform distribution of energy over the frequency spectrum and a normal Gaussian distribution of levels. Equation 9-19 describes the thermal noise–power, P_n, as follows:

$$P_n = KTBW_n \qquad (9\text{-}19)$$

where:

$$K = \text{Boltzman's constant} = 1.380 \times 10^{-23} \text{ J/°K}$$
$$T = 17°C \ (290°K)$$
$$KT = 4 \times 10^{-21} \text{ W/Hz}$$
$$\text{BW}_n = \text{noise bandwidth}$$

A form more frequently used in communications is the noise relative to 1 mW, which is shown in Eq. 9-20.

$$P_n = 10 \log \left(\frac{KT\text{BW}_n}{1 \text{ mW}} \right) \tag{9-20}$$

Equations 9-21 and 9-22 depict voltage and current generators, which are generally more useful in electronic circuits than in optical transmission, where power calculations are more often employed.

$$e_n^2 = KT\text{BW}_n R \tag{9-21}$$

$$I_n^2 = \frac{KT\text{BW}_n}{R} \tag{9-22}$$

Since noise calculations are of little use unless they are referenced to a signal, an important relationship to consider is signal-to-noise ratio (SNR), which is a measure of channel or amplifier quality. SNR is depicted in Eq. 9-23 in log form as a power ratio, which is the most useful. The alternative forms of Eqs. 9-24 and 9-25 in signal ratios are common in electronic calculations, but Eq. 9-23 is ordinarily used in fiber optics.

$$\text{SNR} = 10 \log \left(\frac{P_s}{P_n} \right) \tag{9-23}$$

$$\text{SNR} = 20 \log \left(\frac{e_s}{e_n} \right) \tag{9-24}$$

$$\text{SNR} = 20 \log \left(\frac{I_s}{I_n} \right) \tag{9-25}$$

When using SNR values in optical cabling systems, one must remember that there is a difference from electrical systems. As an example, the optical SNR is 40 dB for a good video display, and this would translate to 80 dB in an electrical system. As an added note, laser fiber-optic transmitters have difficulty in producing an SNR greater than 57 dB.

Another important calculation particularly useful for evaluating amplifiers is noise figure (NF), which is described by Eq. 9-26. In this equation, note that if the noise out of the device, N_o, is equal to the noise into it, N_i, the device is noiseless.

$$NF = \frac{SNR_{in}}{SNR_{out}} = \frac{(S_i/N_i)}{(S_o/N_o)}$$

where S_i is the signal in and S_o, the signal out. Then,

$$NF_{dB} = 10 \log NF = 10 \log \left(\frac{N_o}{(S_o/S_i)N_i} \right)$$

or

$$NF_{dB} = 10 \log \left(\frac{N_o}{KTBW_nG} \right) \tag{9-26}$$

where:

$$G = gain = S_o/S_i$$

An assumption made here is that no intermodulation products are present; i.e., the device is linear.

A noise component attributed to the conversion of analog signals to digital form is called "quantization noise." This type of noise, shown in Fig. 9-11, is due to quantization error.

The quantization noise is usually smaller than the quantization level. Generally speaking, it is less than one-half the amplitude of that level. The quantization noise power, QNP, is calculated as follows:

$$P(n) = \left(\frac{1}{q} \right) - \left(\frac{q}{2} \right) \le n \le \left(\frac{q}{2} \right)$$

where q is the quantization level. (Otherwise, $P(n) = 0$.) Then

$$QNP = \int_{-q/2}^{q/2} \frac{n^2}{qR} \, dn$$

$$= \frac{1}{Rq} \left(\frac{2q^3}{3 \times 8} \right)$$

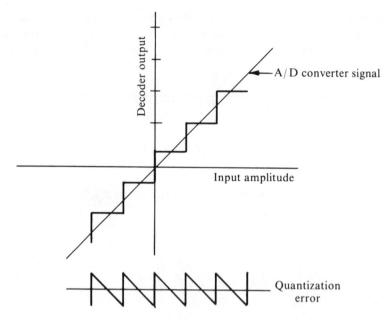

Fig. 9-11. D/A conversion showing quantization error for a single sloping analog input.

or

$$QNP = \frac{q^2}{12R} \tag{9-27}$$

The signal-to-quantization-noise ratio, SN_qR, is calculated as follows:

$$SN_qR = 10 \log \left(\frac{v^2}{q^2/12R} \right) \tag{9-28}$$

or

$$SN_qR = 10.8 + 20 \log \left(\frac{v}{q} \right)$$

where v = the analog signal level

A number of assumptions have been made in Eq. 9-28, as follow:

1. The quantization amplitude is constant, which is not the case for some

types of adaptive A/D conversion. Some of the more advanced techniques reduce step size as the amplitude approaches overload. Overload is a condition when the signal amplitude is larger than the A/D converter can quantize. These adaptive A/D converters decrease step size when signals are low-level. Such an encoding technique increases the dynamic range of the A/D converter.

2. The A/D converter has infinite rise and fall times; i.e., it is not band-limited.

Idle channel noise is present if the noise at the input of an A/D is greater than the quantization level. The effect is a continuous string of pulses depicting the plus and minus values about zero. This type of noise will be noticeable during pauses or in some fricative sounds in speech. Idle noise may be reduced to zero if low-level values are decoded as zeros. By selecting a quantization characteristic that will present a threshold to the analog input, the noise can be suppressed. Another possibility, of course, is to increase the quantization voltage levels so as to reduce sensitivity. The final decision will depend on the application.

Another type of noise, which is due to fiber-optic parameters, is referred to as "quantum noise." The first case we will consider is a digital system where the number of detected photons per second is equal to the number of hole-electron pairs generated by incident light. Equation 9-29 will statistically predict the number of hole pairs generated as follows:

$$P_n(n) = \Lambda^n e^\Lambda / n! \tag{9-29}$$

where:

$$\Lambda = \frac{1}{hf} \int_o^t P_d(t)\, dt = E_d/hf$$

E_d = energy detected over the time interval; h = Planck's constant; and f = frequency.

This is a Poisson distribution. If an assumption is made that no errors occur because of electron-hole pairs of energy E_d, then Eq. 9-28 will have the following value:

$$P_n(o) = e^{-E_d/hf}$$

If, furthermore, Eq. 9-29 is evaluated for a BER of 10^{-10}, then the result is calculated as follows, the average number of photons in the optical pulse being 23.3:

$$\text{BER} = 10^{-10} = e^{-E_d/hf}$$

Thus,

$$E_d = 23.3 \; hf$$

The average minimum power required to maintain this BER is indicated by Eq. 9-29, where the $\frac{1}{2}$ reflects the fact that the digital waveform is assumed to have a 50/50 duty cycle:

$$P_{\min} = 23.3 \times \tfrac{1}{2} \, hf \, \text{BW}_n$$

or

$$P_{\min} = 11.65 \, hf \, \text{BW}_n \tag{9-30}$$

BER is equivalent to SNR in digital systems and is considered a figure of merit. When receiver calculations are made, dark current and thermal noise will also be considered.

To complete the optical noise calculations, the analog signal-to-noise ratio must be considered. The quantum-limited SNR is calculated using Eq. 9-31, with thermal noise and dark current ignored:

$$(\text{SNR})_q = \frac{m^2 \, P_o}{2hf \, \text{BW}_n}$$

where m = the modulation index, $P_o = I_s^2 R_L G$, and $I_s = P_{in} r$. Then, when $G = 1$,

$$(\text{SNR})_q = \frac{m^2 (P_{in} r)^2 R_L}{2 \, hf \, \text{BW}_n} \tag{9-31}$$

where:

r = responsivity in A/W
P_{in} = optical power at the detector
R_L = detector load resistance

This equation represents the absolute limit of optical detection. Its derivation can be found in Personick.[4] If thermal noise and dark current noise can be kept very small, the limit it defines can be approached very closely.

In laser transmission of analog signals, the modulation index must be kept less than 60 percent or the signals will be destroyed. From personal experience, it has been found that a safety factor of 50-percent modulation will usually allow some margin of overdrive. Generally speaking, when overdrive persists for less than 2 μsec, either the laser source will deteriorate to an unusable power-output level or a catastrophic failure will occur.

BACKBONE DESIGN

A major difference between backbone networks and other LANs is that they may connect several different LANs into one large network. The rudimentary design issues involved have already been discussed. The backbone network may have just evolved, or it may be a planned architecture. The typical backbone network shown in Fig. 9-12 has a highly simplified architecture; note that the connection between gateways 2 and 3 could have been a telephone line. For this latter case, a severe limitation would have occurred because of the low transmission rate afforded by telephone facilities. Figure 9-12 is also an example of a system that may have evolved rather than having been planned.

Backbone networks are usually quite outsized and cost-sensitive. The design methodology is usually one of trial and error. One iterative method is as follows: A preliminary design is configured and then tested for connectivity, delay, and other constraints. At this point a cost analysis is completed, and the network is then used as a baseline. The baseline network can be examined for improvements with a connectivity, delay, and cost analysis performed at each iteration. Large networks will require a computer to assist in these analyses.

A number of rather sophisticated backbone design techniques are discussed in Tavenbaum.[5] This subject is scantily treated here because the majority of design applications involve LANs. Large backbone networks are usually designed after some of the facilities have already been installed and will therefore be subject to preconditions determined by these partial installations. Installing a backbone under such conditions constitutes a real challenge to a designer's ingenuity.

NETWORK LAYER (VIRTUAL CIRCUITS)

One of the most important aspects of a local area network with packet switching is that of the virtual circuit. If a single channel within a particular medium is time-shared by multiple users, the time-shared channel between any two users constitutes a virtual circuit. Such circuits must provide the network layer with certain properties. The following paragraphs will present a discussion of each of these properties.

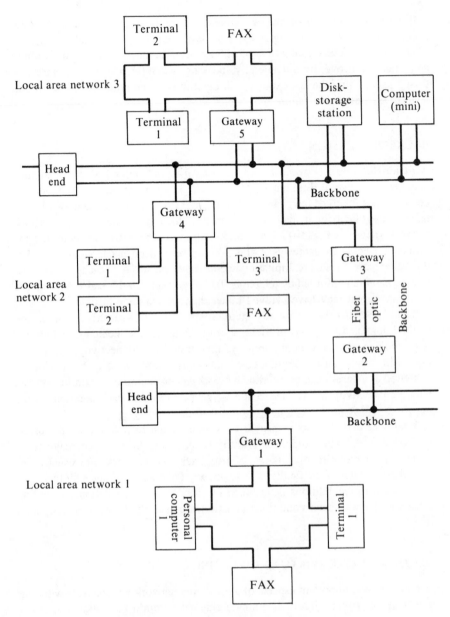

Fig. 9-12. LANs connected through a backbone network.

Data Transfer in Sequence

The data bits delivered to a network for transmission must maintain their order throughout the network. For this to happen, the network facilities must disassemble messages before transmission and reassemble them at the receive end.

Network Transparency

The network must be transparent to users, and no groups (of one's and zero's) can be excluded from the data field of a packet. Data must be delivered errorfree between two users. This condition implies checking for inadvertent flags generated within the data field.

Path Full-Duplex Capability

The path between any two users must appear to be operating full-duplex; i.e., transmission must appear to be simultaneous. Thus, the nodes in the network must have identical network capability; e.g., initiation of a connection and message buffering in one direction must have the same attributes for data in the other direction.

Control Signalling

The signalling must flow freely between users to maintain flow control. Status information must be conveyed between nodes to respond to user inquiries. Signalling can be produced (in-band) as part of the user data stream or (out-of-band) outside the normal data stream.

Data Flow Control

The network must be prevented from congestion during high traffic-density periods. Traffic routing algorithms are one example of flow control.

Error Control

All network transmission must have a negligible amount of errors. The network layer may demand retransmission of data that has errors or perform error correction.

Interface Independence

The network layer must be independent of the physical properties (electrical or fiber-optic) of the node interfaces. It must also be consistent with logical data structures.

Address Flexibility

Network virtual-circuit operation allows information to be exchanged between various user pairs by modifying user address fields within a message.

DATAGRAM

The next network service to consider when the facilities are shared and entire messages are transmitted is the datagram. This service is exemplified by stock quotations, reservations, and point-of-sale terminal-type systems. It is characterized by the following set of characteristics:

Self-Contained Messages

The information within a datagram is self contained; i.e., it does not depend on previous messages nor does it contribute information other than its own message.

Datagram Identification

The datagram contains its own destination address, routing information, and control; it is a complete entity.

Errors in Transmission

A destination node has no previous knowledge of a datagram's being sent. Lost datagrams are not recovered or retransmitted as the result of a request from a destination node. Duplicate datagrams are often sent by the transmitting node to insure that data has been conveyed to the destination node.

Sequencing

Datagrams are not sequenced since they represent entities unto themselves. As a result, messages transmitted to a destination node need not arrive in order.

Uncontrolled Aspect

The originating node of the datagrams will be advised of their progress through the network, but they are uncontrollable after leaving the originating node.

A summary of the differences between datagrams and virtual circuits is given in Table 9-1. Examination of the table should make one aware that datagrams are the primitives of the two network services.

Since many of these local area networks may be required to run for years without problems, not only must they be robust, but the routing algorithm must

TABLE 9-1. A Comparison of Virtual Circuits and Datagrams.

	Virtual Circuit	*Datagram*
Destination address	Only during initial start-up	Needed in every packet
Source address	Only needed at start-up	Not always needed
Error detection	Transparent to upper network layers	Done by the upper layers
Flow control	Provided by the network layer	Not provided by the network
Packet sequencing	Messages passed in order	No order required
Initial network setup	Required	Not possible

be simple and precise. The topology must be flexible; i.e., addition of new implements to the network should not require a change in software. As an example, if the networks of a LAN at a bank use smart terminals as nodes, addition of new terminals should not present a software problem to the terminals already installed. Also, all nodes within the network should be treated equally; i.e., each can access the network with equal probability of success.

ROUTING TECHNIQUES

Two forms of control are used on routing algorithms. The first is a nonadaptive type where traffic is routed according to a fixed table; the second is adaptive and the algorithm will measure traffic flow and route traffic accordingly. The latter technique is also called "dynamic routing."

A method of routing called "flooding" uses a broadcastlike technique. The outgoing packets are sent to all lines except the one they were received on. Fiber-optic star couplers lend themselves well to this type of transmission. Star couplers are made to be rather simple but do not operate well in single mode.

One can imagine what flooding does to traffic flow when several nodes are transmitting simultaneously. Flooding has a limited amount of utility as compared to other routing techniques. Some networks need it to function properly, e.g., networks distributing news or stock market results, tactical military networks (where redundancy is a necessity), and other similar networks in which data must be distributed to all nodes. Of course, it is also possible to flood a network partially, i.e., distribute over only a part of the output lines. If this technique is used to forward data in the direction of the destination node and all other nodes are capable of forwarding the data toward this destination, the shortest route to the destination will be selected. These techniques all increase traffic density however.

Directory routing is the most common technique of packet control. A routing directory is set up by the operator with previously established criteria such as shortest path, least delay, light traffic volume, etc. If several possible routes emerge as candidates to carry the message traffic, then a random number generator can be implemented to select routing.

The discussion to this point has dealt primarily with nonadaptive routing. The next logical step is to examine adaptive routing, which, of course, has a great deal more utility than the nonadaptive kind.

Adaptive routing has one immediate disadvantage over the nonadaptive and that is the complexity of executing and updating a routing matrix in real time. The new generation of microprocessors, however, will certainly overcome the computing power issue. These microprocessors have the computing power of minicomputers and, in certain cases, may be added in parallel to provide parallel processing. Large numbers of controllers that emulate specific network protocols will eventually be available, and these devices will no doubt have adaptive routing capability.

Now let us examine exactly what is involved in adaptive routing. Some form of monitor is required to update various aspects of network status such as traffic density over various routes, delay along all routes, and, lastly, the condition of all nodes along the route. One must note that all these quantities are highly time-dependent. Traffic density may increase and decrease very rapidly because of the bursty nature of the data. Delay due purely to propagation is very predictable, but when a node that processes data lies within the path of transmission, the service time of the node can vary in accordance with the loading. Nodes may become damaged or removed from service, and these contingencies must be reflected in the network status.

The objective of any local area network is to deliver packets of data in a rapid and efficient manner. When the network is lightly loaded, then minimum delay is the logical choice (shortest path), and data must pass through the fewest number of nodes (unless nodes act as repeater).

If the routing table is made adaptive and based on minimum delay, the node transmitting data must select a line to its nearest neighbor. The condition of this line must be known, i.e., the number of bits waiting in the queue, the transmission rate over the line, error rate, and the traffic density. When this information about its neighbor is known, the node can calculate the estimated delay to it. An estimate of nodal delay is made to all nodes other than the transmitting nodes of the nearest neighbor because the nearest node may have longer queueing delays, lower transmission rates, higher bit error rates, or heavier traffic density than some of the more distant nodes.

A transmitting node must exchange delay information with its neighbor. Thus, one must expect the nodes within a network to have a microprocessor or specialized microprocessor-based controller to provide the necessary computing power to accomplish these various tasks and perform delay calculations.

Another approach to adaptive directory routing is the use of a central controller to generate all routing tables. This unit can broadcast updates to all nodes on a special channel or periodically broadcast the data on the primary channel. One of the problems with this technique is that the network will degrade drastically if the central controller fails.

LOCAL AREA NETWORKS

To form a local network, the hardware must be configured in a particular topology whether it be a star, ring, or bus structure. Routing must be implemented by a dedicated microprocessor, microcomputer, minicomputer, etc. The routing algorithm is usually embedded in a network node or a central controller and rarely altered; it may be considered as part of the hardware with firmware embedded (ROM program in a microprocessor). As one might imagine, the network layer can be difficult to divide into software and hardware. As specific networks are examined, the division between the two for each network will become apparent.

Collision-Sensed Networks

The topology of a collision-sensed network can be a ring architecture in which transmission may begin whenever a node has failed to sense a carrier. This is a CSMA (Carrier-Sensed Multiple-Access) network. Rings may also be designed for CSMA/CD (the addition to the acronym being Collision Detect). Such a ring not only listens for carriers but is also able to detect collisions. The advantage is that when a collision occurs, the node will know of it before it receives its own data back with errors. Bus topologies can also use this protocol for determining what the network status is before transmitting. The bus can be a broadband cable with FDM channels dedicated to CSMA or CSMA/CD protocols. For a star topology, the collision-sensed networks are not very practical because each node has multiple transmission paths. The node itself may have a ring topology however, with multiple gateway stations that connect to other ring-type nodes. Thus, star networks cannot be completely ruled out.

Ring CSMA

Ring structures have certain characteristics that are a part of their design philosophy, as follows:

1. Messages are passed in one direction from node to node.
2. Propagation delays are kept small (link distances less than 10 km).

3. Nodes have an ordering inherent in the network design; i.e., they are dependent on their physical placement in the ring.
4. The connectivity of the network is minimal; an N-node network will require $(2N - 1)$ connections.
5. To send a message to all nodes on the ring requires $(N - 1)$ relays of the message.
6. Each node in the ring is an active element; i.e., it amplifies signals, reads addresses of the packet, and processes the data for errors.
7. The ring by its nature limits the number of messages traveling around it at any one time.

Let us now investigate how these contention rings behave according to the design philosophy given.

Before a node begins to transmit, the receiver listens to the ring for traffic; i.e., it monitors the ring interfaces. If no data is present on the ring, the node begins transmitting. At the end of transmission, it may append a token to the last packet, which is the most efficient technique. When another node wishes to transmit, it listens for the token at the end of the packet. The token is changed to a flag (delimiter) by the node waiting to transmit. The waiting node will then begin transmitting and insert a token at the end of its transmission. When data is received by the node that transmitted it, it is removed from the ring and the flag and token at the end. Thus, during heavy traffic loads, the contention ring behaves similarly to a token ring if a token is appended. If the ring is at idle and two stations begin to transmit, a collision of the packets will occur. The receivers of the two transmitting nodes will compare the transmitted data with that which is received, and both nodes will detect errors as the data is taken off the ring. Note that the node can be listening, repeating, or removing data. There are other failures, such as node failure or power failure at the node that can be catastrophic if no provisions are made for them.

Ring network connections are shown in Fig. 9-13(a) for electrically connected nodes (copper cable); Fig. 9-13(b) is a schematic representation of a fiber-optic network with optical switches. Within the interface circuits are embedded the necessary electronics, such as relay solenoids (not shown) and the required drive circuitry. The relay contact networks shown in Fig. 9-13(a) are configured in the power-off state, which allows the nodes to bypass signals through the closed contacts. When the interface is under power, RR' and TT' connect the receiver and transmitter, respectively, to the twisted pair network. Note that the two closed contacts are wired in series since the relays used for these circuits are equipped with four sets of transfer contacts (labeled as form C contacts).

Figure 9-13(b) is a fiber-optic version of the same ring network, being fiber-optic throughout to the transmitters and receivers. A 2 × 2 switch is provided, as shown, that will disconnect the transmitter and receiver of the node and form an optical bypass if a power failure occurs. The solenoid for this switch is again embedded in the interface circuitry and not shown here.

The fiber-optic version has many attributes that make it appealing as a ring network design solution; these will be explored here. Fiber-optic cable plants require no matching networks, and each node is electrically isolated from its neighbor except for the power mains. Bandwidths of the fiber-optic cable plant are in excess of 1 GHz–km for 6-μm core/125-μm clad fiber. For military and industrial applications, the cable plant may be procured with a measure of nuclear radiation hardness. Some of the single-mode waveguides have distance bandwidths of 10 to 100 GHz–km. The losses of single-mode cable are 1 dB/km for 1,300-nm wave guide and 0.9 dB/km for some of the newer 1550-nm waveguide.

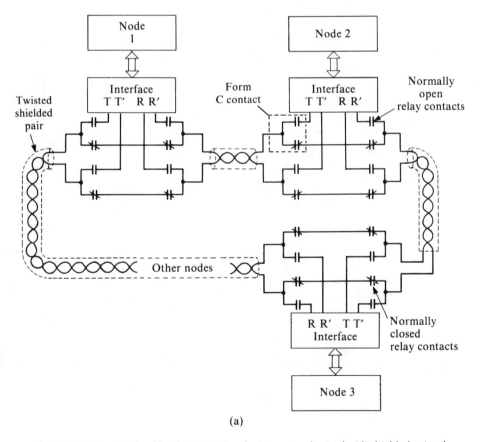

(a)

Fig. 9-13. Ring network cable plants: (a) N-node ring network wired with shielded twisted-pair cable (relay contacts are for failsafe operation), and (b) four-node fiber-optic network.

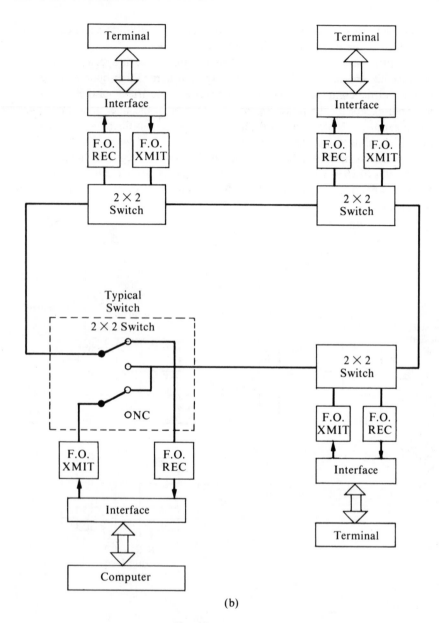

(b)

Fig. 9-13. (*Continued*)

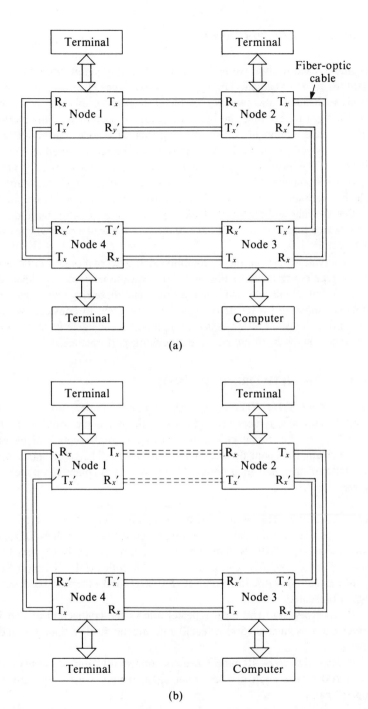

Fig. 9-14. Dual-ring cable plant implemented with fiber optics: (a) Ring with no node failures, and (b) counterrotating ring with a transmitter failure, T_x, at node 1.

A technique used to improve reliability is a dual-ring architecture that allows node failures to be bypassed. The bypass can be either electrical, with relays, or optical, with fiber-optic switches. A block diagram of this ring is shown in Fig. 9-14(a). Let us examine the block diagram for a moment. This particular network is a dual-redundant counter-rotating ring. If a node transmitter failure occurs, such as at T_x of node 1, then the ring will be reconfigured as shown in Fig. 9-14(b). The primary transmission path occurs through the T_x and R_x transceiver components. However, when a failure occurs, the secondary path through the transceiver, R_x' and T_x', will also become active. Note, in Fig. 9-13(b) that the ring will physically look as it did in Fig. 9-14(a), but the cables shown in dashed lines will not be active; therefore, the actual path will resemble Fig. 9-14(b).

Some observations can be made, as follows: If a second node failure occurs before any repairs are made to Fig. 9-14(b), then the network will lose continuity; i.e., it will break into two isolated rings, and these rings will function as two separate rings until a repair is made. Another item to consider, which is obvious, is that the propagation delay through nodes and cable plant has essentially doubled; this will, of course, affect network performance.

CSMA/CD BUS NETWORK (ETHERNET)

An example of a CSMA/CD bus network with collision detection is Ethernet. Let us now examine a typical network using Ethernet implements as depicted in Fig. 9-15. The functional characteristics of the implements will be given before the system aspects of the network are discussed. The cable plants can be designed with off-the-shelf fiber optics. Definitions of the plant components are as follows:

1. *Host computer:* This is a computer that will accept Ethernet controller cards such as an Ethernet-to-Q bus communication controller (DEQNA), an Ethernet-to-UNIBUS communication controller (DEUNA), an Ethernet-to-Professional 350 Communication Controller (DECNA), on Ethernet-to-Local Network Interconnect (DELNI), and others to be discussed when presented.
2. *R (local repeater):* The local repeater allows the connection of two Ethernet cable segments (500-m each) that are no further than 100 meters apart.
3. *R (remote repeater):* Remote repeaters connect all cable segments further than 100 meters apart and less than 1,000 meters apart (these are fiber-optic links.)

The remote and local repeaters are connected to the cable segments via two

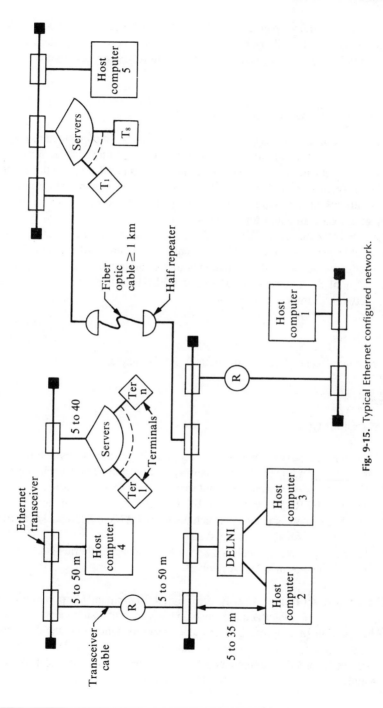

Fig. 9-15. Typical Ethernet configured network.

transceivers and transceiver cables. The remote repeater consists of two local repeaters, each with a fiber-optic interface board. The local repeaters are also called "half repeaters" because two of them are required to configure a remote repeater.

DELNI (Local Network Interconnect)

The DELNI is a low-cost data concentrator that can be implemented as a stand-alone, hierarchical stand-alone, or network-connected unit. In the stand-alone configuration, it can interconnect a maximum of eight systems, and the Ethernet coaxial cable or transceiver will not be required. The hierarchical stand-alone DELNI allows the interconnecting of these units to form a network without Ethernet coaxial cable and transceivers.

The network-connected DELNI allows the network to be expanded without adding cable segments. Each cable segment can then have over 100 implements connected because each DELNI expands the single transceiver connection to it by a factor of 8. A further discussion of DELNIs will be deferred until later when network examples are examined.

Transceivers

These devices were discussed previously in this chapter.

Servers

There are four types of servers, as follows:

1. The terminal server, which connects multiple terminals to one transceiver, thereby expanding the LAN capability without adding new cable segments.
2. The router server, which is a unit that transfers data packets between nodes on an ETHERNET or between ETHERNET LANs or remote nodes.
3. The X.25 gateway (router), which connects LANs to X.25 packet switched data networks and to remote systems.
4. The SNA gateway, which is a server that connects Ethernet LANs to IBM host computers in an SNA network.

The units described previously are Digital Equipment Corporation (DEC) components; other manufacturers also make Ethernet components, however, such as Interlan Inc. Before procurement of components, one should examine the wares of several manufacturers to insure the best cost and also to insure that they are not Ethernetlike components; i.e., that the components are 100 percent compatible.

Observing Fig. 9-15, one will note that four coaxial Ethernet cable segments are connected via repeaters. The segments connected by local repeaters are capable of being interconnected within a building, such as a college campus administration building. The distances between all the implements must be kept within close proximity. In the fiber-optic repeater case, the segments may be located in another building, such as a laboratory. User terminals T_1 through T_8 on the server may thus access not only the local host computer but any other hosts within the network. Sharing of these expensive resources allows more efficient use of them. If, for example, host 5 fails, then the other computers will allow the terminals on that segment access. Performance of the network would suffer, but a complete failure would not occur.

TOKEN PASSING NETWORKS

This technology is used on ring and baseband bus and broadband bus topologies. Token passing allows for a more orderly transfer of data and establishes an upper-bound on waiting time between network access by each station. Token passing networks remain efficient under heavy loading because of their collision-free operation. Token networks are useful for applications requiring guaranteed network access time. Examples of such requirements are: real time processing, process control, factory automation, and packet voice transmission (difficulty is encountered in the latter case; the slotted ring makes a better choice).

Token passing is not without its penalties; CSMA/CD would otherwise become obsolete, which, of course, is not the case. The token-passing techniques are sensitive to the number of nodes in the network. Excessive bandwidth overhead brought about by passing the token to idle nodes causes this condition, especially when network loads are light. These networks are also sensitive to the physical length of the ring or bus structure, which, of course, increases propagation delay.

Token Rings

One of the leading manufacturers using token ring technology is IBM. The rings are local loops connected to a central distribution panel. These local loops are connected to work station terminals or host computer nodes, as shown in Fig. 9-16. As the reader will note, this is a ring-star configuration. Each of the rings operates at a 4-Mbit/sec transmission rate, and the data is Manchester-encoded. The delimiters for the packets (equivalent to flags) are formed by Manchester Code violations. The packets have a four-byte destination address as well as a source address. The first two bytes identify the destination of the ring, and a control byte is provided to maintain ring integrity and indicate whether the traffic is synchronous or asynchronous.

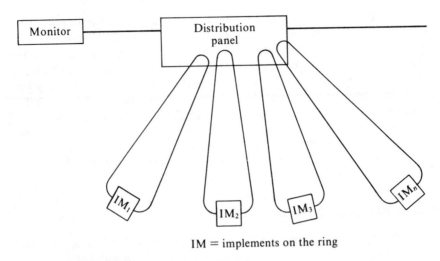

IM = implements on the ring

Fig. 9-16. IBM token ring.

The ring monitor will occasionally declare a period of nothing but synchronous operation. This monitor can refuse new synchronous/asynchronous requests until sufficient bandwidth is available. During synchronous operation, the monitor permits only single-packet transfers between call originator and answering station; i.e., a single packet is transmitted and a reply will be in the return packet. Synchronous operation can be used to implement both telephony and non-IBM devices. IBM and Texas instruments have produced controller ICs to emulate the token-passing ring protocol. Any of the rings previously described in this chapter may be designed as token rings.

Token Bus

During normal operation, the token is passed in a direction toward decreasing node addresses. When the lowest address has been reached, the token is passed to the highest. This is a form of a logical ring, therefore, even though the network has a bus structure. As is the case with all token networks, nodes may transmit only if they have the token; otherwise the node will function as a repeater with error-checking capability.

A station can remove itself from the bus by using a "set-next-node" frame, as shown in Fig. 9-17(a). The station must hold the token to perform this task. First, the station informs its predecessor of the removal and then informs its successor on subsequent cycles. After these two functions have been performed, the node becomes transparent to the network.

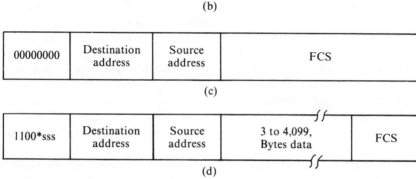

Fig. 9-17. Token bus frames: (a) Set-next-node frame, (b) solicit-successor frame, (c) token frame, and (d) data-transfer frame.

The existence of a logical ring creates a need for maintenance. The routine functions are as follows:

1. Reconfiguration of the logical ring.
2. Adjustment to, and maintenance of, the algorithm, e.g., the maximum time a station may hold the token, larger address field, new control function defined, etc.
3. Implementation of new stations on the ring.

Certain stations may be designated to perform maintenance functions. The first two above are performed in a straightforward manner, whereas the third is not quite so simple because a new station is involved. The process of recognizing a new station on the ring is performed using controlled contention, which is the same process used for a lost or damaged token.

All stations designated as ring maintenance stations initiate controlled contention every n^{th} time they receive the token, where n is set by the system. The maintenance station transmits a solicit-successor frame to start the process, as shown in Fig. 9-17(b).

Maintenance stations with the lowest addresses require two windows. Stations that are to join the network will place a set-next-node frame in the appropriate window.

When a token is lost or damaged, a period of silence on the bus will indicate this fact. The maintenance stations will begin the contention process after the period of silence, and the logical ring will be reestablished. If the maintenance station hears a response after the last token procedure, it will add the new station to the logical ring. If there is no response, the station will conclude the maintenance phase. If a collision occurs, a resolve contention frame will follow it with four window slots. Each demand station selects one of the window slots according to the first two bits of its address and transmits a set-next-node frame. If a collision occurs during this process, the station drops its demand.

If the controlling station detects a second collision, the resolve contention frame is repeated, and the demand nodes respond using the windows with the third and fourth bits of their addresses. This procedure continues until the controlling station receives a valid set-next-node, there are no further demands, or the maximum time-out is exceeded (this prevents thrashing about when damage has occurred to the network).

Transfer of data is a rather straightforward process. A station may not transmit unless it passes the token, which it must capture as the explicit token frame passes through the node. After the token is captured by the node, the node can transfer data for a limited amount of time, determined by a timer that is set at the time of token capture.

Frames to be transmitted have the following prioritized classes of service:

1. Synchronous
2. Urgent asynchronous
3. Asynchronous
4. Time available buses

Frames on the bus will be cleared according to their priority. Descriptions of the explicit token frame and the data transfer frame are given in Figs. 9-17(c) and 9-17(d), respectively.

The attributes of the token bus are similar to those of ring topology; i.e., performance is rather good for lightly loaded networks that have small numbers of nodes and short cable spans. Performance will eventually degrade, but transfer of data will remain orderly and reliable.

When transit delay is critical—such as in telephone systems, voice recognition, and certain types of encryption—token passing is unsuitable. Token passing places an upper bound on transit delay, which is dependent on the number of stations (nodes). Token bus algorithms are extremely complex and very costly to implement. With the advent of the new microprocessor technology and low-cost memories, however, this token bus technology will become more cost effective and appealing to network controller designers.

Microprocessors such as the MC 68000 series (Motorola), 32032 (National Semiconductor), and IAPX 432 (Intel), all of which have the computing power of minicomputers, can make use of the token-passing bus algorithms, which are easily implemented. At the time this book was being written, the cost of these devices was prohibitive, but it will decrease dramatically as the demand increases and manufacturing processes improve.

HYBRID STAR NETWORK

A hybrid fiber-optic star coupler is depicted in Fig. 9-18(a). The input ports can be configured with receivers connected to the outside environment through pigtails and bulkhead connectors. These receivers may have decoding logic design built into the receiver itself, or they may simply be bit drivers that convert the incoming optical signals to electrical. In the latter situation, a gate array may furnish the necessary logic to perform the decoding. Note, in the diagram, that data and clocks from each receiver enter the serial-to-parallel interface. The assumption here is that incoming optical waveforms are self-clocking.

Serial-to-parallel conversion is accomplished with a series of USRTs (Universal Synchronous Receivers Transmitter) or standard shift registers. The USRT requires the data to be in a particular format. For example, the 6852 SSD, a Motorola device, has onboard buffering and operates in a full duplex manner; i.e., it handles both serial input and output simultaneously. The data is processed in the USRT in 8-bit bytes, and sync characters must also be included in the data. This type of serial-to-parallel conversion does not lend itself to packet formats; for this purpose, the standard shift register with buffers is more advantageous.

Since fiber-optic transmitters require serial data, either a parallel-to-serial conversion register or a USRT will be required, depending on data format. Encoders may be embedded within the fiber-optic transmitters, designed in gate arrays, or constructed of discrete components.

The microprocessor, the heart of the system, can be used to perform some of the following functions:

1. Perform error checking
2. Log errors for maintenance purposes
3. Monitor traffic
4. Perform dynamic routing
5. Initiate diagnostics during low-traffic volume, such as local and far-end loopbacks.

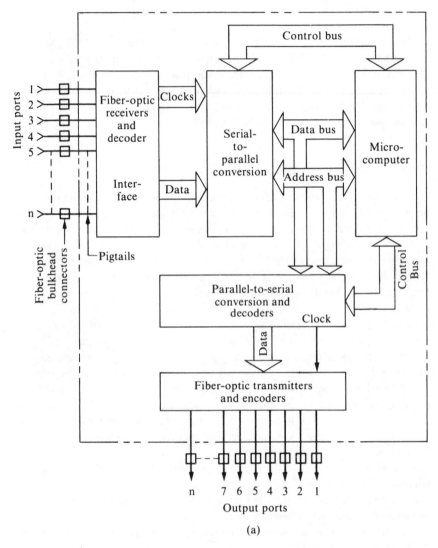

Fig. 9-18. Hybrid star: (a) Block diagram, and (b) flowchart of hybrid-star microprocessor.

6. Report abnormalities to a central maintenance node, which can be a mini-computer.
7. Perform data conversion, i.e., from one format to another.
8. Maintain a directory service, i.e., convert names to address locations or address locations to names.
9. Perform security checks on passwords, e.g., check their validity.

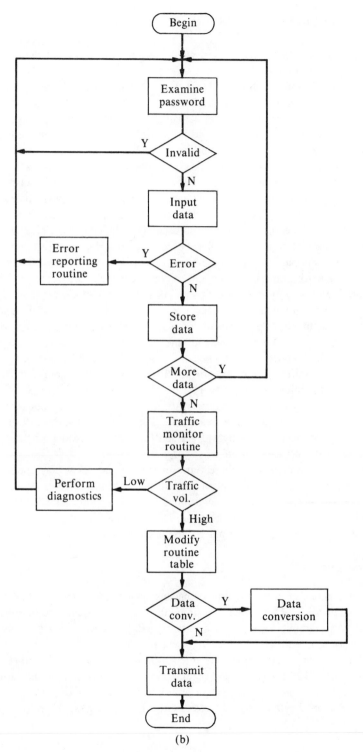

Fig. 9-18. (*Continued*)

(b)

The list of functions shown here would be much expanded if digital voice were also being processed.

For military applications—for example, when networks fail as a result of damage—a simulation program may be embedded to remove nodes and allow the star couplers to circumvent the damage. A flowchart showing the network functionality is shown in Fig. 9-18(b). This diagram is highly simplified; to consider the coupler flowchart would be beyond the scope of this book.

The hybrid star is not without disadvantages and thus requires a trade-off analysis. Microprocessors are rather slow compared to the serial transmission rates of many LANs; consequently, processing time must be kept small or higher speed devices must be used. For LANs in the 1-Mbit/sec range, microcomputers similar to the Intel 8051 may be used. For higher speeds, bit slice technology or discrete components must be considered. All is not lost, however, since GaAs technology is producing some very high-speed devices. Eventually, when processes become more efficient, reduced costs will allow the implementation of high-speed hybrid star couplers. As an example of GaAs products, one manufacturer is producing a 900-MHz bit counter with 40 mW of dissipation. The same manufacturer produces memories, shift registers, multiplexers, dividers, prescalers, and multiplexer/demultiplexers.

Often, high-speed devices may be controlled by low-speed controllers. As an example, if a shift register can process 16-bit words, the input data can be strobed into the microprocessor every 1.6 μsec, for a 10-Mbit data rate. Therefore, with some type of direct memory access (DMA), data may be stored in a large memory, which can serve as a queue. The object of this discussion is not to describe the design of a hybrid star but to alert the reader to the design variations that these devices afford.

One serious problem with these types of stars is the delay they entail. Any processing takes a finite amount of time to execute; therefore, a delay at the star node will occur to a degree determined by the algorithm complexity. If the hybrid star is not produced in large quantities, it cannot become very competitive with other types, such as fiber-optic stars or all-electric star networks.

GATEWAYS

This topic is one of prime importance because gateways are the glue that allows networks that do not have identical protocols or physical plants to be interconnected. As a simple example of a gateway, Ethernet data may be passed to another location with an Ethernet LAN via a telephone line.

Let us examine what is involved in performing this task. The Ethernet transmission rate is 10 Mbits/sec, and the telephone line transmission rate for high-speed modems is 9600 bits/sec; an approximate 1000 : 1 disparity therefore exists between the transmission rates of the two networks. Therefore, when

Ethernet-to-telephone-network transmission occurs, a large buffer must store the data. For the reverse situation—i.e., telephone facilities to Ethernet transmission—a small buffer is needed to assemble the data in packet form before transmission begins. Since Ethernet protocol is not even remotely similar to the protocol of telephone facilities, a conversion process is necessary that will accept either protocol and make them compatible.

Physical plant facilities for Ethernet are vastly different from those of their telephone counterpart. Electrical compatibility is necessary. As one might imagine, the task of producing an adequate gateway is quite formidable. Entire books could be written on this particular subject because of the large numbers of LANs available in the marketplace.

Often a manufacturer may have a network for connecting several implements, and to redesign the system could be very costly. Gateways make it possible for the locally connected equipment to communicate with an outside network.

For connecting networks at two sites, a bridge using leased lines or other communications facilities will be necessary. Each bridge end point will require protocol conversion of some type to allow end-to-end communication. The OSI transport layer will pass packets unchanged and uninterpreted across the bridge network between the two sites. The facilities of the intervening network are usually irrelevant and not available to the bridge user. In some circumstances, however, a user may require access to the network facilities, the following circumstances being among them.

1. Public and private packet switched networks that support X.25, X.28, and X.75 interfaces (CCITT standard)
2. Private switched networks supporting S.21 interfaces (CCITT standard)
3. Private LANs such as SNA, Ethernet, Wangnet, etc.
4. Videotext networks (private and public)
5. Modem interfaces
6. RS 232, 422, and 423 interfaces (EIA standards)

The preceding is only a partial list; many more entries could be added.

Gateways establish a virtual circuit between networks and transfer data across this circuit. Control, log in, and security, however, are not implemented in the gateway; its functions are restricted to the ISO network, the transport layer, and, in some cases, the session layers. It will be difficult to design gateways for interconnecting older networks that do not have the ISO layer approaches with new LAN designs.

The question of datagram versus virtual-circuit service provided by gateways should be addressed before delving into other issues. Datagram service through gateways is simpler to provide than virtual-circuit service. Datagrams can simply be encapsulated with an envelope when they enter the network gateway, and the envelope is then stripped off when exiting. For long-haul networks, virtual circuits are usually employed because they guarantee orderly delivery of message traffic. The IO X.75 protocol is appropriate for this situation.

The next issue to consider is addressing. Let us examine for a moment what is involved in passing a message across a gateway. The "gateway message transfer frame" is shown in Fig. 9-19.

Before data can be transported through the transfer medium, the destination must be known and the path between the originating node and the destination node must be ascertained. Flow control is also a concern of gateways. The destination, internet, and origination networks may all have different data transmission rates; in fact, this is the usual case. Network congestion may lead to gateway rejection or, in some cases, destruction of messages.

Issues of importance that have not been addressed are message fragmentation, privacy, and security. Fragmentation of message traffic occurs when internet, originating, and destination networks have different packet sizes. In situations where origination and destination networks must be kept secure, the gateways will also be required to employ encryption.

Gateways add overhead to the LANs that must be minimized if they are to be effective. Through-put optimization can be accomplished with gateways if they are made to be modular expandable. As an example of this kind of architecture, imagine a single gateway module capable of processing 1000 packets/sec; furthermore, suppose the design requirements for a particular gateway are 10,000 packets/sec. To meet the necessary design criteria, the original design need only be expanded with modules, thus allowing the gateway designer a great deal of flexibility.

A useful technique to reduce loads on large networks is to introduce dedicated subnetworks. Through-put increases because packets not intended for other LANs do not pass through the major gateways. As computers and other devices expand processing power and diversity, dedicated subnetworks become even more important.

As an example of a dedicated subnet, see Fig. 9-20. The high-speed 100-Mbit/sec fiber-optic ring will service the high-speed-devices host computer, laser printer, and disk-storage station. The terminals and telephone gateway

Network 1 interface protocol	Standard internet-working interface protocol	Transfer medium	Standard internet-working interface protocol	Network 2 interface protocol

Fig. 9-19. Gateway-message transfer frame.

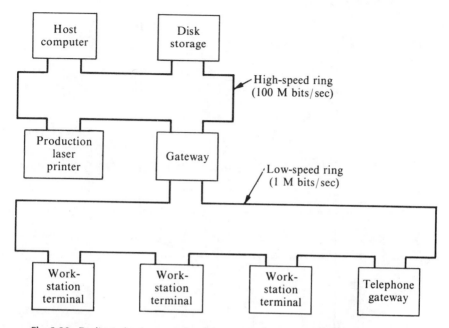

Fig. 9-20. Dedicated subnet configuration with high-speed and low-speed rings.

should not be serviced by the high-speed dedicated subnet. Their slow processing speed will reduce the through-put of the high-speed ring. The gateways will serve to match the speed of the two LANs. Other gateways provide access to the dedicated high-speed network by other low-speed rings.

Further examination of the figure reveals that telephone gateways must further step down/up the data-transmission rate. This would present a problem if the gateways were added to the high-speed ring—i.e., the approximate 10,000 : 1 difference in transmission speed.

BROADBAND FIBER-OPTIC NETWORKS

At present, most of the devices discussed here are available but are either not cost effective or in the laboratory stage. Let us examine the cable plant of Fig. 9-21. It is constructed with single-mode waveguide optimized at 1550 nm with an attenuation of 0.2 dB/km (this is a reasonable assumption, the theoretical limit being 0.17 dB/km). However, it is not the cable and waveguide that present problems with these bus systems but the couplers, because of connector and excess loss.

An assumption will be made that the minimum sensitivity for the up converter

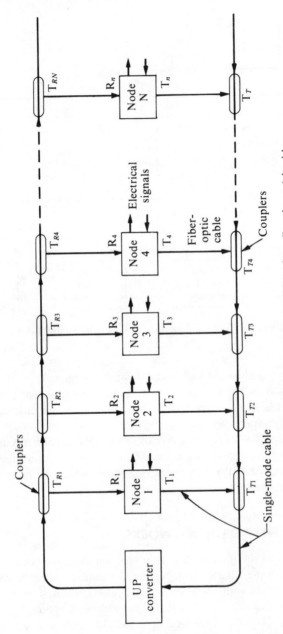

Fig. 9-21. Fiber-optic bus system analogous to broadband coaxial cable.

receiver is −42 dBm, with a dynamic range of 32 dB. Node 1 is 1 kilometer from the up converter; the tap ratio of T_{T1} is 1 percent, with the output signal calculated below:

T_{T1} Optical Power Budget (Without a Spliced Coupler)

Transmitter output	0	dBm
Connector loss	2	dB
Excess loss	0.2	dB
Tap ratio (1 percent)	20	dB
Waveguide loss (1 km)	0.2	dB
Power input to up converter receiver	−22.4	dBm

If each tap were equipped with fiber-optic connectors, any signals passing through 10 or 12 taps would be attenuated below the minimum receiver sensitivity. All is not lost, however. Suppose each tap were spliced; then the loss would be as calculated below:

T_{T1} Optical Power Budget

Transmitter output	−13	dBm
Splice loss out of coupler	−0.1	dB
Excess loss	0.2	dB
Tap ratio (10 percent)	10	dB
Waveguide loss (1 km)	−0.2	dB
Connector loss into coupler	1	dB
Power input to up converter receiver	−23.4	dBm

In the above example, the coupler input is equipped with a connector, but the input from T_{T2} and the output to the up converter are spliced into the bus and each tap (coupler) represents a 0.4-dB loss.

Next, let us calculate the maximum loss between any transmitter and the up converter receiver, which is the difference between the transmitter output power and the receiver maximum sensitivity, or 42 dB. Note that these calculations are not worst-case because some of the devices are experimental. For the node to function correctly, it must meet the criteria in the following equation:

$$5 \text{ dB} < \text{attenuation} \leq 37 \text{ dB}$$

The 5-dB minimum attenuation is required because the receiver dynamic range is only 32 dB. A further assumption is made that the dynamic range does not extend beyond the maximum sensitivity of the receiver.

Let us now examine the nth coupler in the system. Assume that a 1-percent coupler is used; the coupler output will then be -21.5 dBm and the maximum loss from T_{TN} output to the up converter receiver must be less than 20.5 dB. This figure allows 50 taps to be installed, for example, on a 2-km bus without repeaters. If greater than 50 taps are necessary, the same technique for calculating the receiver output power for the up converter may be used to calculate the input power levels for the repeater receiver. Repeaters are not shown in Fig. 9-21 because they are similar to the up converter with the exception of not translating the frequency up.

The up converter translates all incoming signals to the receive band. For example, if T_{T1} through T_{TN} transmit a carrier in the 70- to 100-MHz band, which is FSK modulated, and the nodes use the carrier on a time-division multiplexed basis, the up converter will translate the carrier to the receive band, or 226 to 266 MHz. The repeaters reinforce the signal and are transparent to the nodes on the network.

At the receiver input of node 1, the input at R_1 from T_{R1} must be low enough to prevent saturation. Also, the tap ratios must be extremely large because a part of the signal must enter each tap and be monitored by the receivers. Therefore, each tap induces a loss of signal in addition to splice loss and excess loss.

Let us assume a tap ratio of 0.5 percent. Then the optical power input of the first receiver would be as follows:

Tap Loss Budget

Up converter transmitter output	0	dBm
Waveguide loss (1 km)	0.2	dB
Coupler splice loss	0.1	dB
Excess loss	0.2	dB
Connector loss	1.0	dB
Tap ratio	23.0	dB
Power input at node T_{R1}	-24.5	dB

If the assumption is made that 50 nodes are connected to the bus, then the power at the nth receiver (50th) will be -44.5 dBm, which is too low. One solution would be to increase the tap ratio to 1 percent. As one may readily conjecture, the power levels must be adjusted throughout the bus system to prevent saturation or low power levels.

The couplers used on the transmit side can be programmable; i.e., the tap

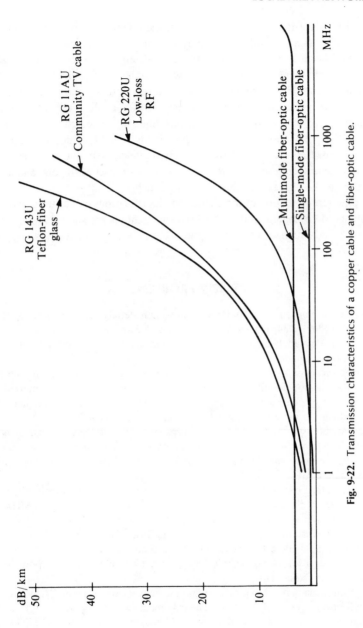

Fig. 9-22. Transmission characteristics of a copper cable and fiber-optic cable.

ratios can be set using voltages to program the channels through the couplers. This type of voltage control applies only to single-mode technology. Fixed tap ratios would be difficult to adjust if a node and length of cable were added to an existing cable plant, for example, between node 1 and node 2.

The major problem reiterated here is if the bus system is to be used for the transmission of video on carriers, such as a CATV application. The SNR for most lasers is 70 to 80 dB at best, but this figure is expected to improve. At present, it does not allow enough margin for high-quality video transmission. At 70 dB, losses must be less than 18 dB or video will not be of sufficient quality. There is such a flurry of research to improve laser SNR, one could even expect a breakthrough before this book is published.

A graphic representation of the characteristics of various cables used in copper cable plants as compared to those of fiber-optic cables are shown in Fig. 9-22. One may quickly realize why fiber-optic cable plants are becoming so popular.

REVIEW PROBLEMS

1. Calculate the propagation delay for the following materials: (a) glass, 1.5; (b) Mylar, 1.64; (c) Teflon, 1.34; (d) water, 1.33; (e) ice, 1.31.

2. Which of the materials listed in the first question has the shortest delay? Is it suitable for optical waveguide?

3. If a message must be received at a node in its entirety before it can be repeated, derive the equation describing the transmission rate for a T1 line. A T1 transmission consists of a synchronization bit, eight bits of quantization code, and 24 channels of audio sampled at the Nyquist rate. Use a 4-kHz audio bandwidth.

4. A transmission system uses one-half its capacity; the total delay is 3 nsec; and the message length is 256 bytes. What is the maximum transmission capacity in bits/sec?

5. Find the maximum capacity of a video channel that has eight levels of coding; assume the standard 6-MHz bandwidth.

6. Given an optical receiver with a modulation index of 50 percent, responsivity of 0.5 A/W, R across the detector diode of 100, and noise bandwidth of 300 MHz, find the quantum signal-to-noise ratio (for -32-dBm optical input power). How does the noise power in the optical system compare with that in a copper cable?

7. Design a single-mode cable plant with 60 taps similar to that in Fig. 9-20: (a) Show the basic budget for the tap nearest the up converter receiver, and (b) show the loss budget for the sixtieth tap.

8. Design the same network but assume that the network will support CATV. Assume an SNR always greater than 52 dB. The maximum laser power will be $+13$ dBm.

9. Design the same cable plant as in question 7 but assume that 1300-nm LEDS will be used and that the maximum signal injected into a coupler after all losses due to connectors and filters is -21 dBm. What is the maximum number of taps?

REFERENCES

1. Tanenbaum, A. S. *Computer Networks.* Englewood Cliffs, N.J.: Prentice-Hall Inc., 1981.
2. Tanenbaum, A. S. *Computer Networks.* Englewood Cliffs, N.J.: Prentice-Hall Inc., 1981.
3. Shannon, C. E. A mathematical theory of communication. *Bell System Technical Journal* **27:** 379–423 (July 1948); 623–656 (Oct. 1948).
4. Personick, S. *Optical Fiber Transmission Systems.* Plenum Press, 1981.
5. Tanenbaum, A. S. *Computer Networks.* Englewood Cliffs, N.J.: Prentice-Hall Inc., 1981.

10 | LONG-HAUL NETWORKS

Perhaps a more appropriate description for long-haul networks would be "switched networks." These networks have long enough runs of wiring for repeaters to be occasionally necessary to reinforce the signals between nodes. The nodes can be end offices, toll offices, CSPs, or other telephone switching centers. The objective here is not to discuss the entire switching hierarchy—telephone plant facilities are much too large in scope—but only its fiber-optics applications. For the reader interested in communication systems based on switching circuits, see Reference 1.

A subscriber-to-subscriber connection is shown in Fig. 10-1. The end office is where all subscribers are connected (although only one to an office is shown here) by a two-wire loop, which is the common connection found in most homes. The transmission is bidirectional on this loop and separated at the subscriber instrument with two- to four-wire converters. The trunk between the end office and toll office may be either two-wire bidirectional or four-wire unidirectional (two wires in each direction). The long-haul transmission between two toll offices is usually a four-wire connection.

The two-wire subscriber loop usually consists of twisted pairs, and replacement of these facilities with fiber-optic waveguides is not very feasible at this time. Toll trunks and intertoll trunks, however, are of paramount interest. These require a high-capacity transmission; i.e., each of the lines may carry a multiplicity of subscriber transmissions at the same time over FDM channels, TDM channels, or combinations of both. As one may observe, Fig. 10-1 is somewhat simplified, but it illustrates where the trunks are located and how they relate to the subscriber loops.

In 1962, the first commercial digital PCM baseband system was installed for distances less than 50 miles. This system, designed for paired cables, was the T1 system. The T1 is composed of a TDM digital frame with 24 voice channels. The voice sampling rate is 8000 frames/sec and the total number of bits per frame is 8 bits/sample times 24 samples/frame, plus one framing bit, for a total of 193 bits/frame. The channel transmission rate is 1.544 Mbits/sec.

Before discussing the T1 format, a comparison should be made to determine how fiber optics can be used in place of wire pairs. Standard multimode fiber-optic cable has a bandwidth of 600 to 800 MHz-km. Therefore, multimode cable has a bandwidth of 7 to 10 MHz at 50 miles (80 km), which is adequate to meet the bandwidth requirement. The system may be attenuation-limited. The

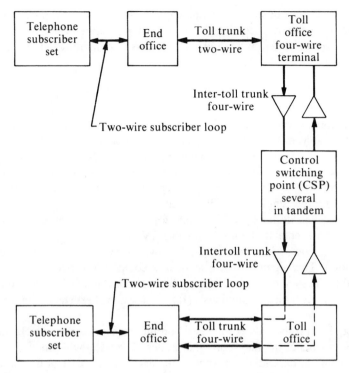

Fig. 10-1. Longhaul subscriber-to-subscriber connections showing various offices en-route.

lowest loss multimode waveguide is approximately 1 dB/km. A system constructed with multimode waveguide would require a loss budget of approximately 80 dB, a figure very difficult to achieve. Single-mode waveguides, on the other hand, have bandwidths of 2 to 40 GHz-km and losses of 0.15 to 0.5 dB/ km for certain wavelengths. Their use will reduce the loss budget to 12 to 16 dB, which can easily be accommodated with single-mode transmitter and receiver pairs. Later in the chapter, a complete loss budget will be generated for the T1 carrier system. This will give the reader some insight into why fiber optics is gaining such popularity.

A single-mode waveguide has sufficient bandwidth to accommodate a multiplicity of T1 carriers. The optical waveguide is physically much smaller than the wire pair and sometimes seems to offer nothing but advantages; in every bowl of cherries, however, there are a few pits. Water vapor is particularly damaging to optical cable. It increases loss brought about by OH^- and produces stress corrosion in the waveguide. Care must be taken when cabling fiber-optic waveguides to prevent stress from opening microcracks; these can result in large losses for temperature excursions below room temperature. The advantages and disadvantages of optical waveguides are discussed in more detail at the end of this chapter.

Modems

In some applications, analog FDM channels may be necessary to use FSK modulation at various channel frequencies to expand the use of single-wavelength systems. At the time this book was written, SNR was not suitable for carrier-type broadcast TV. However, some research facilities reported using SNRs of 80- to 100-dB range for video-carrier systems. One may well expect broadcast TV on single-mode fiber-optic cable within the next three years. Therefore, when designing cable plants for analog use, it should be assumed that they will eventually be used for cable TV.

The word *modem* is a combination of two words—"modulator" and "demodulator." This signifies that a modem is composed of two parts—a modulator, which is shown in Fig. 10-2(a), and a demodulator, which is depicted in Fig. 10-2(b). As indicated previously, this is only one of several design solutions possible.

In Fig. 10-2(a) the data is impressed on the channel carrier, which, in turn, intensity-modulates the fiber-optic transmitter. The VCO will produce either of two frequencies, depending on whether the data is a logical 1 or 0. Let us now examine how the modulator functions:

f_C = carrier frequency
f_K = deviation from the carrier frequency

Let f_V be the free-running frequency of the VCO without any signals impressed on the input or with the input in a high-impedance state. Then,

$$\Delta f = f_C \pm f_V$$

If f_M = modulation frequency and f_S = signal frequency, then

$$f_V - f_S = f_M - f_V = f_P$$

where f_P is a dummy variable. The passband of the filter can then be defined as

$$f_C \pm (\Delta f + f_P) = f_C \pm f_K$$

Note that if the VCO had been $f_V = 0$ with no output, then $f_C \pm \Delta f$ would

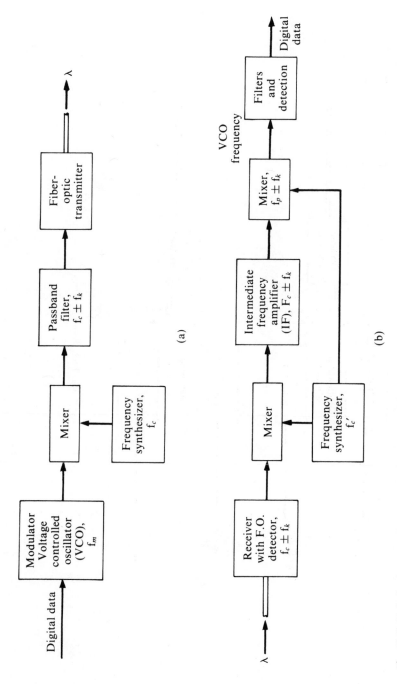

Fig. 10-2. Full duplex communication using FDM or WDM and a single-mode fiber-optic waveguide: (a) Analog channel showing modulator detail (transmitter), (b) analog channel showing demodulator detail (receiver), (c) full duplex fiber-optic link using WDM, and (d) full duplex fiber-optic link using WDM and filters.

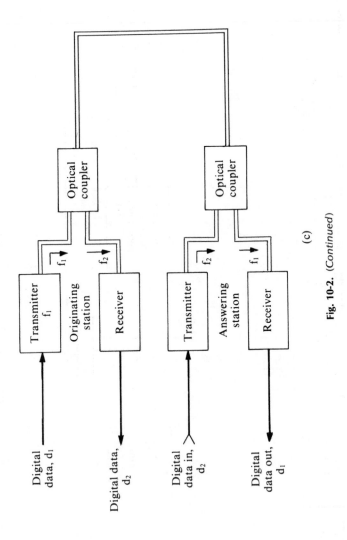

Fig. 10-2. (*Continued*)

(c)

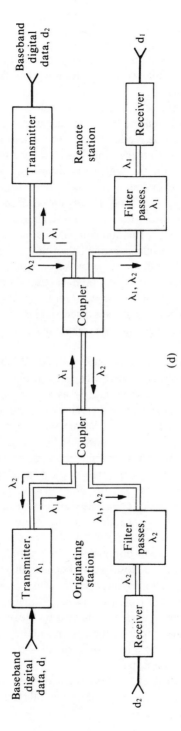

Fig. 10-2. (*Continued*)

have been filter passband, but the objective here was to produce a general solution. The output of the filter will intensity-modulate the fiber-optic transmitter (laser, LED, or YAG).

The receiver at the remote end of the link will detect and amplify the optical signals. Both receivers and transmitters were discussed in previous chapters, and this information will not be repeated here. The receiver must have a passband limited to $f_C \pm f_K$. An IF stage is included to keep the solution general. If multiple frequencies are on the optical medium—i.e., several carriers are present such as would occur in common FDM systems—then the IF stage will improve selectivity and image rejection. If only two carriers are present and the frequencies are chosen carefully—i.e., they are neither harmonics nor subharmonics of each other—and if modulation components of their harmonics do not appear in the passband of the other channel, then the IF stage is not required. The VCO idle frequency, f_V, will appear at the second mixer output if no digital data is present. If a 1 is present at the digital data output, then f_M appears at the mixer output, and for the condition when a zero appears at the digital data output, f_S appears at the mixer output. Not discussed here is the fact that mixers are nonlinear devices and that filtering takes place in Figs. 10-2(a) and 10-2(b).

A typical application showing a full-duplex FDM scheme is shown in Fig. 10-2(c). The data, d, is impressed on carrier f_1 [as in Fig. 10-2(a)] and passed through a $\Delta\beta$ coupler. The receive signal, f_2, is separated out at the coupler. The originating station transmits on f_1 and receives an f_2 from the remote station.

If wavelength division multiplexing (WDM) were to be implemented, the implementation would be quite similar to that of Fig. 10-2(c). Instead of f_1 and f_2, however, wavelengths λ_1 and λ_2 would be used, and the receiver would require an optical filter to filter out the transmitting wavelength of the station. Figure 10-2(d) is a diagram of the system. Note that the data is transmitted at the baseband, which reduces the complexity of the transmitter and receiver design. The couplers will allow a certain amount of crosstalk from the transmitter to enter the receiver at 20 to 60 dB. If the crosstalk can be kept small enough, the receiver filters may be eliminated. One must also keep in mind that if the transmitters are spaced too close, very short length attenuators should be added to the link. The reason for the attenuators is as follows: When the links are short, the crosstalk from the receiver wavelength that enters the originating and remote transmitters—the λ_1 and λ_2 indicated by broken lines—can destroy the laser sources. As one may observe, the design solution is heavily dependent on the intended applications, and a trade-off-analysis should be performed prior to selecting a transmission philosophy.

Another alternative to FDM and WDM is to use two separate waveguides with TDM channels—one waveguide to transmit and the other to receive, much as in a four-wire trunk application.

The VCO shown in Fig. 10-2(a) could have been implemented with two oscillators that are switched depending on the nature of the digital data, whether 1 or 0. This type of FSK (frequency-shift keying) modulation is fairly easy to implement compared to the VCO, but the transition boundaries between 1, 0 and 0, 1 cause abrupt changes in waveform, and these result in large harmonic content.

Phase-shift and frequency-shift keying are often used in modem design. Amplitude modulation is also used, but it will not be discussed because PSK and FSK are more common. Note that in the PSK waveform shown in Fig. 10-3(a) the carrier frequency remains the same with only a 180° shift in phase. During the phase transition, a sharp peak occurs that will result in approximately a 7-percent second and 3-percent third harmonic. These will not cause a problem if the adjacent channels are not operating at a frequency that is a harmonic of this carrier, which is usually the case. Figure 10-3(b) is a power spectral density plot, where P_c represents the peak power in the carrier and P_d represents the peak power in the sidebands. For a phase difference between the carrier and each phase of 90° ($\beta = \pi/2$), all of the power resides in the sidebands. This concentration makes the system efficient because all of the signal power carries information. If this value of β is substituted into the peak values of P_c and P_d, the carrier power becomes zero and the information-carrying peak power, P_d, is $A^2/2$. The curve shown in Fig. 10-3(b) is of the form, $[\sin x/x]^2$. The following equation expresses the actual values of the power spectral density, $F_S(f)$, where T_b is the period of the binary data rate, $R_B[R_B = 1/T_b$ (bits/sec)]:

$$F_S(f) = \left(\frac{A^2 T_b \sin \beta}{2}\right) \left\{\frac{\sin\left[2\pi(f - f_c) T_b\right]}{2\pi(f - f_c) T_b}\right\}^2 \qquad (10\text{-}1)$$

If $\beta = \pi/2$, Fig. 10-3(c) provides a phaser plot of PSK information-carrying vectors as compared to the carrier vectors. The dotted lines around the $\sin \omega_c t$ and $-\sin \omega_c t$ vectors indicate the phase noise effects on these values. If β has any angular value other than $\pi/2$, then the $\cos \omega_c t$ term will have some amplitude and the noise will have a greater effect (see problem 10-4).

FSK modulation that uses a VCO modulator is shown in Fig. 10-3(d). Note that the transition between a binary zero and one and vice versa is smooth and does not exhibit sharp transitions in the waveforms; the harmonic content of these waveforms is therefore rather low. Figure 10-3(e) shows a switched oscillator version of FSK, i.e., one in which two free-running oscillators are gated to form a modulator. The diagram shows sharp discontinuities that produce a large number of harmonics. The transition between "0" and "1" or "1" and "0" need not occur at the waveform crossovers as shown but may also occur in the middle of the peak amplitude since both oscillators are free-running. For

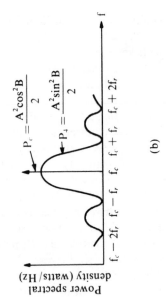

Phase-Shift Keying (PSK)

(a)

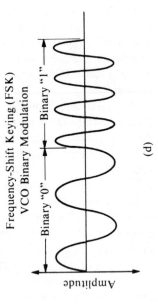

(b)

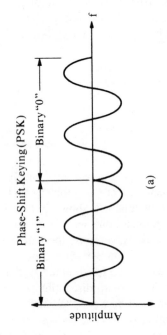

$$\sqrt{E_b}\,d = 2\sqrt{E_b} = A\sqrt{2\,T_b}$$

$$E_b = \frac{A^2\,T_b}{2}$$

(c)

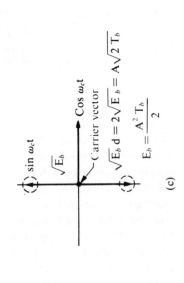

Frequency-Shift Keying (FSK)
VCO Binary Modulation

(d)

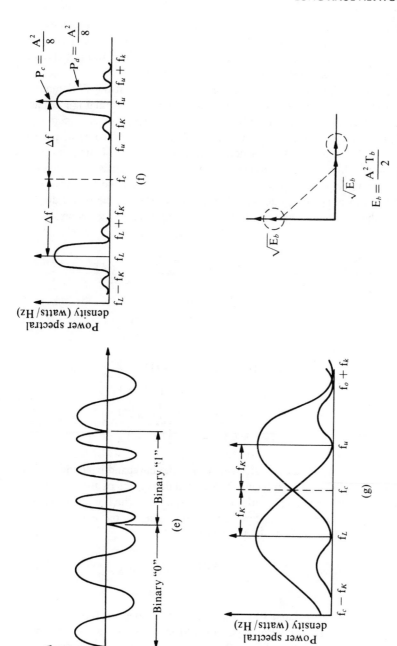

Fig. 10-3. Waveforms and spectral representation for various analog modulation schemes: (a) Phase-shift-keying waveforms (PSK), (b) PSK spectral density, (c) PSK phasor representation, (d) frequency-shift-keying (FSK) waveform, (e) FSK power-density plot, (f) orthogonal tone-spacing spectral power-density plot, (g) phasor representation of orthogonal tone spacing, and (h) switched oscillator FSK.

conditions shown, the harmonics are more predictable and may be analyzed. The analysis will be left to the reader, who is advised to use a Fourier series analysis and $u(t)$ function to obtain the sharp discontinuity waveforms that occur other than at sinusoidal crossovers.

A diagram depicting the power density for tone spacing of the FSK tones is shown in Fig. 10-3(f). Since this is a generalized tone spacing that ignores any reference to their phase, tones must be separated sufficiently or they cannot be resolved. Also, for the switch oscillator case, other sidebands would be present and the spectral power would be located above and below; these are not shown in the figure to allow a more simplified approach to the analysis.

Figure 10-3(f) depicts the upper tone as $f_u = (f_c + \Delta f)$ and the lower tone as $f_L = (f_c - \Delta f)$. The data rate is f_K. For the curves shown, the power density for each of the tones is of the form $(\sin x)/x$. The power density is derived as follows:

$$f(\omega) = \frac{A^2}{8} \left\{ \frac{\sin \left[(\omega_L - \omega_K)(T/2) \right]}{(\omega_L - \omega_K)(T/2)} + \frac{\sin \left[(\omega_U - \omega_K)(T/2) \right]}{(\omega_U - \omega_K)(T/2)} \right\}$$

where $\omega_L = \omega_c - \Delta\omega$, $\omega_U = \omega_c + \Delta\omega$, and $\omega_K = 2\pi/T_b$.
Then,

$$f(\omega) = \frac{A^2}{8} \left\{ \frac{\sin \left\{ \left[\omega_S - (\Delta\omega + \omega_K) \right](T_b/2) \right\}}{\left[\omega_c - (\Delta\omega + \omega_K) \right](T_b/2)} \right. $$
$$\left. + \frac{\sin \left\{ \left[\omega_c + (\Delta\omega + \omega_K) \right](T_b/2) \right\}}{\left[\omega_c + (\Delta\omega + \omega_K) \right](T_b/2)} \right\} \tag{10-2}$$

For the situation when $\Delta f \gg B_K$, where B is the baseband bandwidth, then the FSK bandwidth is approximately $2\Delta f$ and independent of the base bandwidth. This situation described is for large tone spacing, which may or may not involve orthogonal tones. When tones are closely spaced, the bandwidth will approach $2f_K$, which is that required for AM.

$$\beta = \frac{\Delta f}{B_K} \tag{10-3}$$

where $\beta \ll 1$ for narrowband FM and $\beta \gg 1$ for wideband FM.

Equation 10-3 is the modulation index for FM systems. As one can observe, narrowband FM assumes that $\Delta f < B_K$, which makes detection difficult, but for $\beta = 0.01$, only a single sideband is necessary to retain fidelity. For the other condition, $\beta = 2$, for example, four side bands are necessary to maintain fidelity of the binary pulses or $8\,f_K$.

When $\beta = 1$, the amplitude distribution of the information waveform has approximately 90 percent of its energy concentrated within the first three side-bands, and the bandwidth requirement becomes $6\,f_K$, not $2f_K$ as one may erroneously assume. If orthogonal tones can be used, however, then detection only depends on resolving the two power density peaks as shown in Fig. 10-3(g), but the tone phases shown in Fig. 10-3(h) can also be used.

This discussion, which has addressed modulation FSK and PSK, will leave AM to the reader (see Schwartz[2]).

Demodulation for FSK or PSK will be examined in either of two categories: coherent or noncoherent. Coherent demodulators have superior performance in noisy environments but are also complex and expensive. Detection often requires phase-lock loops with external sweep circuits that drive the PLLs toward lock. Phase jitter is sometimes a problem. Other problems include false loss of lock and long lock acquisition time.

Noncoherent demodulation does not rely on PLL, and demodulation usually requires envelope detection only, which is simply constructed. Another type of detection, somewhat unique, is discriminators for FM; they can often be obtained in IC form and are relatively inexpensive. Their performance is rather close to that of coherent demodulation. Another type not to be covered in detail is differential coherent demodulation. Performance of these devices is similar to that of noncoherent demodulators; i.e., no PLLs are required because they are not too complex. They perform well when channel fades or line hits cause irregularities in phase over two-bit periods.

Coherent demodulation can be accomplished using several different techniques for PSK or FSK. Some of these schemes will be investigated in the following paragraphs.

A PSK demodulation scheme is shown in Fig. 10-4(a). The incoming signal is represented by $S_i(t)$, where $\beta = \pi/2$; this represents residual carrier PSK demodulation.

$$S_i(t) = A \cos \beta \sin \omega_c t + Ad_i(t) \sin \beta \cos \omega_c t$$

The PLL locks in on the residual carrier, $A \cos \beta \sin \omega_c t$, of the $s_i(t)$ input. The output, S_o' is shown below:

$$S_o'(t) = A \cos \beta \sin \omega_c t \cos \omega_c t + Ad_i(t) \sin \beta \cos^2 \omega_c t$$

$$= \left(\frac{A}{2}\right) \cos \beta \sin 2\omega_c t + \left(\frac{A}{2}\right) d_i(t) \sin \beta + \left(\frac{A}{2}\right) d_i(t) \sin \beta \cos 2\omega_c t$$

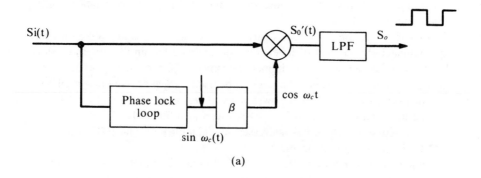

(a)

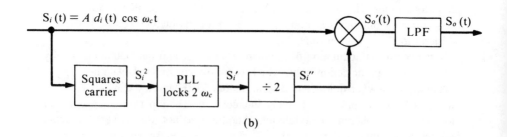

(b)

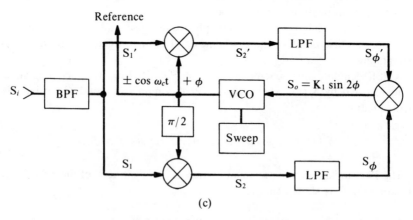

(c)

Fig. 10-4. Various types of demodulation: (a) Coherent demodulation PSK, (b) suppressed-carrier PSK demodulation, (c) Costas loop demodulation, (d) coherent FSK demodulation, (e) noncoherent FSK demodulation, and (f) discriminator demodulation for FSK.

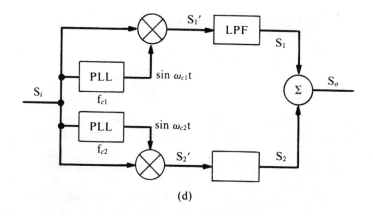

(d)

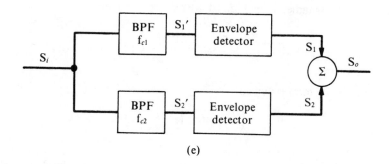

(e)

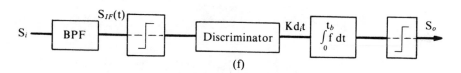

(f)

Fig. 10-4. (*Continued*)

or

$$S_o'(t) = \left(\frac{A}{2}\right) d_i(t) \sin \beta/2 \qquad (10\text{-}5)$$

Equation 10-5 is the output of the demodulator after the second harmonics of the carrier have been filtered by low-pass filter LPF. The power output is represented by Eq. 10-6:

$$P_o = \left(\frac{A^2 \sin^2 \beta}{4}\right) d_i(t) = 1, 0 \qquad (10\text{-}6)$$

Note that if β is exactly $\pi/2$, then the peak output has noise on the recovered digital signal, as shown in Fig. 10-4(a), but this may be minimized by adjusting β phase to compensate.

The next PSK demodulation scheme is suppressed carrier with $\beta = \pi/2$. A diagram of the demodulator is shown in Fig. 10-4(b). Note that, for this type of demodulation, the second harmonic of the carrier is used for PLL locking, which results in the need for higher frequency components in the demodulator; this may or may not result in a problem depending on the application. The values of the parameters of Fig. 10-4(b) are shown below:

$$S_i^2 = A^2 d_i^2(t) \cos^2 \omega t = \frac{A^2 d_i^2(t)}{2}(1 - \cos 2\omega_c t)$$

or, with the dc term removed,

$$S_i^2 = \frac{A^2 d_i^2(t)}{2} \cos \omega_c t$$

Making the amplitude adjustment and dividing the frequency by 2, we get

$$S_i = \pm\cos \omega_c t$$

Likewise,

$$S_D = \pm A d_i(t) \cos^2 \omega_c t = \pm\left[\left(\frac{A D_i(t)}{2}\right) - \left(\frac{A d_i(t)}{2}\right) \cos \omega_c t\right]$$

Filtering out the second harmonic of the carrier, we get

$$S_o = \frac{A d_i(t)}{2}$$

The PLL of Fig. 10-4(b) may be replaced by a Costas loop, as shown in Fig. 10-4(c). The signals for Fig. 10-4(c) (before and after filtering) are shown below:

$$S_i = A d_i(t) \cos \omega_c t$$

The band-pass filter is used to reduce the noise bandwidth and filter out any stray harmonics from this channel or others. Then,

$$S_2 = Ad_i(t) \cos \omega_c t \sin(\omega_c t + \phi)$$

$$S_2' = Ad_i(t) \cos \omega_c t \cos(\omega_c t + \phi)$$

Using trigonometric expansions for S_2, we have

$$S_2 = Ad_i(t) [\cos \omega_c t \sin \omega_c t \cos \phi - \cos \omega_c t \cos \omega_c t \sin \phi]$$

$$S_2 = Ad_i(t) [\tfrac{1}{2} \cos \phi \sin 2\omega_c t - \tfrac{1}{2} \sin \phi(1 + \cos 2\omega_c t)]$$

or, after filtering,

$$S_2 = -\left(\frac{Ad_i(t)}{2}\right) \sin \phi$$

Using a similar trigonometric argument for S_2', we have, after low-pass filtering,

$$S_2' = \left(\frac{Ad_i(t)}{2}\right) \cos \phi$$

The two values for ϕ are then multiplied together, and the result is shown as Eq. 10-7:

$$S_o = -\left(\frac{A^2 d_i^2(t)}{4}\right) \sin \phi \cos \phi = -\left(\frac{A^2 d_i^2(t)}{8}\right) \sin 2\phi$$

or

$$S_o = K \sin 2\phi \tag{10-7}$$

The Costas loop is locked when the ϕ error is approximately zero. A sweep generator is shown that drives the loop toward lock. Two advantages of the Costas loop is its ability to lock for large differences between VCO and carrier, and phase error to drive the VCO further toward lock is 20 degrees. For further details, see Hedin et al.[3]

Coherent FSK demodulation is shown in Fig. 10-4(d), where f_{c1} and f_{c2} are the two FSK tones. The incoming signal, S_i, is as follows:

$$S_i = Ad_i(t) \sin \omega_{c1} + A\overline{d_i(t)} \sin \omega_{c2} t$$

The equation for S_1' and S_2' before the LPFs are as follows:

$$S_1' = Ad_i(t) \sin \omega_{c1} t \sin \omega_{c1} + A\overline{d_i(t)} \sin \omega_{c2} t \sin \omega_{c2} t$$

$$S_2' = Ad_i(t) \sin \omega_{c1} t \sin \omega_{c2} t + A\overline{d_i(t)} \sin \omega_{c2} t \sin \omega_{c2} t$$

$$d_i(t) \neq \overline{d_i(t)}$$

The equations for S_1' and S_2' can be rewritten, after the necessary trigonometric expansions, as follows:

$$S_1' = \left[\frac{Ad_i(t)}{2} \right] [1 - \cos 2\omega_{c1} t] + \left[\frac{A\overline{d_i(t)}}{2} \right]$$

$$\left[\cos (\omega_{c1} - \omega_{c2}) t - \cos (\omega_{c1} + \omega_{c2}) t \right]$$

$$S_2' = \left[\frac{Ad_i(t)}{2} \right] \left[\cos (\omega_{c1} - \omega_{c2}) t - \cos (\omega_{c1} + \omega_{c2}) t \right]$$

$$+ \left[\frac{A\overline{d_i(t)}}{2} \right] [1 - \cos 2\omega_{c2} t]$$

The frequency terms are filtered out by the LPFs, and the resulting output after the summer is S_o, as shown in Eq. 10-8.

$$S_o = \left(\frac{Ad_i(t)}{2} \right) + \left(\frac{A\overline{d_i(t)}}{2} \right) \tag{10-8}$$

Note that, for the values shown in Eq. 10-8, bipolar data may be detected, i.e., NRZ. This detection scheme has a restriction that must not be violated. The value, $\Delta\omega = \omega_{c1} - \omega_{c2}$, must be greater than R_B, the binary data rate. If this restriction is not observed, a low-frequency component will appear at the summer output, and ambiguity will result. The LPF bandwidth must be sufficient to reject $\Delta\omega$, or a component will appear at the summer output. For the orthogonal set of tones, the calculations should be performed (see problem 10-7 at the end of the chapter).

Let us now consider noncoherent FSK demodulation. The block diagram for this detection scheme is shown in Fig. 10-4(e). The input S_i is the same as that shown previously for the coherent FSK detector, with the exception that the tones are restricted as shown in Eq. 10-9. The associated inputs to the summer

$$\Delta\omega \geq R_B \tag{10-9}$$

$$S_1'd = Ad_i(t) \tag{10-10}$$

$$S_2'd = A\overline{d_i(t)} \tag{10-11}$$

$$S_o = Ad_i(t) + A\overline{d_i(t)} \tag{10-12}$$

and the output are given by Eqs. 10-10, 10-11, and 10-12, respectively.

If common discriminators are used to recover data from FSK using FM demodulation techniques, the demodulator can be represented by the block diagram in Fig. 10-4(f). The input is given by Eq. 10-13:

$$S_i(t) = A \sin \left[\omega_c + \Delta\omega d_i(t) \right] t \tag{10-13}$$

in which the term $\Delta\omega$ in the deviation that represents a frequency proportional to the data signal amplitude. The optimum bandwidth for the BPF, B_W is 2.4 R_B for NRZ data and 2.8 R_B for Manchester-coded data. The optimum deviation is $\Delta\omega(2\pi\Delta f)$, where Δf is 0.35 R_B for NRZ and 0.8 R_B for Manchester codes. The K in the discriminator is made up of insertion loss from the BPF and discriminator characteristics. The last section—data detection—recovers the data and reshapes the waveform.

No other singular-modulation techniques will be discussed in detail, but the reader should be aware that there are a number of others. Differential coherent demodulation is one of them, in which the reference is a delayed version of the input. Commonly called "differential phase-shift keying" (DPSK), it has the advantages of noncoherent detection (no PLLs are necessary), performs well where channel fades or other irregularities occur, and has no data ambiguity. Another type is Δ-PSK, a technique that requires decoding differential phase-coded PSK signals. There are also PCM μ law, PCM A law, PCM linear adaptive differential pulse code modulation (ADPCM), continuous slope delta modulation (CVSD), linear predictive coding (LPC), and a popular compression technique, LPC 10, which compresses normal speech to a 2400-bit/sec rate. The latter technique doesn't operate well in noisy environments, and its speech reproduction is rather mechanical sounding. To analyze coding in terms of all these modulation schemes would obviously require a lengthy text. Only the common PSK and FSK and certain AM techniques will be considered.

M-ARY CODES

The following paragraphs will address M-ary (multisymbol coding), which is a technique of coding that reduces transmission bandwidth by increasing the information content. This discussion will delve into coding schemes using multiphase, multiamplitude, multifrequency, and combinations of these techniques. To some extent, QAM and QPSK are special cases of M-ary amplitude and phase-shift keying modulation; both schemes will be examined. The FSK equivalent is MFSK, i.e., multiple tones.

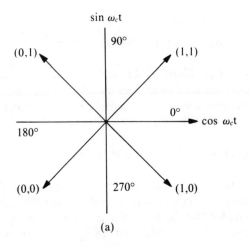

(a)

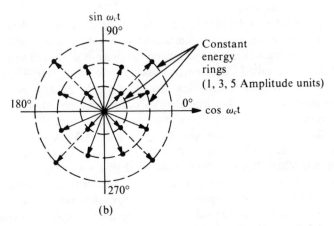

(b)

Fig. 10-5. M-ary phasor representations: (a) QPSK phasor diagram; (b) 16-QAM, 3 amplitudes, 12 phases, and 16 levels of code; and (c) 16-QAM, 3 amplitudes, 8 phases, and 16 levels of code.

To exemplify M-ary modulation, let us examine each type of modulation individually, i.e., QAM, QPSK, and MFSK. Then we will look at an example of M-ary modulation that combines all three.

Note in Fig. 10-5(a) that each phase represents a 2-bit code and that all the combinations of that code are represented. The code is arranged so that only one bit changes at a time as the phasor diagram progresses around 360 degrees. The Gray code/phase reduction is shown in Table 10-1. All phases are in

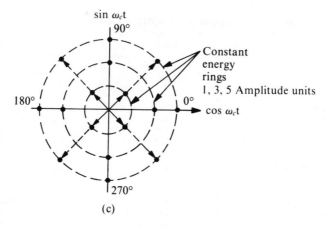

Fig. 10-5. (*Continued*)

quadrature (90 degrees). It is only natural to divide the code up in this manner since only four phases are needed.

QAM is a combination of phase and amplitude modulation; 16-QAM is one of the most popular M-ary codes for 9600-bit/sec rates. The phasor diagram in Fig. 10-5(b) represents 16-QAM coding. The code for Fig. 10-5(b)—a binary Gray code—is given in Table 10-2.

This is only an example of one coding technique; the binary-to-phase/amplitude relationship could have been presented differently. Table 10-2 illustrates a point. If the transmission rate of the waveform is 2400 bits/sec for each variation in waveform and the phase or amplitude conveys 4-bit words, the equivalent transmission rate is 9600 bits/sec. This technique is very efficient, but it is also sensitive to phase jitter and amplitude variation. The telephone lines must be conditioned so that there is no amplitude or phase variation across the bandwidth of 300 Hz to 3 kHz. This does not present a problem for fiber-optic mediums, but telephone lines must be conditioned to obtain these bandwidth characteristics.

Figure 10-5(c) shows another technique for coding waveforms, and Table

TABLE 10-1   Quadrature Phase-Shift Keying.

Phase	Binary Representation
45°	11
135°	01
235°	00
315°	10

TABLE 10-2   QAM 3-Amplitude 12-Phase Code.

Binary Representation	Phase	Relative Amplitude
0000	15°	3
0001	60°	3
0011	105°	3
0010	150°	3
0110	195°	3
0111	240°	3
0101	285°	3
0100	330°	3
1100	45°	1
1101	135°	1
1111	225°	1
1110	315°	1
1010	45°	5
1011	135°	5
1001	225°	5
1000	315°	5

10-3 provides the code information. This technique for 16-QAM provides at least 45° of phase between each phasor and 90° between some of them. Also, one will note that the large amplitudes are used more efficiently and should thus provide better BERs because the SNR is larger for larger amplitude signals. This code requires rather flat passbands, but phase errors are a little less strin-

TABLE 10-3   16-QAM 3-Amplitude 8-Phase Code.

Binary Representation	Phase	Relative Amplitude
0000	0°	3
0001	90°	3
0011	180°	3
0010	270°	3
0110	45°	1
0111	135°	1
0101	225°	1
0100	315°	1
1100	0°	5
1101	45°	5
1111	90°	5
1110	135°	5
1010	180°	5
1011	225°	5
1001	270°	5
1000	315°	5

gent. The 16-QAM coding scheme has been adopted by the CCITT, and VLSI circuits are available.

If several QAM signals are used to modulate a multitude of tones, then a hybrid form of M-ary results. For example, if two tones are used and 16-QAM techniques modulate each tone, then the data rate will be 19.2 Kbits/sec for a rate of 2400 symbol/sec (the word "symbol" must be used because each waveform variation represents a 4-bit symbol, and each tone carries a 2400-symbol/sec rate). See problem 10-8 at the end of the chapter for the development of a symbol table for M-ary codes with MFSK included.

A 16-QAM modulator is shown in Fig. 10-6(a). The general equation for a QAM signal, S_i, is given by Eq. 10-14, where r_i is the amplitude and θ_i the phase modulation:

$$S_i = r_i \cos \left(\omega_c t + \theta_i \right) \qquad (10\text{-}14)$$

The general equation for M-ary amplitude, phase, and frequency modulation is given by Eq. 10-15, where ω_k represents the various tones:

$$S_i = r_i \cos \left(\omega_K t + \theta_i \right) \qquad (10\text{-}15)$$

The equations for S_o and S_e will both have four levels of amplitude and phase. The sine and cosine will determine the quadrant. When these two signals are summed, they produce 16-signal-state 16-QAM. The vector components are calculated using four levels of amplitude at S_o and S_e. The amplitudes of $A \sin \omega_c t$ and $A \cos \omega_c t$ are multiplied by the respective S_o and S_e values for vectors corresponding to one of the 16 signal states. Therefore, the two 2-bit coders need only be D/A converters.

The technique described is similar to changing both variables in Eq. 10-14. If this technique were extended to two tones and 32 signal states, then the serial-to-parallel conversion would require two 16-QAM coded tones, the bit rate to each would be $R_B/2$, and the bit rate to the 16-QAM vector summers would be $R_B/4$. To go to more complex modems would be a matter of extending the number of FSK tones or utilizing all of the available amplitudes and phases more efficiently.

Receiving and decoding 16-QAM is rather more difficult than the modulation scheme. To examine a 16-QAM demodulator, refer to Fig. 10-6(b). The 16-QAM general equation is the same as Eq. 10-14, reiterated here as

$$S_{QAM} = r_i \cos \left(\omega_c t + \theta_i \right)$$

S_o'' and S_e'' are represented by Eqs. 10-16 and 10-17, respectively.

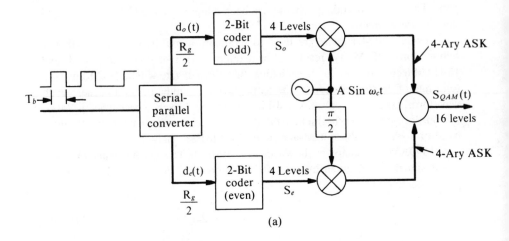

(a)

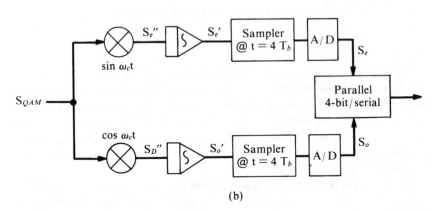

(b)

Fig. 10-6. 16-QAM modulation/demodulation: (a) Modulator block diagram, and (b) demodulator block diagram.

$$S''_e = r_i \cos (\omega_c t + \theta_i) \sin \omega_c t \qquad (10\text{-}16)$$

$$S''_o = r_i \cos (\omega_c t + \theta_i) \cos \omega_c t \qquad (10\text{-}17)$$

where $S''_e = S_{\text{QAM}} \sin \omega_c t$ and $S''_o = S_{\text{QAM}} \cos \omega_c t$. Using trigonometric identities on these two equations, we have Eqs. 10-17 and 10-18:

$$S''_e = \tfrac{1}{2} r_i [\sin 2\omega_c t + \sin \theta_i] \qquad (10\text{-}17)$$

$$S''_o = \tfrac{1}{2} r_i [\cos 2\omega_c t + \cos \theta_i] \qquad (10\text{-}18)$$

After these signals pass through the integrator, we have the equations for S'_e and S'_o, which are the components necessary to present the sampler:

$$S'_e = 2T_b r_i \sin \theta_i$$

$$S'_o = 2T_b r_i \cos \theta_i$$

The samplers will sample and hold the values of S'_e and S'_o, which are A/D decoded to reconstitute the original 2-bit digital values at the transmitter. These 2-bit samples from S_o and S_e reconstitute the original 4-bit value of the transmitter in the parallel-to-serial register.

One question the designer should ask is: How do these modulation techniques compare in performance? Figures 10-7(a) through 10-7(d) will give some performance criteria to make an assessment of the techniques. Often one will find that a designer may trade-off bandwidth for SNR; the correct choice will depend on the application.

Figure 10-7(a) is the representation of the ratio expressed in Eq. 10-19 to the probability of symbol error:

$$\frac{\text{Symbol Energy}}{\text{Noise Energy}} = 10 \log \left(\frac{E_b}{N_o} \right) \qquad (10\text{-}19)$$

The conversion from symbol error to bit error is represented by two equations. The first, Eq. 10-20, is an approximation for the more linear part of the curves in Fig. 10-7(a), and the second, Eq. 10-21, is the actual value:

$$\text{Binary bit error (approx.)} = \left(\frac{1}{\log_2 M} \right) P_M \qquad (10\text{-}20)$$

$$\text{Binary bit error} = \left(\frac{1}{\log_2 M} \right) \text{erfc} \sqrt{\frac{\log_2 ME_b}{N_o}} \left(\sin \frac{\pi}{M} \right) \qquad (10\text{-}21)$$

If one observes where the curves cross the symbol/noise ratio, it can be seen that MPSK for phases 2 and 4 allows the designer almost something for nothing at approximately the same energy ratio. When $M = 8$, the penalty for doubling the phase is 4 dB. When $M = 16$, the penalty becomes 9 dB (4 + 5). When M is greater than 16, an additional 6-dB penalty occurs for every doubling of phase. As one may observe, 8-phase QAM requires only approximately 4 dB more symbol energy to maintain a P_M of 10^{-4} to 10^{-5} whereas 12-phase QAM requires between 6 and 9 dB more symbol energy.

Let us next examine the MFSK performance represented by the curves in Fig.

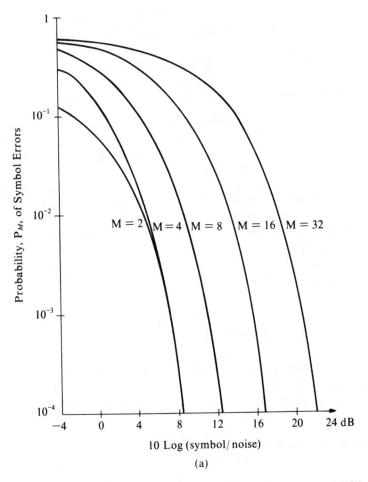

Fig. 10-7. Performance of various modulation/demodulation schemes: (a) MPSK symbol error versus SNR for 2, 4, 8, 16, and 32 level codes; (b) MFSK symbol error versus SNR for 2, 4, 8, 16, and 32 level codes; (c) bandwidth comparison of MPSK, MFSK, MASK, and QAM; and (d) BPS/Hz per 10 log (symbol/noise) comparison for different modulation schemes.

10-7(b). The curves depict M levels of FSK and the relationship of symbol energy represented by $10 \log E_b/N_o$ to symbol error probability, P_M. Note that, as the symbols are distributed between a larger number of frequencies, M, the energy required to produce lower values of P_M is less. From the curves in Fig. 10-7(b), one may ask the question: Why bother with other types of modulation

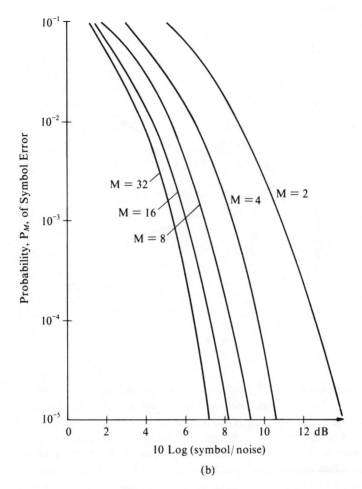

Fig. 10-7. (Continued)

when the superiority of MFSK is so apparent? The answer is a logical one: more tones require greater bandwidth. If a large amount of bandwidth is available, a large tone library can be used. For a more comprehensive treatment of MPSK, MFSK, and other modulation combinations, see Chap. 6 of Schwartz.[2]

A comparison of MFSK, MPSK, MASK, and QAM is given in Fig. 10-7(c). This provides the answer to the question posed in the paragraph above. As one may observe, QAM, MPSK, and MASK offer no increases in bandwidth

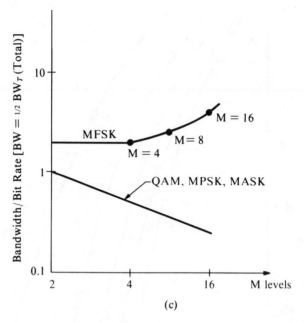

Fig. 10-7. (*Continued*)

as M levels increase, but previous curves have indicated that the energy level must be increased. If four tones are used four times, the same bit rate is needed for the required bandwidth. Also, if the number of tones is 4 or less, the bandwidth requirement remains constant, and the required symbol energy increases. By examining this curve, one can observe again how MFSK makes a trade-off of modulation technique for bandwidth and minimum symbol energy possible.

The curve in Fig. 10-7(d) describes the relationship between bits per Hertz versus the symbol energy ratio. The Shannon limit shown is calculated using Eq. 10-22 or Eq. 10-23, as follows:

$$C = \text{BW} \log_2 \left(1 + \frac{S}{N} \right) \tag{10-22}$$

$$C = \text{BW} \log_2 \left(1 + \frac{P_R}{N_o B} \right) \tag{10-23}$$

where C is the channel capacity in bits/sec. This is the absolute limit of the channel capacity, and the more closely a modulation technique can approach it,

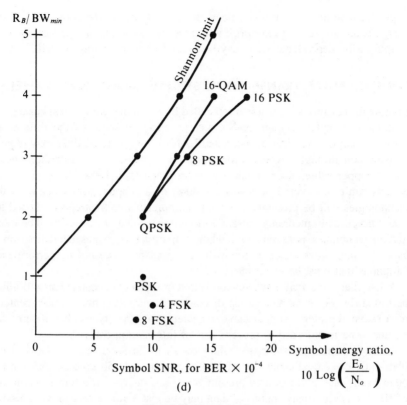

Fig. 10-7. (Continued)

the higher its efficiency. If capacity C is equivalent to the bit rate, R_B, then Eq. 10-24 which calculates the limit in bits/Hz versus the symbol energy ratio results, as follows:

$$\frac{R_B}{BW} = \log_2 \left(1 + \frac{P_R}{N_o B}\right) \qquad (10\text{-}24)$$

This equation is plotted as the Shannon limit in Fig. 10-7(d). Observing the curve for 16-QAM, we note that the efficiency is very good for the bandwidth utilized. We can also observe that the QPSK is rather efficient. Efficiency may not always be desirable, however, if the designer has the option of trading-off bandwidth, energy, complexity of receiver design, and other attributes. The designer must always be aware of trade-offs when making a modulation technique decision to make sure he selects the correct one for the application. Each design is an engineering enterprise with multiple solutions, and the correct one is application-dependent; there may be several correct designs as well.

LONG-DISTANCE TRANSMISSION SYSTEMS (OVER 10 KILOMETERS)

Long-haul transmission (as it is often called by telephone companies) has a great deal more complexity than LANs. Let us examine some of the differences between long-distance transmission and LANs. Long-distance links may require transmission through several communication mediums, e.g., satellite, microwave, copper cable, fiber optics, and short-range laser links. Since all design aspects can't be covered in a single section of a chapter, an overview of the technologies will be presented and more comprehensive treatments referred to. The differences in mediums present a more diverse class of problem. The modulation techniques required, say, when converting from baseband fiber-optic cable to microwave transmission will also present a number of performance variations that must be taken into account.

A long-haul link may have several different types of repeater that will affect performance. For example, a long-distance link may have fiber-optic repeaters, microwave repeaters, copper-cable repeaters, or other types that require the signals to be re-enforced to prevent loss of information content.

Delay is another consideration when long-haul links are at issue. Delay for a fiber-optic or coaxial cable link of 10 km may be only 50 μsec or less. On the other hand, satellite links may present $\frac{1}{4}$- to $\frac{1}{2}$-sec delays. At a transmission rate of 10 Mbits/sec, many frames of data may be lost before errors in transmission are discovered. For long-haul fiber optic or coaxial cable, delays may be 5 msec for 100-km links that may only represent a frame. A laser link of 1 km will have a delay of 3 μsec (assuming that the speed of light in the atmosphere is the same as in a vacuum—3×10^5 km/sec). As one can imagine, these accumulative delays in a system, including repeater signal-processing delays, can result in a multiplicity of problems. The next section will consider fiber-optic mediums only, which will scale down the problems by about a magnitude.

Let us concentrate on a specific long-haul link composed of several different transmission mediums and various signal formats, as shown in Fig. 10-8. Terminals T_1 and T_2 will communicate; transmission links LAN_1 and LAN_2 are token rings. The data in LAN_1 and LAN_2 have the same format for the purpose of illustration. Gateways that translate the data to an analog format suitable for transmission over conditioned telephone lines are equipped with queues and data buffers to absorb high-speed data onto token rings when receiving from telephone lines. These gateways are provided with modems that operate at 9600 bits/sec. Gateways may also be equipped to perform data compression, for example, if digital speech were to be transmitted, but token rings would not be

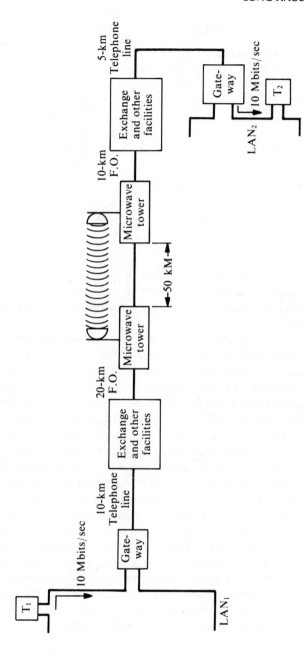

Fig. 10-8. Typical long-haul transmission system.

TABLE 10-4   Delay Budget.

LAN$_1$ maximum delay	50	nsec
Gateway queuing delay (1 kbyte packet)	0.8	msec
Telephone twisted pair (10-km)	67	nsec
Exchange delays	0.9	msec
fiber-optic link (20-km)	100	nsec
Microwave processing	10	nsec
Air medium	150	nsec
Microwave processing	10	nsec
fiber-optic link (10-km)	50	nsec
Exchange delay	0.5	msec
Telephone twisted pair (20-km)	33.75	nsec
Gateway queuing delay	0.8	msec
LAN$_2$ maximum delay	30	nsec
TOTAL end-to-end delay	2.5	msec

used; slotted rings would be required because of the critical timing necessary for each voice sample. The modems located at each end of the link would disassemble and assemble data packets due to the differences in the modem formats themselves.

Data transmitted over the link would have various delays depending on the medium it is being transmitted in. For example, the delays in LANs can be assumed small except for that resulting from the queue in the gateways. The telephone line to exchange may be a shielded twisted pair. Intertoll trunks and toll trunks are lumped together and labeled telephone exchange along with other facilities. The delays in this section can be quite high. The final items to be considered are the delay in the fiber-optic link that remotes the microwave towers and, of course, the microwave transmission between the two towers. A rather superficial analysis of all the delays can be made by using an itemized delay budget like the one shown in Table 10-4. The labels are meant as a typical delay budget to find exact numbers for all the values shown. The exchange and facilities must be known in detail. Details of long-haul network design may be found in Refs. 4 and 5. The same procedure can be followed for a loss budget; however, most readers will not be dealing with this issue unless they are designing a telephone network. Since some large corporations such as General Motors, Grumman, IBM, and others are designing their own telephone facilities, engineers may soon be involved more heavily in such designs. The design of these more massive networks is beyond the scope of the book. A satellite version of the link in Fig. 10-8 is presented in review problem 10-12.

LONG-DISTANCE CABLE PLANT AND THE LOSS BUDGET

The previous section presented an overview of long-haul systems. In this section, long lengths similar to trunks will be investigated. Here we are interested in the exchange and exchange facilities from the fiber-optic cable-plant point of view. The examination of a transatlantic cable installation will be examined from purely a feasibility point of view.

The first item for consideration will be the loss budget, to determine if the system is attenuation-limited, and then the treatment of the bandwidth-limiting parameters, to determine if the link is bandwidth-limited. After these two items have been investigated, repeater spacing can be determined.

To begin with, the loss budget for LANs can be appropriated for long-haul network use, and additional items such as repeater performance figures can be added to make the LANs table appropriate. The table to determine the rise-time budget for LANs can also be extended to accommodate long-haul networks. Most of the LANs discussed thus far have not made use of repeaters. If repeaters process the data—i.e., examine it for errors and make corrections—then delays will occur and a delay budget will be necessary for summing up the total delay, end-to-end.

If repeaters are necessary, then power must be supplied, either via long transmission lines or manufactured locally with salt-water batteries or other techniques for generating electrical power mechanically or chemically.

The loss budget for end-to-end transmission lengths can be calculated using the loss budget in Table 8-4, but any repeaters required must also be taken into consideration. Lets configure a transatlantic cable installation (see Table 10.5). The first item to consider is the loss budget for a single segment—from the initial transmitting station to the first repeater (use Table 8-4). Note that the terminations are splices because the cable plants are installed along the sea bottom and connections are never disconnected. For a total cable-plant installation, 3600 kilometers of cable are required. The installation requires 30 repeaters along the cable. Each repeater will consume power that is supplied by sea batteries with a power-distribution cable or manufactured in some other manner. First, the power consumption of each repeater must be calculated. If all the circuitry is designed using VLSI technology with CMOS, the power consumption per repeater is approximately 5 W. The reader can verify this estimate by constructing a repeater and summing up the power consumption of each component (one will usually find that the laser diode, drivers, and on-board voltage regulator consume most of the power). The most cost-effective solution at present is to route power lines for the repeaters with the cable. If all the power were supplied from one end of the cable to the repeaters and number 8 wire were used for power distribution, the line drop between the first repeater station on the cable and the first land station would be 187 V for 1 A (number 8 wire has a resistance of $1/2 \ \Omega/1000$ ft). As one can imagine, the power-distribution system is not trivial to design and will not be attempted here.

For a cable installation along a land route, repeaters can be powered locally or with solar power. Power grounding problems can easily be eliminated for locally supplied power because each repeater is electrically isolated from its

TABLE 10-5 Loss Budget for Transatlantic Cable Plant.

X_1	Transmitter output power $+10$ dBm
X_2	Transmitter termination loss $\underline{0.1}$ dB (splice)
X_3	Cable attenuation $\underline{0.3}$ dB/Km
X_4	Cable length $\underline{120}$ Km
X_5	Cable attenuation $\underline{36}$ dB (total)
.	.
.	.
.	.
X_{10}	Number of splices $\underline{25}$ N_{sp}
X_{11}	Splice loss $\underline{0.1}$ dB/each
X_{12}	Total splice loss $(X_{10} \times X_{11})$ $\underline{2.5}$ dB
X_{16}	Receiver termination loss $\underline{0.1}$ dB (splice)
X_{17}	Receiver minimum sensitivity $\underline{-42}$ dBm
X_{18}	Temperature variation allowance $\underline{2}$ dB
X_{19}	Format NRZ $\underline{-3}$ dBm
X_{20}	Aging loss $\underline{2}$ dB
X_{21}	Transmitter power, end of life $\underline{-1}$ dB
X_{22}	Required SNR $\underline{21.6}$ dB or BER $\underline{10}^{-9}$ dB
X_{23}	Source output halved, to increase life time $\underline{-3}$ dB
X_{24}	Miscellaneous loss $\underline{0}$ dB
X_{25}	Total losses (coupling insertion, etc.) $\underline{42.7}$ dB
X_{26}	Adjusted power output $\underline{+3}$ dBm
X_{27}	Total available power margin $\underline{45}$ dB
X_{28}	Power margin $\underline{2.3}$ dB

neighbor as a result of the nonconductive character of the optical cable being used.

Next, the bandwidth of the system must be considered. Most cable is available in single-mode form with bandwidths exceeding 1 GHz/km. Bandwidths of greater than 10 GHz/km are also available at an extra cost. Because of cable limitation, the bandwidth for 1300-nm cable will be about 8 to 80 MHz, depending on the cost sensitivity of the cable plant. Other limitations on bandwidth must now be considered.

A great deal of research is being conducted to increase the bandwidth of fiber-optic waveguides. At present, special waveguides still in laboratory stages have bandwidths of over 100 GHz. The vast improvement in performance expected within the next few years is a primary reason the performance of cable plants is difficult to deal with. The basic equation will continue to hold in most cases, but as purity of materials improves, the models will begin to break down when some of the terms become insignificant.

The rise-time budget given in Chap. 8 for LANs will be sufficient to use for the long-haul. If bandwidth of the cable becomes a limiting factor, spacing the repeaters more closely will alleviate the problem, as it will when attenuation-limiting occurs. For the latter situation, however, lossy items such as connectors, couplers, switches, etc., can sometimes be removed to eliminate the problem.

Another problem arising in the analog systems of commercial television is the SNR required to reproduce commercial TV at the receiver. The maximum SNR at the transmitter is approximately 70 dB, with a bandwidth of 6 MHz. The requirement at the receiver is 58 to 64 dB. To meet even the minimum require-ment, losses must clearly be less than 12 dB. For single-mode, long-wavelength cable—i.e., 1550 nm—this loss represents a distance of less than 24 kilometers with no couplers, switches, etc. in the transmission path. Some researchers are claiming increases in SNR to 120 dB. If it becomes possible to produce this SNR in commercial lasers, commercial TV will one day be transmitted via fiber-optic cable.

HIGH-SENSITIVITY RECEIVERS

The problem with most fiber-optic receivers is not optical noise but electrical noise. Generally speaking, thermal noise is the limiting noise component. The objective is to reduce all noise until only quantum noise (due to the optical components) limits the receiver preamplifiers. The SNR equation for receivers is repeated here as Eq. 10-25, with the noise terms divided into quantum noise, both optical and electrical:

$$\text{SNR} = \frac{i_s^2}{i_{\text{optic}}^2 + i_{\text{elect}}^2} \tag{10-25}$$

$$i_s^2 = \frac{1}{2}\left[\frac{\eta e G m P_i}{h\nu}\right]^2 \tag{10-26}$$

$$i_{\text{optic}}^2 = \frac{2e^2\eta}{h\nu} P_i G^2 F_d B_W + 2\left(\frac{e^2\eta}{h\nu}\right) P_g G^2 F_d B_W \tag{10-27}$$

$$+ 2\left(\frac{eG\eta P_i}{h\nu}\right)^2 \frac{B_W}{JW}\left(1 - \frac{B_W}{2W}\right)$$

$$i_{\text{elect}}^2 = \frac{4KTB_W F_t}{R_{eq}} + 2eI_d G^2 F_d B_W + 2eI_L B_W + i_{\text{device}}^2 \tag{10-28}$$

$$i_{\text{device}}^2 = \frac{3.75(\pi C)^2 B_W^3 KT}{g_m} \quad (\text{FET channel noise}) \tag{10-29}$$

or

$$i^2_{\text{device}} = \frac{8\pi KTC_T J_2 J_3}{t_d^2 \sqrt{\beta}} \text{ (Bipolar)} \tag{10-30}$$

Note that all quantum terms have photon energy embedded in them. Replacing all the $\eta e/h\nu$ terms with γ, the responsivity, is more useful because manufacturers of detectors give this value on specification sheets. Then the equations for signal and noise currents are given as Eqs. 10-31 and 10-32:

$$i^2_s = \frac{1}{2}\left[\gamma GmP_i\right]^2 \tag{10-31}$$

$$i^2_{\text{optic}} = 2\gamma eP_i G^2 F_d B_W + 2\gamma eP_g G^2 F_d B_W$$

$$+ 2G^2\gamma^2 P_i^2 \frac{B_W}{JW}\left(1 - \frac{B_W}{W}\right) \tag{10-32}$$

The electrical noise is unaffected by this change in variable. An intuitive approach to the increase in SNR is to increase the equivalent resistance, R_{eq}, which reduces the thermal noise, the largest electrical component. Larger R_{eq} results in larger RC time, which reduces BW and further reduces noise. The decrease in bandwidth is a nonlinear function of R_{eq}.

Let us now evaluate the SNR for a low-noise application and examine the terms to observe which of them are insignificant. Suppose that the following parameters are used to evaluate the equation:

γ = responsivity = 0.5 W/A
P_i = input power = 1 μW
e = charge on electron = 1.6 $\times$ 10^{-19} C
G = gain = 1 (assume PiN diode detector)
B_W = 6 MHz
F_d = noise figure of amplifier = 2
m = modulation = 0.5
R_{eq} = equivalent resistance = 1 MΩ
T = room temperature = 290°K
J = 1 (single-mode)
W = spectral width = 0.2 nm
I_d = dark current = I = 0.1 nA
g_m = gain term in FETs = 30 m mhos
i^2_s = 3.125 $\times$ 10^{-14} amps^2
i^2_{optic} = 1.912 $\times$ 10^{-21} amps^2
i^2_{elect} = 7.3 $\times$ 10^{-15} amps^2

For these values,

$$SNR = 6.23 \text{ dB}$$

The noise contribution due to thermal noise is the largest, and one must therefore try to reduce it when designing the preamplifier stage. If the electrical noise can be made equivalent to, or less than, the optical noise, the amplifier load would be at an optimum. The inequality reflecting this condition is shown by the following equation:

$$i^2_{\text{optic}} \geq i^2_{\text{elect}} \qquad (10\text{-}31)$$

Only the dominant noise terms are included below:

$$2\gamma e P_i G F_d B_W \geq \frac{4KTB_W F_t}{R_{eq}} + 2eI_d G^2 F_d B_W + \frac{3.75(\pi C)^2 B_W^3 KT}{g_m}$$

Solving for R_{eq} after setting both sides of the above equation equal, we have Eq. 10-32:

$$R_{eq} = \frac{4KTF_t}{2\gamma e P_i G F_d - 2eI_d G^2 F_d - \dfrac{3.75(\pi C)^2 B_W^2 KT}{g_m}} \qquad (10\text{-}32)$$

If the values of the original calculation are inserted into Eq. 10-32, the value of R_{eq} is approximately 7,000 MΩ. For an R_{eq} of this value, SNR is 42 dB, which implies that it is almost impossible to attain this condition. Increases in P_i, G, F_d, I_d, g_m, and B_W are all parameters under the control of the designer. Since each one will affect the value of R_{eq}, the designer has a great many tradeoffs to consider. We can also observe that increasing the power by a factor of 2 increases the optical noise component by 2 but increases the signal level by the square. When the electrical noise and optical noise get closer to magnitude, an increase will have less effect because of the electrical noise term.

Some of the parameters in these equations will be determined by the constraints of the application such as bandwidth. The resistor load across the detector will often be determined by the bias conditions of the amplifier; e.g., a FET preamplifier stage may have a low noise figure at a particular operating point. The operating point will determine, to some extent, the load resistance across the FET bias circuit.

Avalanche detectors usually do not operate as well as PIN FETS for low-noise applications, but some APD are beginning to become more competetive as low-noise detectors. Write-ups in trade journals indicate that many new PIN FETS have closely approached the quantum limit of detection, claiming that i_{elect}^2 terms have almost been eliminated. At the time this book was being written, none of the research material was available.

The difference between LANs and long-haul receiver designs is that the performance of the latter must be a great deal better. Repeaters are costly and require power supplies that can be complex if power has to be distributed along with the signal, as in undersea applications. The design of long-hauls is similar to that of LANs; in this case, however, the cost involved to squeeze maximum amount of sensitivity out of the receiver and power out of the transmitter is justified. Also, high reliability is a necessity because most of these plants must operate for 20 to 30 years with little or no maintenance. Therefore, cost will be secondary to reliability and performance.

Long-haul applications often require redundancy to increase reliability. In certain situations minimum performance can be achieved with a single unit, and the back-up may be used to enhance performance. As an example, two microprocessors may be used in a repeater to allow it to operate at peak performance, whereas a single processor is sufficient to provide minimum performance. This is a typical decision one must make when designating LANs with back-up computing power.

REVIEW PROBLEMS

1. Trace the frequencies through Fig. 10-2(a) for the following parameters: $F_c = 10$ MHz; $f_v = 10$ kHz; $f_m = f_x = \pm 2$ kHz.
2. Trace the frequencies through Fig. 10-2(b) assuming that the IF is 500 kHz.
3. Device a two-channel system with carrier frequencies of 10 and 12.5 MHz; $f_v = 100$ kHz; $f_m = f_s = \pm 10$ MHz; and IF = 500 kHz.
4. Assume that the carrier vector in Fig. 10-3(c) is 10 percent of $\sqrt{E_b}$ and that noise produces a circular plot at the end of the vector of 5 percent. Find the value of the information-carrying component.
5. The noise peak power and information-carrying component peak power are known; find the apparent SNR.
6. Assume values of 10, 20, 30, and 40 percent for the carrier magnitude in problem 10-4 and plot the apparent SNR in problem 10-5.
7. Assume that the tones of an FSK modulator are orthogonal. Determine output S_o in Fig. 10-4(d). Are there any restrictions?
8. Derive a symbol table similar to Tables 10-1 and 10-2, but include f_1 and f_2 for M-ary codes with MFSK.
9. Draw a block diagram of a M-ary 32-level modem using two 16-QAM tones.
10. Given a P_M of 10^{-5}, what the approximate is P_e for PSK when $M = 2$, $M = 4$, and $M = 32$?
11. In Fig. 10-8, replace the two microwave towers with satellite dishes. Assume that the satellite is 55,000 miles above the earth. Recalculate the delay budget.

12. Using a loss budget, determine the maximum transmission distance for commercial TV. Assume that the transmission system is equipped with a four-port star (two couplers), that it is single-mode and has connectors at each component in the system with 0.5-dB loss, that each splice in the system has a 0.1-dB loss, and that the laser has a bandwidth of 6 MHz and an SNR of 71 dB. What is the noise level at the receiver if the initial laser output is 1 mW?

13. How can the transmission distance be increased in problem 10-13?

REFERENCES

1. Bell Telephone Laboratories, Inc. *Transmission Systems for Communications*, 1982.
2. Schwartz, M. *Information Transmission, Modulation, and Noise*, pp. 209–227. New York: McGraw-Hill, 1980.
3. Hedin, G. L.; Holmes, J. K.; Lindsey, W. C.; and Woo, K. T. *Theory of False Lock in Costas Loops. IEEE Transactions on Communications* COM-26, No. 1, 1978.
4. ITT Corp. *Reference Data for Radio Engineers*. Indianapolis: Howard Sams & Co., 1982.
5. Freedman, Roger L. *Data Communication*. New York: Wylie, 1979.

11 | FIBER-OPTIC SENSORS

There has been a great deal of interest in fiber-optic sensors for some time. These devices can replace several of the less reliable mechanical ones, and they usually require a small (physically small) form factor. This chapter will present a discussion of the principles of operation of a number of sensors and several equations that govern their behavior in general. The devices to be examined are interferometers, rate sensors, accelerometers, strain gauges, pressure sensors, and mode conversion. Several of them use similar means to measure the phenomena they were designed for. For example, strain and temperature sensors may be designed to use interferometers. The scope of this chapter must be kept limited because it represents a vast field of study.

SAGNAC EFFECT

One of the principles used for rotational sensors is the *Sagnac Effect*.[1] For many years the primary instrument for measuring angular rotation was the gyroscope, which depended on a spinning ball or wheel to generate high angular momentum. The disturbances on the wheel or ball could then be measured electrically or mechanically to determine the angular movement. These mechanical components naturally required high-precision bearings and balancing. A warm-up period was needed to allow the moving parts to stabilize. One important drawback to mechanical gyros was their inability to withstand heavy gravitational loads (g loads). Fiber-optic gyros, on the other hand, have no moving parts; they can thus withstand 100-g forces without damage. Missiles, high-speed aircraft, and other pieces of equipment requiring rugged sensors are good candidates for fiber-optic sensors.

The Sagnac Effect is the optical path difference, ΔL, that is brought about by, and proportional to, an angular rotation rate, ω_s. Figure 11-1(a) shows two beams of light traveling around the periphery of a loop of mean radius R that is rotating about the loop axis at angular velocity ω_s. The light beams travel around the loop in opposite directions. Equation 11-1 represents the optical path difference as derived in References 2 and 3:

$$\Delta L = \left(\frac{4A}{c}\right) \omega_s \qquad (11\text{-}1)$$

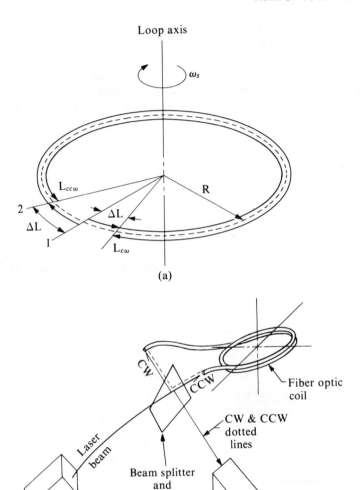

Fig. 11-1. Fiber-optic gyro: (a) Single-turn fiber-optic loop, and (b) Sagnal effect gyro.

where A is the area of the loop, c is the speed of light in a vacuum, and ω_s is the rotation rate about the loop axis, as shown in Fig. 11-1(a). Note that R is the *mean* radius of the loop.

The intuitive approach to Eq. 11-1 is to imagine two beams launched from the same point in the loop, i.e., from point 1. The beams are launched in clockwise and counterclockwise directions. If $\omega_s = 0$, then both beams traveling at the speed of light will arrive at point 1 at the same time, which is calculated as follows: $t = 2\pi R/c$.

If rotation occurs (i.e., $\omega_s \neq 0$), then the distances L_{ccw} (counterclockwise distance) and L_{cw} (clockwise distance) will be different. It is clear from Eqs. 11-2 and 11-3 that $L_{cw} > L_{ccw}$. Both are alternatively expressed by Eqs. 11-4 and 11-5:

$$L_{cw} = 2\pi R + R\omega_s t_{cw} \tag{11-2}$$

$$L_{ccw} = 2\pi R - R\omega_s t_{ccw} \tag{11-3}$$

$$L_{cw} = c_{cw} t_{cw} \tag{11-4}$$

$$L_{ccw} = c_{ccw} t_{ccw} \tag{11-5}$$

Then the time of arrival difference, Δt, between clockwise and counterclockwise beams can be expressed by Eq. 11-6, derived as follows:

$$\Delta t = t_{cw} - t_{ccw}$$

$$= \left(\frac{2\pi R + R\omega_s t_{cw}}{c_{cw}}\right) - \left(\frac{2\pi R - R\omega_s t_{ccw}}{c_{ccw}}\right)$$

$$\Delta t = \frac{2R\omega_s t}{c}$$

where $t = (t_{cw} + t_{ccw})/2$ and $c_{cw} = c_{ccw} = c$. Then,

$$\Delta t = \frac{4\pi R^2 \omega_s}{c^2} = \frac{4A\omega_s}{c^2} \tag{11-6}$$

Equation 11-1 can be derived directly from Eq. 11-6 because the path length is the difference in time, Δt, multiplied by the velocity, which is merely that of light in a vacuum. The results are shown in Eq. 11-7.

$$\Delta L = \Delta t \times c = \left(\frac{4A\omega_s}{c^2}\right)c = \frac{4A\omega_s}{c} \tag{11-7}$$

Next, the Sagnac Effect in a medium must be considered. The velocities, c_{cw} and c_{ccw}, were derived in References 1 and 2 as given in Eqs. 11-8 and 11-9:

$$c_{cw} = \frac{(c/n) + R\omega_s}{1 + (R\omega_s/nc^2)} \tag{11-8}$$

$$c_{ccw} = \frac{(c/n) - R\omega_s}{1 - (R\omega_s/nc^2)} \tag{11-9}$$

These equations can in turn be expanded into a series as shown by Eqs. 11-10 and 11-11:

$$c_{cw} = \frac{c}{n} + R\omega_s\left(1 - \frac{1}{n^2}\right) + \cdots \tag{11-10}$$

$$c_{ccw} = \frac{c}{n} - R\omega_s\left(1 - \frac{1}{n^2}\right) + \cdots \tag{11-11}$$

Now that C_{cw} and c_{ccw} are known, Δt can be calculated as was previously done for a vacuum, i.e., $\Delta t = t_{cw} - t_{ccw}$. The derivation of Δt for the nonvacuum case as a function of A, C_o, and ω_s, is as follows:

$$\Delta t = \frac{2\pi R + R\omega_s t_{cw}}{c_{cw}} - \frac{2\pi R - R\omega_s t_{ccw}}{c_{ccw}}$$

$$\Delta t = \frac{2\pi c_{ccw} - 2\pi R c_{cw} + R\omega_s t_{cw} c_{ccw} + R\omega_s t_{ccw} c_{cw}}{c_{cw} - c_{ccw}} \tag{11-12}$$

but

$$t_{cw} = \frac{2\pi R}{c_{cw} - R\omega_s} \quad \text{and} \quad t_{ccw} = \frac{2\pi R}{c_{ccw} + R\omega_s}$$

Substituting these values for t_{cw} and t_{ccw} into Eq. 11-12 and simplifying terms, we have Eq. 11-13:

$$\Delta t = 2\pi R\left[\frac{2R\omega_s - (c_{cw} - c_{ccw})}{c_{cw}c_{ccw}}\right] \tag{11-13}$$

Substituting the value of Eq. 11-14 into Eq. 11-13, we have, for $c_{cw} = c_{ccw} \approx c/n$, Eq. 11-15:

$$c_{cw} - c_{ccw} = 2R\omega_s\left(1 - \frac{1}{n^2}\right) \qquad (11\text{-}14)$$

$$\Delta t = \frac{4\pi R^2 \omega_s}{n^2 c_{cw} c_{ccw}} \approx \frac{4A\omega_s}{c^2} \qquad (11\text{-}15)$$

The result is identical to that for Δt in a vacuum. If A is a multiple of the area enclosed by N loops, then the result is as follows:

$$\Delta t = \frac{4NA\omega_s}{c^2} \qquad (11\text{-}16)$$

To find the nonreciprocal phase shift, $\Delta\theta$, the derivation is as follows:

$$\Delta\theta = 2\pi f_o \Delta t = \left(\frac{2\pi c}{\lambda}\right)\Delta t$$

Then,

$$\Delta\theta = \frac{8\pi NA\omega_s}{\lambda c} \qquad (11\text{-}17)$$

Another form of Eq. 11-17 showing more of the geometric parameters of the coils is indicated by Eq. 11-18, where $A = \pi D^2/4$ and $N = L/\pi D$:

$$\Delta\theta = \frac{2\pi LD\omega_s}{\lambda c} \qquad (11\text{-}18)$$

The Sagnac Effect gyro shown in Fig. 11-1(b) has a 50/50 beam-splitter and combiner mirror. The mirror is assumed lossless. As the laser beam impinges on the mirror, 50 percent of it is reflected into the ccw end of the fiber optic waveguide and the remainder passes through to the cw end. As the beams emerge from the two ends of the waveguide, they are combined and either constructively or destructively interfere. The actual result, of course, is dependent on the beam splitter used. If any delay occurs in the beam splitter, it will be reflected in the resulting beam that impinges on the detector. Any rotation will result in a phase shift, which can be detected by the photodiode detector.

One may immediately surmise that turns may be added to keep increasing the sensitivity, but the designer will soon discover that the input amplifier noise will limit detection sensitivity. The dominant noise is due to the electrical parameters and not the optical. As the optical attenuation is increased as a result of the increase in turns, the optical power impinging on the detector will soon become insufficient to overcome the noise of the electrical parameters of the amplifier. Only the basic principles for the various sensors can be discussed here.

The equation governing the intensity of the signal is as follows:

$$I = \frac{1}{2} I_o (1 - \cos \Delta\theta)$$

This equation may be expanded into a cosine series, as shown by Eq. 11-19:

$$I = \frac{1}{2} I_o \left(1 - \frac{\Delta\theta^2}{2} + \frac{\Delta\theta^4}{8} \cdots \right) \qquad (11\text{-}19)$$

Note that Eq. 11-19 has no linear terms (i.e., only $\Delta\theta^2/2$, $\Delta\theta^4/8$, etc.). If one of the signals, either λ_{cw} or λ_{ccw}, can be phase-shifted 90°, then Eq. 11-20, which is linear, is the result.

$$I = \frac{1}{2} I_o (1 - \sin \Delta\theta) \qquad (11\text{-}20)$$

$$I = \frac{1}{2} I_o (1 - \Delta\theta + \cdots) \approx \frac{1}{2} I_o (1 - \Delta\theta)$$

An example of a typical gyro calculation is given below. Assume that $r = 0.01$m, $\lambda = 800$ nm, and $\omega_s = 2$ deg/hr; then,

$$\Delta\theta = 2\left(\frac{I_1 - I_2}{I_o} \right) = 3.06 \ \mu\text{rads} \qquad (11\text{-}21)$$

where

$$I_1 = \frac{1}{2} I_o \text{ maximum radiation intensity}$$

$$I_2 = \frac{1}{2} I_o (1 - \Delta\theta)$$

The example shows the relationship between intensity and $\Delta\theta$, i.e., Eq. 11-21. This example does not appear to have any limit—i.e., it would seem that $\Delta\theta$ can be as small as possible; this, of course, cannot be true. The recombination rate of the photons at the detector limits the minimum detectable phase angle; this is expressed by Eq. 11-22:

$$\Delta\theta_m = \frac{I_1 - I_2}{I_o} = \frac{1}{\sqrt{\eta_{Ph}}} \qquad (11\text{-}22)$$

where $\Delta\theta_m$ and η_{Ph} are the minimum phase angle and photon recombination rate, respectively. A typical value is $\Delta\theta_m = 2.2$ μrads for $\eta_{Ph} = 2 \times 10^{11}$ photons/sec.

The Sagnac Effect sensors use a single fiber. Another type of sensor uses two fibers of equal length and physical characteristics. Known as an "interferometer," its operation is similar to that of the Sagnac sensor, but a signal is passed down each leg and all are summed at the output end, where they will add either constructively or destructively. One of the legs will be used as a reference, as shown in Fig. 11-2; the other will be the sensing leg. If the sensing leg is exposed to temperature variation, microbending, or other effects that change its transmission characteristics, the summed signals will vary. These devices are used as strain gauges and for temperature, pressure, fluid level, radiation, etc., measurement. Some of them will be discussed in the following paragraphs.

For the present, only a silica single-mode interferometer will be considered. The phased, $d\theta$, will be affected by changes in length, dL, and in index of refraction, dn. An equation derived in reference 4 relates these parameters as follows:

$$\frac{d\theta}{L} = Kn\left(\frac{dL}{L}\right) + Kdn \qquad (11\text{-}23)$$

If Eq. 11-23 is divided by a change in pressure, dP, Eq. 11-24 results:

$$\frac{d\theta}{LdP} = Kn\frac{dL}{LdP} + K\frac{dn}{dP} \qquad (11\text{-}24)$$

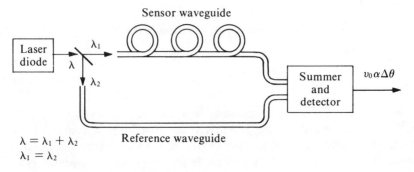

$$\lambda = \lambda_1 + \lambda_2$$
$$\lambda_1 = \lambda_2$$

Fig. 11-2. Interferometer sensor.

Then, substituting Eqs. 11-25 and 11-26, as follows:

$$\frac{dL}{LdP} = -\frac{1 - 2\gamma}{y} \tag{11-25}$$

$$\frac{dn}{dP} = \left(\frac{1 - 2\gamma}{y}\right)\left(\frac{n^3}{2}\right)(P_{11} + 2P_{12}) \tag{11-26}$$

into Eq. 11-24, we have Eq. 11-27, which reflects changes in phase per unit pressure for a specific length:

$$\frac{d\theta}{LdP} = Kn\left(\frac{1 - 2\gamma}{y}\right)\left[\left(\frac{n^2}{2}\right)(P_{11} + 2P_{12}) - 1\right] \tag{11-27}$$

The variables and constants are as follows: y (Young's modulus) = 6.5×10^{10} N/m^2; n (index of refraction) = 1.46; P_{11} and P_{12} (photoelastic coefficients for silica) = 0.12 and 0.27, respectively; ν (Poisson's ratio of silica) = 0.16; K (free-space propagation of light) = $2\pi/\lambda$; and wavelength = 633 nm. The evaluation of the right-hand side of Eq. 11-27 yields -4.5×10^{-5} rad/Pa–m. For an L of 1 km, the $d\theta/dP$ is thus -4.5×10^{-2} rad/Pa.

TEMPERATURE SENSITIVITY

If Eq. 11-23 is divided by dT, the result is an equation for temperature sensitivity, as shown below:

$$\frac{d\theta}{LdT} = Kn\left(\frac{dL}{LdT}\right) + K\left(\frac{dn}{dT}\right) \tag{11-28}$$

where, for silicon:

$dL/LdT = 5 \times 10^{-7}/°C$

$dn/dT = -10^{-5}/°C$

Then,

$$\frac{d\theta}{LdT} = \left[\frac{2\pi \times 10^9}{633}\right][1.46(5 \times 10^{-7} - 10^{-5})] = 107 \text{ rad}/°C - m$$

The value of this equation for pure silicon is theoretical, and measured values may deviate from the theoretical, depending on doping and construction, which may induce strains in the structure and thus affect the results. Temperature effects are quite significant and must be considered when designing sensors.

One current sensor uses a coating on the fiber that generates heat so that the current squared may be made proportional to temperature. Another technique is to coat the fiber with magnetostrictive substances, which will produce strain proportional to the current. Combinations of coatings and fiber design can obviously enhance various desired properties in a sensor.

Another principle that may be exploited is to use unclad fiber that changes its transmission properties when surrounded by a medium with variations in index of refraction. For example, if a coil of optical waveguide is surrounded by air ($n = 1.0$) and a fluid surrounds the waveguide such as water, ice, nitrogen (liquid, solid, gas), changes in transmission properties will occur. When these changes can be measured, the device can function as a sensor. For liquids, the sensor may be used as a level indicator, such as to measure levels in storage tanks. Gas measurement becomes possible if the refractive index changes sufficiently with changes in concentration or density. Water and ice have differences in refractive index, for example, and it may be possible to detect that difference.

Examples of three types of level sensors are shown in Figs. 11-3(a), (b), and (c). The first of these, in Fig. 11-3(a), shows a prism that has total internal reflections at the prism–air interface (i.e., no liquid in the container). When liquid is present in the container, the critical angle is no longer exceeded, and most of the light passes into the liquid. This particular sensor has the following shortcomings: If the index of the liquid and prism are not closely matched, there will be residual reflections in the prism. The matching may be difficult for some liquids and impossible for others. Failure of the liquid to wet the prism surface will cause internal reflections (i.e., air bubbles).

The difference between the wet and dry state may be in the 10- to 20-dB range. Connectors, couplers, and other devices that induce loss may cause the sensor to lose calibration periodically. If the liquid-level measurements are located in fixed-plant facilities, elastometric splices may be used to reduce transmission losses and increase measurement accuracy by removing connector loss uncertainty.

The second sensor, in Fig. 11-3(b), is very similar to the first, with the exception that the prism is replaced with a coil of unclad fiber having sufficient coils and bend radius to produce radiation from the coil. A wet coil will radiate more energy than a dry one; the difference in loss represents the liquid measurement. This device has an additional advantage: Its coil may extend down the length of the container. It also permits an analog sensor output whose accuracy is dependent on the smallest change in loss that can be measured.

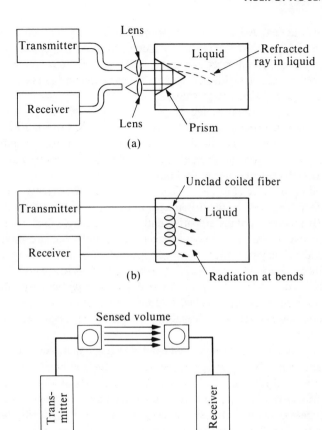

Fig. 11-3. Transducer elements: (a) Prism type liquid sensor, (b) coil liquid-level sensor, and (c) proposed level sensor.

Figure 11-3(c) depicts a sensor that depends on the absorption of liquid at one of two closely spaced wavelengths. Water, for example, has a high absorption at 980 nm due to the OH^-, and hydrocarbons have a corresponding absorption for C–H. Spherical lenses collimate the beam passing through the volume. The sensitivity of this device is dependent on differential absorption between the two wavelengths. A certain amount of noise is summed out of the system, and this results in a high SNR. This type of measurement will require a source for each different liquid, which may or may not be a problem. Care must be taken to select fiber-optic waveguides that do not have absorption bands in the wavelength of interest. A LED of broad spectral range may be used as a source (low-cost method), or two lasers may be, depending on the application. If a source of sufficiently broad spectral range is used, only the receiver filters need be changed to accommodate different liquids. These receiver filters can be either dichronic mirrors or diffraction gratings.

Dichronic mirrors are constructed so that one band of wavelengths will pass through the mirror and the other will be reflected. If the mirror has been made carefully, one detector may be used to capture the reflected wavelength, while the other detects light passing through. Only a change of mirrors may be necessary to accommodate sensing other liquids.

The grating technique can be used in a similar manner. Wavelength separation is accomplished by means of the differences in reflection caused by wavelength differences. If the wavelengths are sufficiently close, a single lens may be used to image the two beams on separate detectors. Liquids that are mixtures of various substances may be difficult to devise a sensor for. What's more, changes in liquid concentration of particularly absorbent material can cause erroneous measurements. For example, sewage effluent would be particularly difficult to sense because of its heavy metal ions. Cadmium used in plating is occasionally dumped into sewage systems with resulting changes in the absorption of one or both wavelengths.

A sensor for liquid-level measurement can be designed that uses a combination of optical-radar and fiber-optics principles to measure the position of a liquid surface from a fixed reference point above the surface. The distance measurement is essentially absolute, using optical radar that requires a phase comparison of an amplitude-modulated lightwave reflected from a target to the original phase of a launched beam. The detection system will simultaneously determine the position of the free surface and the height of a containment (boundary layer between two liquids). Optical waveguides are used for both input and signal beams to allow the multiplexing of several tank measurements. One large advantage of this system is that it requires no electrical connection to the tanks; the measurement technique is therefore inherently safe for use with volatile liquids such as gasoline or naptha.

Distance measurements can be made by measuring the transit time of a short pulse of light reflected from a remote target (optical radar). If the target is relatively close to the emitter, then this time measurement related to distance is very small. A target $\frac{1}{2}$ meter away from the emitter requires a 3.3-nsec resolution (in air) and a measurement accuracy of 6.6 psec to yield a distance resolution of 10^{-3} meters. These time differences and accuracy measurements are made possible by using methods developed for nuclear physics, but the equipment needed is quite costly at present.

An alternative approach to the measurement that can produce similar results is based on experiments by Barr.[5] This is a simple but accurate method of determining the velocity of light in transparent media by modulating a light sinusoidally and using a technique similar to the continuous wave (CW) ranging used in conventional microwave radar.

Figure 11-4(a) is a diagram of the measurement apparatus. The source can be LED, edge-emitting LED, or laser. A modulation frequency, ω_1, is impressed on the source and collimated by the lens to produce an unequally power divided beam at the beam splitter. Photodetector PD1 is the optical phase reference. The signal beam is detected at PD2. The photodetector currents are represented by Eqs. 11-29 and 11-30. If the modulating signal, f_{mod}, is

$$f_{mod} = A_D \sin \omega_1 t$$

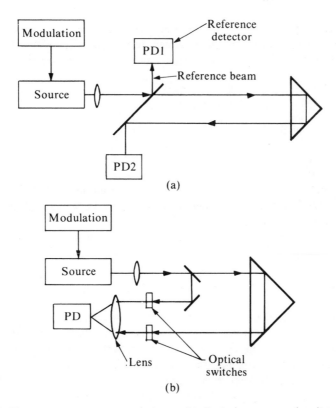

(a)

(b)

Fig. 11-4. Distance measurement techniques: (a) Optical apparatus for distance measurements, and (b) distance measurement using optical switches to implement a sampling technique of the signal and reference beams.

where A_D is the peak amplitude of the modulating signal, then

$$f_{ref} = A_{Ref} \sin \left(\omega_1 t + \phi_E + \phi_{f1} + \phi_{D1} + \phi_{fc} \right) \qquad (11\text{-}29)$$

$$f_{sig} = A_{sig} \sin \left(\omega_1 t + \phi_E + \phi_{f2} + \phi_{D2} + \phi_{fc} \right) \qquad (11\text{-}30)$$

where:

ϕ_E = delays in emission
ϕ_{fc} = delays in transit over the same optical path.
ϕ_{f1}, ϕ_{f2} = increases in phase due to the time of flight of the light between the beam splitter and PD1 and PD2, respectively.
ϕ_{D1}, ϕ_{D2} = increases in phase due to detection process and amplification in each channel

The phase terms, ϕ_E, ϕ_{D1}, and ϕ_{D2} are unpredictable. They vary with the amplitude of the incident light and ambient temperature. Equations 11-31 and 11-32 express the relationship between ϕ_{f1} and ϕ_{f2} and path length:

$$\phi_{f1} = \left(\frac{2\pi}{\lambda_e} \right) l_r \qquad (11\text{-}31)$$

$$\phi_{f2} = \left(\frac{2\pi}{\lambda_e} \right) (2l_f + k) \qquad (11\text{-}32)$$

where the path lengths l_r and l_f are the distances from the reference and liquid surfaces to the beam splitter, respectively, and λ_e is the wavelength of light in air (it is approximately equal to $2\pi c / \omega$).

A comparison of the two photodetector phase outputs will produce Eq. 11-33:

$$\Delta\phi = (\phi_{f2} - \phi_{f1}) + (\phi_{D2} - \phi_{D2}) \qquad (11\text{-}33)$$

The second term of Eq. 11-33 constitutes an irreducible error if the system shown in Fig. 11-4(a) is implemented as shown. These last two phase variables can be summed out differentially at the detector, however, if the scheme shown in Fig. 11-4(b) is implemented instead. The switches shown in Fig. 11-4(b) function as shutters. Reference and signal beams are sequentially sampled by the photodetector and summed differentially, thereby producing the result shown in Eq. 11-33. If the intensity is adjusted by an optical attenuator, however, the second term of Eq. 11-33 may be eliminated, thus producing a signal, $\Delta\phi$, that is proportional to the liquid surface changes. The accuracy of the measurement will depend on how well the system can resolve this phase difference. Note that errors can be caused by reflection resulting from connector mismatches in the path, variations in waveguide geometry, material variations, etc. The errors in the electronics and a small phase error will always be present. The errors shown were the larger, optically induced type.

The accuracy of the measurement will be determined on the basis of a 0.1-degree resolution in phase angle, which is attainable with commercial instrumentation. If the distance accuracy is 10^{-3} meters, then the wavelength of the light source is 7.2 meters (assuming that the increase in optical path l_f is twice the distance from the reference point to target surface). The modulation frequency of the laser calculated from the equation, $\lambda_e = (2\pi c)/\omega_1$, is 41.6 MHz. The unambiguous range of the level gauge is 3.6 meters. Since most storage tanks are typically 10 to 15 meters in depth, it will be necessary to use longer wavelengths (values for λ_e) to increase the depth resolution and provide unambiguous measurements. The optical depth-measuring system must have the source modulated at two frequencies for the necessary measurements to be performed with the required accuracy. The limiting condition to ensure unambiguous measurements is that the depth of measurement be less than $\lambda_e/2$ (i.e., a 15-m measurement requires a 7.5-m light source).

An experimental configuration for an optical measurement system is given in Reference 6. This particular reference discusses the details of design, such as light-gathering efficiency, ripple effects on measurements, and signal processing aspects of the system with several optical configurations. This reference is also useful for other sensors not discussed in this chapter and has many practical applications.

A transducer that can be used for pressure-sensing is the microbend sensor. It utilizes induced benching loss in optical waveguides. Light loss in the waveguide depends upon the induced coupling of propagating modes to radiation modes. When the waveguide distortion produces wave numbers equal to the difference between propagating and radiating modes, a large, easily measurable loss occurs.

A diagram of the microbend sensor and its measurement system are shown in Figs. 11-5(a) and 11-5(b), respectively. The two periodically ridged plates shown in the former will be optimally constructed, depending on the modal properties of the sensor waveguide (the sensors are typically constructed with periodicity in the millimeter range). Changes in pressure on the upper plate— i.e., displacement of the pressure plate—affect the amplitude of the bends, which results in amplitude modulation of the intensity. The modulation index, m_i, formulated in Eq. 11-34 is composed of two components, the first of which is a function of the optical properties of the sensor and the second, a function of the mechanical parameters of the sensor design.

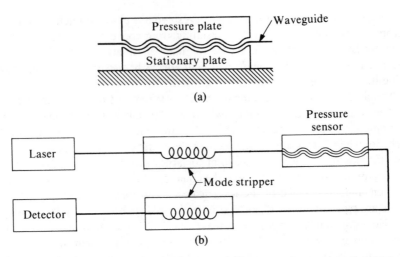

Fig. 11-5. Pressure sensor: (a) Transducer, and (b) pressure measurement system.

$$m_i = \left(\frac{dt_r}{dx}\right)_{op} \left(\frac{dx}{dP}\right)_{mech} \qquad (11\text{-}34)$$

where x, p, and t_r represent displacement, pressure, and transmission, respectively.

Consider the parameters that affect the optical modulation index. The propagation constants β_o and β_f are the initial and final propagation constants, respectively. The latter is the final constant as a result of microbending. The periodic distortion wavelength that produces this difference is depicted by Eq. 11-35, where Λ represents the mechanical wavelength of the distortion:

$$\Delta\beta = \beta_o - \beta_f = \pm\frac{2\pi}{\Lambda} \qquad (11\text{-}35)$$

Note that the result is rather intuitive: the more ripples in the sensor (smaller wavelength and higher frequency), the larger the difference between β_o and β_f. Moreover, $\Delta\beta$ is also a function of α (constant due to waveguide profile), M (the total number of modes), m (the mode label), and Δc (fraction difference between core and clad; assume multimode fiber-optic waveguide). Equation 11-36 depicts this relationship; it was developed by Gloge and Marcatillic using the WKB (Wentzel-Kramer-Brillouin) approximation[7]:

$$\Delta\beta = \beta_{m+1} - \beta_m = \left(\frac{\alpha}{\alpha+2}\right)^{1/2} \frac{2\sqrt{\Delta}}{a} \left(\frac{m}{M}\right)^{\alpha-2/\alpha+2} \qquad (11\text{-}36)$$

For a parabolic index waveguide, $\alpha = 2$ and Eq. 11-36 reduces to Eq. 11-37, where a is the waveguide core radius. Equation 11-37 indicates that $\Delta\beta$ is independent of m and that all modes are equally spaced. Thus, for parabolic-index waveguide, a critical distortion wavelength, Λ_c, exists for optimum coupling between adjacent modes, as shown in Eq. 11-38.

$$\Delta\beta = \frac{\sqrt{2\Delta}}{a} \tag{11-37}$$

$$\Lambda_c = \frac{2\pi a}{\sqrt{2\Delta}} \tag{11-38}$$

For step-index waveguide, $\alpha = \infty$ and Eq. 11-36 reduces to Eq. 11-39. Note that Eq. 11-39 cannot result in a simple Λ for critical coupling.

$$\Delta\beta = \beta_{m+1} - \beta_m = \frac{2\sqrt{\Delta}}{a}\left(\frac{m}{M}\right) \tag{11-39}$$

Spacing between the higher order modes is farther apart than for lower order modes, and small mechanical Λ wavelengths are required to couple higher modes.

A plot of dt_r/dx versus Λ for both stepped- and graded index fiber-optic waveguides is shown in Fig. 11-6. The stepped-index fiber is not readily avail-

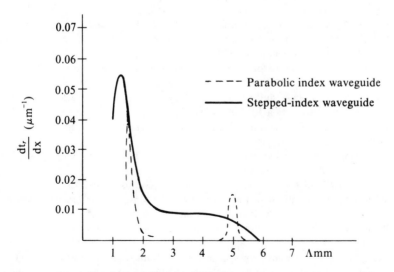

Fig. 11-6. Plot of Λ versus dt_r/dt for parabolic and stepped-index waveguides.

able because the present trend is toward graded-index waveguides owing to their superior bandwidth.

The dx/dP values are a function of the mechanical design of the device, i.e., the displacement of the plates of the sensor per unit pressure. See review problem 13 for an example of determining psi.

The minimum detectable pressure is expressed by Eq. 11-40. Where the threshold pressure is dark-current-limited (I_d).

$$P_{min} = \frac{1}{m_i} \frac{(2eI_dBW)^{1/2}}{qI_o} \qquad (11\text{-}40)$$

See problem 14 in the review problems for an example of these calculations.

The reader is urged to seek more in-depth study should he or she be involved in sensor design. The principles of operation don't change, but implementation of the various applications do. For example, GaAs digital circuitry will become commonplace by the 1990s by making instrumentation for high-accuracy phase measurements cost effective enough to be readily available. Also, the push for integrated optics may allow low-cost optical measurement techniques to become abundant. Therefore, the reader who wishes to become technically astute in this area should always search the journals for new and innovative sensor designs.

REVIEW PROBLEMS

1. Show the series expansions for Eqs. 11-8 and 11-9.
2. Show the derivation for the exact expression for Eq. 11-15.
3. Derive an expression for ΔL as a function of L, D, W_s, λ, and C.
4. Calculate Δt, ΔL, and $\Delta \phi$ for a single loop gyro for $A = 100$ cm^2, $\omega_s = 0.015°/$hr (7×10^{-8} rad/sec), and $\lambda = 1300$ nm.
5. Calculate Δt, ΔL, and $\Delta \theta$ for a multiple-turn coil with $A = 100$ cm, $N = 2000$ (i.e., $D = 11.3$ and $L \approx 710$ cm), $\omega_s = 1$ rad/sec, and $\lambda = 1300$ nm.
6. Calculate the $\Delta \theta$ phase shift in problem 5 for $\omega_s = 2\pi$ rad$/24$ hr (earth's rotation); navigational gyros have a 10^{-7}-m radius of $\Delta \theta_o$.
7. Find the phase difference for a Sagnac Effect gyro for $\omega_s = 1°/$hr, $\lambda = 1$ μm, $r = 0.01$ m, and $N = 10$ turns.
8. Is it within the detectable limit for η_{Ph} to be 2×10^{11} photon/sec? If not, what alternative is there to bring it within limits?
9. Calculate $d\theta/LdP$ for $\lambda_1 = 1.55$ μm, $\lambda_2 = 1.3$ μm, and $n = 1.46$.
10. What is the sensitivity of an interferometer if the sensor and reference legs have 1 kilometer of fiber, the source (laser) operates at 1.3 μm, and $n = 1.493$.
11. What is the temperature sensitivity of an interferometer for $\lambda = 1.55$ μm, 1.3 μm and 0.8 μm ($n = 1.493$)?
12. Design a surface measurement system using a 3-μm source to measure liquid film on a surface. What is the modulating frequency for a 0.1-μm measurement accuracy and angular resolutions of 0.01 and 0.1 degrees?

13. Find the optimum M_i for a graded-index waveguide if the pressure per unit displacement is 1 Pa per 1 μm.

14. Use the result of the M_i calculation to find the detection threshold for the sensor for $I_o = 1$ mW, $q = 0.4$ A/W, $I_d = 0.1$ nA, BW $= 1$ KHz, $e =$ electronic charge, and $I_o = 1$ mW (output optical power).

REFERENCES

1. Sagnac G. *Compt. Rend.* **157**:708 (1913).
2. Post, E. J. *Rev. Modern Physics* **39**:475 (1967).
3. Arditty, H. J., and LeFevre, H. C. *Optics Letters* **6**:401 (1981).
4. Hacker, G. B. *Applied Optics* **18(9)**:1445–1448 (1979).
5. Barr, R. An inexpensive apparatus for the measurement of light in transparent media. *Journal of Physics Scientific Instruments* **5**:1124 (1972).
6. First International Conference on Optical Fibre Sensors, 26–28 April 1983. Published by Multiplex Techniques Ltd., St. Mary Cray, Kent, U.K. (1983).
7. Gloge, D., and Marcatillic, E. A. J. Multimode theory of graded-core fibers. *Bell System Technical Journal* **52**:1563–1578 (1973).

12 | ANGULAR DIVISION MULTIPLEXER (ADM)

The object of the Angular Division Multiplexer (ADM) is to provide several channels for data transmission over a single-stepped index multimode fiber. The acceptance angle of the fiber is divided into small angular segments, each of which represents a channel. At the receive end is a series of annular cones, each also representing a channel. ADM has another inherent quality that results from the differential excitation required by each channel. Since the modal groups minimize intermode dispersion, each channel has fewer modes than it would in a normal stepped-index transmission. The resultant bandwidth is much improved and similar to graded-index transmission.

This transmission technique has a distance limitation of approximately 1 kilometer. Mode coupling caused by core/cladding interface roughness, a nonuniform refractive index, and microbending are responsible for this restriction. Lasers or edge-emitter diodes are a desirable choice for use as transmitter sources because fewer modes are supported and therefore less mode coupling can occur.

The system distance can be increased if WDM or FDM are combined with ADM. The transmission distance can be increased at the expense of greater transmitter and receiver complexity. Once a system has been designed for ADM, however, wavelength division multiplexing may be added later, usually by replacing each source with a different wavelength. Expansion in transmission distances at a later date may thus become a rather simple task if both ADM and WDM are used.

In this chapter, a few simple LAN cable plants using ADM techniques will be investigated. The transmitters and receivers of these LANs are complex enough to warrant this examination. They are much more complex than the simple monomode or multimode types.

MODAL ANALYSIS

Guided modes in a step-index waveguide have azimuthal quantum number k and radial quantum number q related by the equation,

$$k + 2q = m \qquad (12-1)$$

where m is a compound group number shared by a group of degenerate guided modes. These modes are plane waves that propagate down the waveguide axis at an angle, θ_q. Maximum values for q and m are as follows:

$$Q \cong \frac{\pi^2 q^2}{8} \qquad M = \frac{2V}{\pi} \qquad V = \left(\frac{2\pi a}{\lambda}\right) NA$$

$$Q \cong \frac{V^2}{2} = \left(\frac{2\pi A_c}{\lambda^2}\right) (NA)^2 \qquad (12\text{-}2)$$

$$M = \left(\frac{4a}{\lambda}\right) NA$$

where $A_c = \pi a^2$

The propagation angle of all modes with quantum numbers (k, q) can be calculated by rearranging Eq. 12-2 as follows:

$$\frac{M\lambda}{4a} = NA$$

Since

$$NA = n_1 \sin \theta$$

then,

$$\frac{M\lambda}{4an_1} = \sin \theta$$

and, for small angles,

$$\theta_m = \frac{M\lambda}{4an_1} \qquad (12\text{-}3)$$

$$N_{ch} = \frac{\theta_c}{\Delta\theta}$$

where N_{ch} = the channel capacity and $\Delta\theta = \theta_{m+K} = \theta_m = \lambda/4a$, as developed in Ref. 1.

Thus,

$$N_{ch} = \frac{4\theta_c a}{\lambda} \qquad (12\text{-}4)$$

The result of Eq. 12-3 can be used to calculate the maximum number of modes the waveguide can propagate at the same angle (see Eq. 12-4).

For a typical application, a stepped-index waveguide with the following characteristics is selected.

Core diameter $= 100 \ \mu m \ (a = 50 \ \mu m)$
Wavelength of operation $= 800 \ nm = 0.8 \ \mu m$
Numerical aperture $= 0.2 \ rad$

Calculations:

$$N_{ch} = \frac{4 \times 0.2 \ rad \times 50 \ \mu m}{0.8 \ \mu m}$$
$$= 50 \ ADM \ channels$$

A number of trade-offs depend on the application; for example, if bandwidth is not a concern, larger core waveguides may be used. Thus, the value of core radius can be doubled; θ_c can be larger as well; and the channel capacity, N_{ch}, will increase by a factor of 3.

Since each channel has fewer transmitted modes, than it would in a normal multimode application, the bandwidth is also increased.

The following equations will develop a relationship that shows how rise time is related to N_m, the number of modes supported by a normal stepped$=$index waveguide application with or without ADM:

$$\frac{\Delta \tau}{l} = \frac{n_1 \Delta}{c}$$

where $\Delta = (n_1 - n_2)/n_1$.

$$N_m = \frac{1}{2} V^2$$

where $V = (2\pi a / \lambda) NA$. Therefore,

$$N_m = \left(\frac{2\pi^2 a^2}{\lambda^2} \right) (n_1^2 - n_2^2)$$

and

$$N_m \cong \left(\frac{2\pi^2 a^2}{\lambda^2} \right) \Delta 2 n_1^2$$

Solving for Δ,

$$\Delta = \frac{N_m \lambda^2}{4a^2 \pi^2 n_1^2}$$

The relationship between rise time and N_m is then

$$\frac{\Delta\tau}{l} = \frac{N_m \lambda^2}{4a^2 \pi^2 nc} \qquad (12\text{--}4)$$

The percentage of improvement in $\Delta\tau$ is $[1 - m/N_m] \times 100$. The rise time is related to bandwidth, as discussed previously in relation to cable plant design in Chap. 2.

A series of curves that show the relationship between various ADM parameters when trade-offs exist is depicted in Fig. 12-1. As shown in the diagram, large-core diameters will allow a larger number of channels if bandwidth is

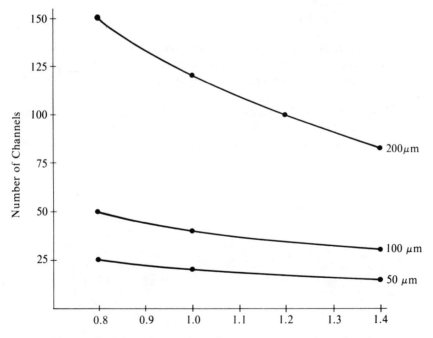

Fig. 12-1. Number of ADM channels versus operational wavelength.

initially large for the fiber. In previous discussions, long wavelengths were a desirable feature because losses were lower. The curves in this figure, however, indicate a square-law type of decrease in the channel capacity.

Figure 12-2 illustrates ADM transmission over a fiber-optic link. Angles θ_c, θ_1, and θ_2 are called the critical angle and channel one and channel two angles, respectively. Note that the beam enters at the channel angle and that an annular cone is produced at the output. The annular cone results from a uniform scattering of light at azimuthal angles in the waveguide. The far field—at the end of the waveguide—produces a cone with an inner opening of twice the entry angle. The exit angles at the receiver end of the waveguide are θ_c, θ_1, and θ_2.

As transmission of optical signals progress down the waveguide, mode coupling occurs as a result of geometric variation in the core, core/cladding interface smoothness, refractive index uniformity, and effects of microbending.

The length dependence of mode coupling can be described by the following analysis: Mode coupling causes a change in the power of mode m with length Z. The coupling coefficient between mode groups m and n is δ_{mn}. The equation describing the change in optical power per unit length is as follows:

$$\frac{dP_m}{dZ} = -\alpha_{mn} P + \sum_{n=1}^{M} \delta_{mn} (P_n - P_m) \qquad (12\text{-}5)$$

The first term in the equation is due to radiation, and the second describes the coupling between mode group m and other mode groups. Equation 12-5 does not address crosstalk, however, and this is the parameter that affects performance.

The strongest crosstalk component will come from adjacent channel signals. Adjacent channel crosstalk has been measured at -32 dB for a 12-channel ADM link. Equation 12-6 is a differential equation relating power coupled between modes.

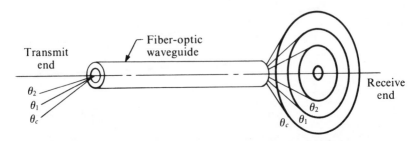

Fig. 12-2. Diagram of waveguide using ADM.

$$\frac{\partial P(\theta, z)}{\partial z} = \alpha(\theta) P(\theta, z) + \frac{1}{\theta} \frac{\partial}{\partial \theta} \left[\theta f_d(\theta) \frac{\partial P(\theta, z)}{\partial \theta} \right] \qquad (12\text{-}6)$$

The term on the left-side of the equal sign is the power coupled per unit axial length; the coupled power is a function of Z and channel angle θ. On the right side of the equal sign, $\alpha(\theta)$ is the loss as a function of channel angle, and $P(\theta, z)$ is the power coupled between modes as a function of z down the waveguide. The term in brackets is the channel angle, with $f_d(\theta)$ the coupling coefficient.

CROSSTALK

As the signal propagates down the waveguide, angular broadening occurs, thus increasing crosstalk. Angular broadening also increases the number of propagating modes, which results not only in crosstalk but intermodal dispersion; as a result, bandwidth also decreases.

Crosstalk analysis is dependent on finding solutions to Eq. 12-6 for various boundary values. If the variables are separable, then the solution is fairly straightforward, as shown in Eq. 12-7:

$$P(\theta, z) = F(\theta) G(Z) \qquad (12\text{-}7)$$

For the nonseparable case, however, the solutions become more difficult to determine.

The crosstalk may be calculated as a function of axial length for a solution of Eq. 12-6 in the form shown in Eq. 12-10. Let the total power distributed at the output of channel i be $E(\theta, z) = P_i(\theta, Z)$.

$$\theta \frac{\partial P_i(\theta, z)}{\partial z} = \alpha(\theta) \theta P(\theta, z)$$
$$+ \frac{1}{\theta} \frac{\partial}{\partial \theta} \left\{ \theta f_\alpha(\theta) \left[P(\theta, z) + \theta \frac{\partial(\theta, z)}{\partial \theta} \right] \right\} \qquad (12\text{-}8)$$

Divide Eq. 12-8 by θ and transpose the first term on the right-hand side of the equals sign; the result will be Eq. 12-9:

$$\frac{\partial P_i(\theta, z)}{\partial z} - \alpha(\theta) P(\theta, z)$$
$$= \frac{1}{\theta^2} \frac{\partial}{\partial \theta} \left\{ \theta f_d(\theta) \left[P(\theta_0 z) + \theta \frac{\partial P(\theta, z)}{\partial \theta} \right] \right\} \qquad (12\text{-}9)$$

The right-hand side can be replaced by its equivalent from Eq. 12-6 and the expression manipulated to produce Eq. 12-10 (the limits of integration are the inner and outer angle of the annular channel at which $E(\theta, z)$ is a maximum):

$$E_i(z) = \int_{\theta_{im}}^{\theta_{cm}} E_i(\theta, z) \, d\theta \qquad (12\text{–}10)$$

Crosstalk due to any channel j in channel i is calculated using Eq. 12-11. The total crosstalk can be written as the sum of the contributions from all other channels, as given by Eq. 12-12.

$$CT = \text{crosstalk} = 10 \log E_j/E_i \qquad (12\text{–}11)$$

$$CT_{\text{total}} = \sum_{j \neq i}^{N} 10 \log E^j/E_i \qquad (12\text{–}12)$$

ADM BANDWIDTH CONSIDERATIONS

One of the most important parameters to consider, because of its limiting effect on link performance, is bandwidth. Experimentation has indicated that ADM transmission will extend bandwidth by a factor of three or four.[1] This increase in bandwidth is dominant for the core axial channel. For the gaussian impulse response, the pulse width of the output waveform, τ_e, is defined as the time between the rise time and the fall time ($1/e$ points), as follows:

$$\tau_e = 2\sqrt{2}\,\sigma = 1.2\tau \qquad (12\text{–}13)$$

where σ, the RMS Gaussian response, is one-half the height value of the pulse width.

At $1/e$ points,

$$f_{3\text{dB}} = \frac{0.375}{\tau_e} \text{ Hz} \qquad (12\text{–}14)$$

At half height,

$$F_{3\text{dB}} = \frac{0.3125}{\tau} \text{ Hz} \qquad (12\text{–}15)$$

Equations 12-14 and 12-15 represent the 3-dB electrical bandwidth as a function of Gaussian output response.

A simplified version of an ADM link is shown in Fig. 12-3. At the transmit

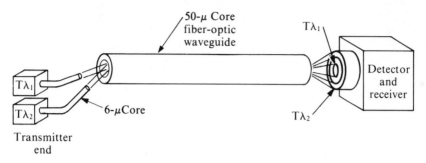

Fig. 12-3. ADM implementation.

end, each is shown operating at a different wavelength. The summation of the two wavelengths is easily accomplished on the surface of the 50-μm core fiber-optic waveguide. At the receive end, note that the rings allow easy separation of the two wavelengths without the use of optical filters. If the two wavelengths were not angular division multiplexed, a single optical beam would be detected at the receiver.

Another technique to separate the ADM received signals—instead of using different wavelengths (WDM)—is to modulate with each transmitter at a different frequency but with the same wavelength. As one can readily observe, there are a number of different methods of using ADM, WDM, and FDM. Table 12-1 describes various techniques with comments.

LOSS MECHANISMS

If transmitters are angular-multiplexed, the loss caused by radial misalignment can be calculated using Eq. 12-16:

$$\text{Loss (radial misalignment)} = 10 \log \left[\frac{1 - n\theta_n}{\pi n_1 (2\Delta)^{1/2}} \right] \quad (12\text{-}16)$$

where:

n = refractive index of the medium filling the gap between the source and waveguide

n_1 = refractive index of the core

Δ = refractive index difference between core and clad

θ_n = the channel angle, in radians

TABLE 12-1   ADM Configurations.

ADM	FDM	WDM	Technique	Comments
$\theta_1, \theta_2 \ldots \theta_n$	f_1	λ_1	ADM only	Transmitters all identical with angular spacing critical; detection requires masking unwanted rings to prevent large crosstalk components at the detector
$\theta_1, \theta_2 \ldots \theta_n$	$f_1,$ $f_2 \ldots f_n$	λ_1	ADM with each channel at a different frequency	Rings do not need to be masked; the channels are separated with tuning at f_1, $f_2 \ldots f_n$; technique lacks any large advantages over standard FDM.
$\theta_1, \theta_2 \ldots \theta_n$	f_1	$\lambda_1, \lambda_2 \ldots \lambda_n$	ADM with each channel at a different θ.	Wavelengths separated with masking; no optical filters needed
$\theta_1, \theta_2 \ldots \theta_n$	$f_1, f_2 \ldots f_n$	$\lambda_1, \lambda_2 \ldots \lambda_n$	ADM, FDM, and WDM	Not a very practical technique because of the complexity of the receiver

The loss for lateral separation is given by Eq. 12-17:

$$\text{Loss (lateral misalignment)} = 10 \log \left[\left(1 - \frac{d}{4a_1} \right) \left(\frac{n_1}{n} \right) \left(2\Delta \right)^{1/2} \right]$$

where d is the separation between source and the end of the waveguide.

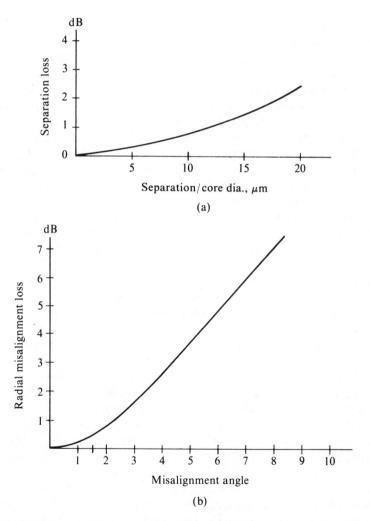

Fig. 12-4. Loss mechanisms for cable: (a) Separation loss versus separation/core diameter, and (b) radial misalignment loss versus misalignment angle.

A set of curves showing these equations graphically is provided in Figs. 12-4(a) and 12-4(b). These plots show loss versus separation (normalized d/a_1) and loss versus misalignment, respectively. As ADM channel angles increase, radial loss will also increase; therefore, when WDM is used with ADM, longer wavelength transmitters should be implemented at the larger channel angles. A more comprehensive treatment of misalignment is given in Chap. 2.

NOISE

In Chaps. 3 and 4, noise components were discussed for single-mode systems, but noise is somewhat different for ADM applications because multimode waveguides are necessary. Crosstalk will be the dominant term and affect both maximum transmission distances and receiver sensitivity.

Modal noise will be more of a problem in ADM systems because of the use of lasers in conjunction with multimode stepped-index waveguide. Previously, noise of a statistical nature at detectors was assumed to be shot noise. When lasers are used, the familiar speckle pattern of near-field radiation can be observed. In the far field (long distances from the laser source), dispersion causes propagation delays between modes to exceed the coherence time of light in each mode. The speckle pattern, which is due to constructive and destructive interference at a given cross section of the waveguide near the source (near field), will merge into a uniform background radiation. This noise component will appear at the detector as random noise that is similar to quantum noise. Modal noise can be minimized if certain design precautions are taken to reduce its effect.

The speckle pattern is not in itself modal noise, but when disturbances in the pattern are mode selective and this mode selectivity is temporary, a random change in optical power will occur at the detector, and it is this that constitutes modal noise.

Mechanical disturbances such as connector misalignment, star couplers, poor splices, switches, etc., can cause modal noise. Poorly aligned connectors are one of the most common sources of this noise component, especially those close to the transmitter. Star couplers are very mode-sensitive because coupling of higher order modes to the output ports occurs. The star is therefore a very mode-selective device. Modal noise is not a problem with LEDs, but for ADM, lasers are the most desirable. If large bandwidths are unnecessary, high-dispersion waveguide will reduce the modal noise. Another technique is to use lasers that have larger spectral widths because these devices also result in higher dispersion.

Some background will be presented here to reveal how detector response is related to other time-related parameters. There is a time period in which the light remains in phase, or coherent τ_c. Coherent phase time, τ_c, is long compared to the optical period. The relationship can be expressed by the inequality in Eq. 12-18:

$$\tau_{\text{optical}} \ll \tau_c \ll \tau_d \tag{12-18}$$

where the line width of the source, τ_c, is

$$\tau_c \simeq \frac{1}{2\pi\Delta f}$$

For the optical case,

$$\gamma = \frac{\Delta f}{f_0} = \frac{\Delta\lambda}{\lambda_0}$$

Then,

$$\tau_c = \frac{\lambda_0}{2\pi\left(\dfrac{\Delta\lambda}{\lambda_0}\right)c} = \frac{\lambda_0}{2\pi\gamma c} \tag{12-19}$$

In Eq. 12-19, λ_0 is the nominal optical wavelength and $\Delta\lambda$ is the spectral width about the wavelength (half-power points). From Eq. 12-19, one would assume the narrower the spectral width, the better, but, as calculated in Chap. 2, dispersion also decreases, and this will further increase modal noise. Thus, there is a practical limitation to increasing or decreasing spectral width that must be determined by the application.

ADM TRANSMITTERS

Design of ADM transmitters is fairly straightforward with two exceptions. First, ADM transmitters must maintain the channel angle mechanically; this is different from previous situations where it was necessary to minimize misalignment (radial). The second difference is that the desirability of narrow spectral width lasers to reduce dispersion will not always be the case for ADM applications. Figure 12-5 is a rather simplified illustration of an ADM transmitter. In this figure, transmitters T_1 through T_4 are the electronic drivers and sources that produce an optical version of the electronic input signal. The sources are connected to an optical summer through single-mode optical pigtails. These pigtails are fixed within the optical summer at their channel angles. Although the optical

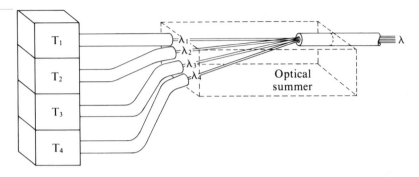

Fig. 12-5. Four-channel transmitter.

summer block in the diagram is rather large for purely illustrative purposes, in actuality it will be very small—not more than a cubic centimeter in volume.

The transmitters can be selected from commercially available devices or designed with discrete components; they may be analog or digital. Their design has been discussed in Chap. 3. In this section, the emphasis will be on variation in design that results from the ADM technique.

One most obvious problem is keeping the separation between the end of the transmitter pigtail and the multimode waveguide small so as to prevent large losses in optical signals. For the situation where large numbers of channels are to be implemented, the loss may become great; thus, the limit on the number of channels may be mechanical rather than optical.

A diagram of optical summer detail is shown in Figs. 12-6(a) and (b). In Fig. 12-6(a), the light beam will increase in width. The relationship between beam angles and channel angle is shown in the following inequality:

$$\theta_1 < \theta_n < \theta_2 \tag{12-20}$$

If $\theta_n \approx \theta_c$, then $\theta_c < \theta_2$. Unless $\theta_2 = \theta_c$, part of the beam will be lost. Another technique is to collimate the beam with a lens that will make the beam angles identical. This will result in an improvement in coupling and a slight reduction in crosstalk. The dispersion in the beam of Fig. 12-6(a) is not to scale. For beams coming out of the single-mode fiber, $I(\theta) = I_o (\cos \theta)^n$, where $n = 10$ to 20 for far-field radiation. Note that some epoxy must be used to affix the pigtail to the multimode-fiber core. In both figures, the refractive index of the epoxy, n_{ex} is 1.3, which is a rather common value.

To analyze the power transfer in Fig. 12-6(a), we will first consider the brightness equation for a laser, as follows:

$$B(\theta) = B_o(\cos \theta)^n \tag{12-21}$$

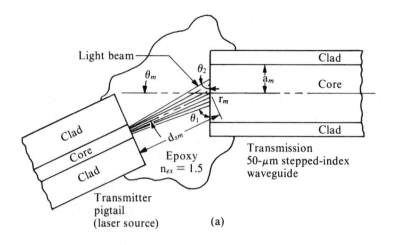

(a)

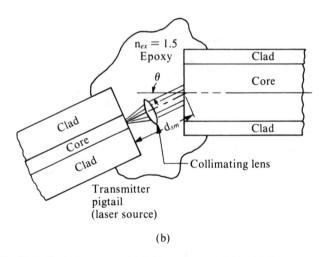

(b)

Fig. 12-6. Optical summer: (a) Without lens, and (b) with lens.

The assumption here is that the pigtails in Fig. 12-6(a) are short and that this is the near-field representation. The units for brightness are watts per cm^2-steradian (sr). An assumption is also made that the radiation is uniform across the source.

Equation 12-22 is an expression for the power transfer from A_s of a single-mode pigtail to the multimode waveguide, where dA_s is the differential pigtail area and $d\Omega_m$ is the differential solid acceptance angle of the multimode waveguide:

$$dP_{sm} = B(\theta_s) \, dA_s \, d\Omega_m \qquad (12\text{-}22)$$

$$d\Omega_m = \frac{\cos\theta_m \, dA_m}{d_{sm}^2} \qquad (12\text{-}23)$$

The result of substituting Eq. 12-23 into Eq. 12-22 and integrating is shown in Eq. 12-24:

$$P_{sm} = \int_{A_s} \int_{A_m} \frac{B(\theta)\cos\theta_m \, dA_m \, dA_s}{r_{sm}^2} \qquad (12\text{-}24)$$

This equation can be simplified somewhat. If the source is considered cylindrically symmetric—which happens to be the case—then Eq. 12-24 reduces to Eq. 12-25, as follows:

$$P_{sm} = \pi a_s^2 \int_{A_m} \frac{B(\theta_m)\cos\theta_m \, dA_m}{r_{sm}^2} \qquad (12\text{-}25)$$

where channel angle $\theta_m \leq \theta_c$. To solve this equation is rather a vigorous task. A reasonable value for $B(\theta_B)$ is

$$B(\theta_B) = B_0(\cos\theta_B)^{20}$$

The 3-dB beam width and the beam radius at the multimode waveguide can be calculated as shown below:

$$\frac{B}{B_0} = \left(\frac{1}{\sqrt{2}}\right)^{1/n} = \cos\theta_B$$

$$\frac{r_m}{r_{sm}} = \tan\theta_B$$

$$r_m = r_{sm}\tan\theta_B$$

where r_m is the minor axis of an ellipse projected on the surface of the multimode waveguide. Then $r_{maj} = r_m/\cos\theta_m$ and

$$A_m = r_m r_{maj}\pi = \frac{r_m^2\pi}{\cos\theta_m}$$

$$dA_m = \left(\frac{2\,r_m\pi}{\cos\theta_m}\right)dr_m + r_m^2\pi\left(\frac{\sin\theta_m}{\cos^2\theta_m}\right)d\theta_m$$

Substituting this value of dA_m into Eq. 12-25, the result is Eq. 12-26, as follows:

$$P_{sm} = a_s^2\pi \int 2B(\theta_B)\, \pi\, \frac{r_m}{r_{sm}^2}\, dr_m + a_s^2\, \pi^2 \int B(\theta_B)\, \tan^2\theta_B \tan\theta_m d\theta_m$$

$$(12\text{-}26)$$

Then, substituting the following values,

$$B(\theta_B) = B_o \cos^n \theta_B$$

$$dr_m = r_{sm} \sec^2\theta_B d\theta_B$$

into Eq. 12-26, we have

$$P_{sm} = 2a_s^2\pi^2\left[\int B_o(\cos\theta_B)^{n-3} \sin\theta_B d\theta_B\right.$$

$$+ \left.\int B_o(\cos\theta_B)^{n-2} \sin^2\theta_B \tan\theta_m\, d\theta_m\right]$$

$$= 2a_s^2\pi^2\left[\frac{B_o(\cos\theta_B)^{n-2}}{n-2} + B_o(\cos\theta_B)^{n-2}\sin^2\theta_B \ln\cos\theta_m\right]$$

$$= \frac{2a_s^2\pi^2 B_0(\cos\theta_B)^{n-2}}{n-2}\left[1 - (n-2)\sin^2\theta_B \ln\cos\theta_m\right]$$

where:

$$\ln\cos\theta_m = -\frac{\cos^2\theta_m}{2} - \frac{\cos^4\theta_m}{12} - \frac{\cos^6\theta_m}{45} - \cdots$$

Then,

$$P_{sm} \cong \frac{2a_s^2\pi^2 B_o(\cos\theta_B)^{n-2}}{2}$$

$$\cdot \left[\sin^2\theta_B\left(\cos^2\theta_m + \frac{\cos^4\theta_m}{6}\right) + 2\right]$$

or

$$P_{sm} \cong \frac{2a_s^2\pi^2 B_o (\cos \theta_B)^{n-2}}{2}$$

$$\cdot \left\{ \frac{1}{4}[\sin (\theta_m + \theta_B) + \sin (\theta_m - \theta_B)]^2 + 2 \right\} \qquad (12\text{-}27)$$

Alternatively,

$$P_{sm} \cong P_1 \frac{(\cos \theta_B)^{n-2}}{2} \left\{ \frac{1}{4}[\sin (\theta_m + \theta_B) + \sin (\theta_m - \theta_B)]^2 + 2 \right\} \qquad (12\text{-}28)$$

Equation 12-27 represents the power coupled into the elliptical area on the surface of the multimode waveguide in Fig. 12-6(a). The launched power at the single mode pigtail is P_l. Note that the epoxy in the figure has good matching characteristics. If the beam were transmitted through air, Eqs. 12-27 and 12-28 would need modification to allow for differences in the index of refraction.

Equation 12-28 reveals an important restriction on beam width, as shown below:

$$\theta_m + \theta_B \le \theta_c$$

If this condition is not observed, then the part of the beam greater than the critical angle will be lost.

If the beam width is small or less than 10 degrees and channel angles are less than 15 degrees, then Eq. 12-28 will be approximated by the following:

$$P_{sm} \cong P_1 (\cos \theta_B)^{n-2} \qquad (12\text{-}29)$$

Equation 12-29 reflects a larger beam width, and the source is fairly coherent. The beam widths for the pigtail and the multimode waveguide are calculated as follows. First, the pigtail beam width is

$$2\theta_{B(3 \text{ dB})} = 2 \cos^{-1} \left(\frac{1}{\sqrt{2}} \right)^{1/20}$$

$$= 21.3°$$

Second, the beamwidth of light entering the multimode waveguide is

$$2\theta_{B(3dB)} = 2 \cos^{-1} \left(\frac{1}{\sqrt{18}} \right)^{1/18}$$

$$= 22.4°$$

Figure 12-7 depicts the pigtail beam entering the multimode waveguide. It

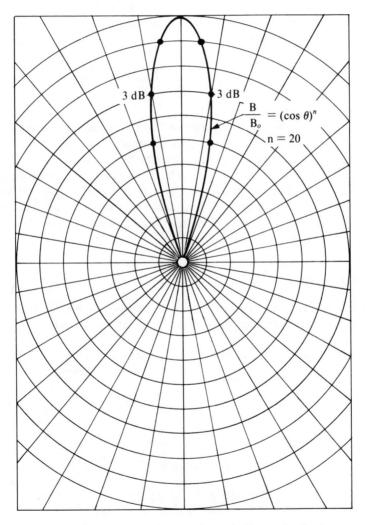

Fig. 12-7. Beam pattern for a single-mode fiber-optic pigtail.

will have a similar shape except that the width is approximately 1 degree wider.

In Fig. 12-6(b), the beam is collimated, which implies a large n. If a laser with a small value of n is used, the colliminating lens improves performance. The full analysis of Fig. 12-6(b) will be left to the reader.

Variation from the analysis can be caused by a multitude of variations in parameters: pigtail-to-epoxy and epoxy-to-multimode core matching, which can result in larger beam widths and low n values, as previously mentioned. Large values of r_s will cause part of the beam projected on the surface of the multimode waveguide to enter the cladding, where it is then lost. The restriction to prevent this from happening is as follows:

$$r_{sm} \leq a_m / \tan \theta_{B(max)}$$

where $\theta_B \neq 0$ degrees.

For a 100-μm multimode waveguide and a 11.2-deg maximum beam angle, r_{sm} is 250 μm ($\frac{1}{4}$ mm). Since dimensions are small, tolerances are also close, which provide another source of variation from the analysis. For the maximum value of r_{sm}, only four or five pigtails can be physically packed within the sector of a sphere formed by the intersection of a θ_c cone and a spherical surface formed by $r_{sm(max)}$. Therefore, physical rather than optical restrictions on the coupling can be a limiting factor on design. Several iterations of the design may be required to optimize the finished product.

ADM RECEIVERS

These devices require an analysis of the coupling similar to that for transmitters. Once the optical signals enter the detector, the analysis is similar to what was presented in Chap. 4.

Figure 12-8 depicts the receiver coupling geometry. Equation 12-30 is the differential of the optical power in an annular ring, as follows:

$$dP_o = I(\theta) \, d\Omega = I_o \cos \theta_{ch} \delta \Omega \tag{12-30}$$

where:

$$\delta \Omega = 2\pi \sin \theta_{ch} d\theta_{ch} \tag{12-31}$$

Equation 12-31 is the differential of the solid angle in the ring. Substituting Eq. 12-31 into Eq. 12-30 and integrating over the ring, we have

$$dP_o = \int_{\theta_{B1}}^{\theta_{B2}} I_o \cos \theta_{ch} 2\pi \sin \theta_{ch} d\theta_{ch} \tag{12-32}$$

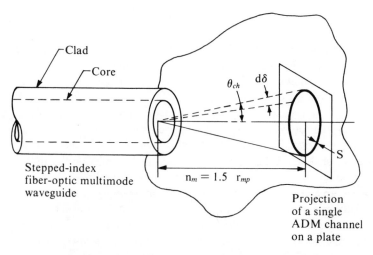

Fig. 12-8. Coupling geometry for a single ADM channel.

Then,

$$P_o = I_o \pi \left[\frac{1}{2} \cos^2 \theta_{ch} \right]_{\theta_{B1}}^{\theta_{B2}}$$

$$= \left(\frac{I_o \pi}{2} \right) [\cos^2 \theta_{B1} - \cos^2 \theta_{B2}]$$

$$= \left(\frac{I_o \pi}{2} \right) \sin(\theta_{B1} + \theta_{B2}) \sin(\theta_{B1} - \theta_{B2}) \qquad (12\text{-}33)$$

Then, for $\Delta\theta = \theta_{B1} - \theta_{B2}$,

$$P_o \cong \left(\frac{I_o \pi}{2} \right) \sin(\theta_{B1} + \theta_{B2}) 2\Delta\theta \qquad (12\text{-}34)$$

Beam angles θ_{B1} and θ_{B2} are related to the channel angle by the following expression:

$$\theta_B = \theta_{ch} \pm \Delta\theta$$

The total beam width is therefore $2\Delta\theta$. Equation 12-34 is an approximation that is valid when beam width is small, which is usually the case. The relationship between projection plate to waveguide surface is given by Eq. 12-35:

$$S = 2\Delta\theta r_{mp} \qquad (12-35)$$

For the transmitted channel in Fig. 12-6(a), the beam width, $2\Delta\theta$, is 22.4 degrees for $r_{mp} = 250$ μm and $S = 95.26$ μm. Note that the beam would be wider for an unmatched medium, i.e., when the refractive index of the core and the medium have large differences in n.

A problem that prevents long-distance transmission is when multiple channels must be separated at the receiver. Tee or star couplers are normally implemented to perform this task, but the mode mixing that occurs in these devices will increase crosstalk and bring about a reduction in transmission distance.

The large numbers of ADM channels that result from optical power splitting will induce great amounts of attenuation. As an example, a 50-channel system will initially have an attenuation of 17 dB, which excludes connectors and excess loss of the couplers. When ADM systems are examined for their feasibility, the designer must not only assess the system loss budget but also any mechanical limitations involved in coupling channels to the cable plant and separating the channels at the receive end.

CABLE PLANT DESIGN

A simple example of a cable plant design will now be presented. The transmitters and receive end will be considered as part of the optical cable plant. One of the first tasks is to produce a system block diagram and describe with some detail the functionality of the various blocks. The first iteration is shown in Fig. 12-9(a). Figure 12-9(b) shows how an angular division MUX and DEMUX can be coupled for a single-fiber application; the link is bidirectional. Figure 12-9(c) is a detail of the MUX internal structure.

The enclosures shown in the diagram can be tactical shelters connected to a command post or building on a college campus connected to the main administration building. Each block within the enclosure is a transceiver; i.e., it has two channels for full duplex operation. These are labeled n and n'. The ADM multiplexer and demultiplexer are where the optical signals are summed and separated, respectively; this block is entirely optical. The power splitter is a coupler with a 50/50 power-split ratio; i.e., transmitted power is split equally between any two enclosures.

The channels are dedicated as follows:

1. Channels 1, 1': These channels have an audio bandwidth that is used for intercom service; they can be used with a 2400-baud modem for data service.
2. Channels 2, 2': These channels operate at a high data rate 10 (Mbits/sec)

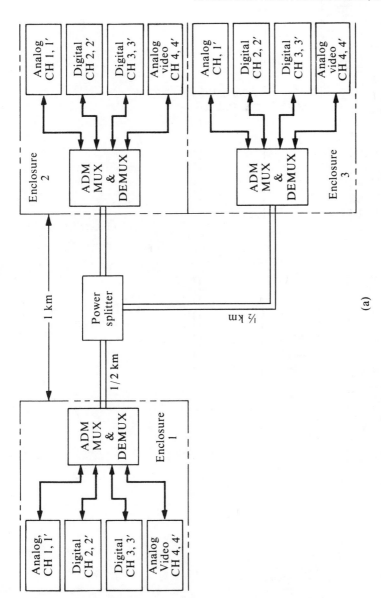

Fig. 12-9. ADM design considerations: (a) First design iteration, (b) MUX/DEMUX block diagram, and (c) MUX with eight single-mode waveguides.

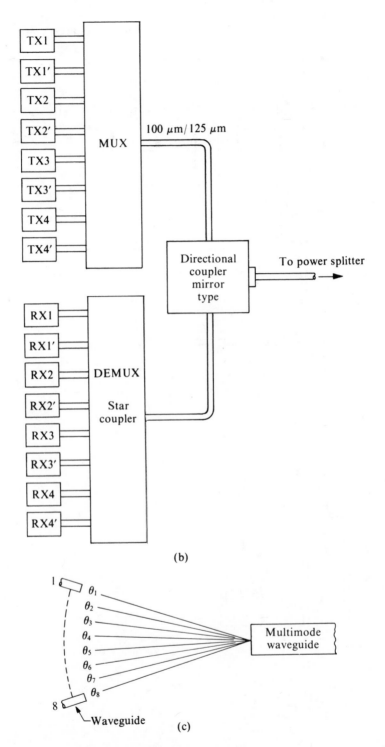

Fig. 12-9. (*Continued*)

and are used for communication between minicomputers.

3. Channels 3, 3′: These channels are used for communication between low-speed devices such as terminals. The data transmission rate is 1 Mbit/sec. The data is packet switched, which allows 15 64-kbit/sec TDM channels.

4. Channels 4, 4′: These channels, which have a 6-MHz bandwidth, are dedicated to video conferencing or video distribution. Which have a 6-MHz bandwidth, are dedicated to video conferencing or video distribution. Analog carrier frequencies are 55.2, 61.2, 66.2, or 72.2 MHz.

The next step is to refine the block diagram and show all the optical components in a symbolic form. The electronics will not be shown because most of the information has been presented previously. Note that two couplers are present but not shown in the first block diagram; these devices represent large losses.

The number of ADM channels required for this system is eight. Evaluating Eq. 12-4 for the multimode waveguide, the maximum number of channels is 50; therefore, the design is within the theoretical limit. Next, a rough estimate of the loss budget must be made to determine if the losses will be too excessive to proceed with the design. This loss budget is shown in Table 12-2.

From this rough approximation of the power gain margin, the video channels cannot be implemented. For adequate display, the SNR must be greater than 46 dB. Some methods may be employed to improve the system and get a suitable SNR. Examine the transmitter and receiver first. One limiting factor on video SNR is the laser transmitter, which has a maximum SNR of 57 dB. The total losses within the signal path cannot be greater than 11 dB. A receiver with

TABLE 12-2   Loss Budget Estimate.

Laser transmitter output power	+ 5 dBm
Video channel with 50% mod index	− 3 dB
Connector losses, four at 1 dB each	− 4 dB
Power splitter includes excess loss	− 4 dB
Directional coupler loss	− 2 dB
Star coupler with excess loss	−13 dB
Radial loss due to channel (angle this for approximately 10 degrees)	− 8 dB
Fiber-optic cable loss (1 km max°; 6 dB/km)	− 6 dB
Receiver sensitivity	−42 dB
Total power gain margin	47 dB
Total estimated losses	40 dB
Power gain margin after losses	7 dB

increased sensitivity will help somewhat; i.e., a figure of -52 dBm will increase power gain after losses to 17 dB, which is still far from the 46-dB requirement. Next to consider is the use of angular channels close to the axis of the multimode waveguide, which will increase the gain margin to 23 dB. If the video channels are separated from the incoming signal before the star coupler with a tee coupler, and the coupler is spliced into the system to reduce coupler loss, the gain margin will again be increased to 42 dB. The link performance can be increased with improved cable quality; i.e., lower-loss cable can be provided (2 dB/km) and the number of connectors reduced (replaced with splices); this will allow the channel to surpass the 46-dB SNR requirement.

The bypass of the star coupler will cause an extra 4 dB of loss in the data channels, which will reduce the power gain margin after losses to 3 dB.

Another parameter to consider during cable plant design is mode coupling. Since crosstalk increases with distance as a result of this effect, some estimate of coupling should be made by either empirical techniques or simplified analytical estimates.

Mode coupling is caused by nonuniformity in the dielectric waveguide. This phenomenon not only causes crosstalk but also attenuates the signal. Coupling that occurs between the signal and unguided modes (leaky modes) rapidly attenuates the signal.

Nonuniformity in waveguides is due to splice mismatches, poor connector alignment, couplers, fiber-optic switches, etc. These disruptions in the waveguides cause other modes to be generated that can rapidly attenuate or be launched into the cladding and become leaky modes. Some of these leaky modes can propagate for up to a kilometer, and, of course, modes that couple between signals cause large crosstalk components. Disruptions in the waveguide that occur close to the transmitter will have the greatest impact.

Crosstalk between channels can be measured in the following manner:

1. Transmit a continuous carrier on each channel at different frequencies.
2. Observe each detected angular channel output on a spectrum analyzer.
3. The fundamental frequency of the monitored channel will appear, and small crosstalk components of the other channels will also appear.
4. Check each crosstalk component by switching off the suspected channel producing it.
5. When selecting channel frequencies for the test, do not choose frequencies that are harmonics of each other as stated below:

$$f_{ch\,2} \neq 2 f_{ch\,1}, 3 f_{ch\,1}, \cdots$$

6. If the receiver is slightly nonlinear, intermodulation IM products can also appear. This aberration is discussed in Chap. 4. One method of determining which transmitter is causing these products would be to remove the suspected transmitter or transmitters from the link. The IM product will disappear when the correct transmitter has been removed.

The coupling constants and angular channel attenuation factor α can be estimated to solve Eqs. 12-6 and 12-7 for crosstalk. Note that in Eq. 12-6, if $f_d(\theta)$ can be evaluated for a particular channel—i.e., $f_d(\theta_{ch}) = k_1$ and $\alpha(\theta_{ch}) = k_2$— then the partial differential equation can lead to a separation of variable solution.

The receiver detectors required that each projected channel ring must be separated (refer to Fig. 12-2). The separation can be accomplished with lenses and a masking technique. The channel rings may first be magnified and all the undesired channels masked. The remaining channel, which is the desired one, is then demagnified and focused on the detector. An illustration of the techniques is given in Fig. 12-10(a). The channel chosen to be the video channel, as previously mentioned, should be channel θ_3; no lens is needed for separation and the losses can be kept lower.

The mask shown in Fig. 12-10(a) may require slots that are extremely small. These may be chemically etched for good precision. The magnification must be sufficient to allow them to be made with low-cost fabrication techniques. The method of Fig. 12-10(a) is rather simple but will increase attenuation. Other techniques are possible, of course, such as a tapered cone to expand the rings with a mask as shown in Fig. 12-10(b). This second technique will exhibit lower loss because the tapered end provides index matching. The mask may be metallic, or the end of the taper may be roughened by chemical etching so as to leave only a clear window for the desired channel.

Channel beams may be separated even further by changing the refractive index of the glass within the cone. As one can imagine, a large number of variations exist, including the use of mirrors, polarizers, etc.; a book might be written on the subject. Another technique with a great deal of potential is making masks of liquid crystal which may also be used to select channels in a way similar to tuning. For example, if the rings can be selected electronically, then the channel presented to the detector can be selected through logic circuitry. This technique has another advantage: If all the channels are not used simultaneously, then star couplers will not be needed, thus reducing the loss during channel separation.

For low data rates, several channels can be demultiplexed by switching a liquid crystal mask and connecting the receiver electronically to the electronic channel corresponding to the angular channel. Data may then be demultiplexed from several ADM channels with a single receiver.

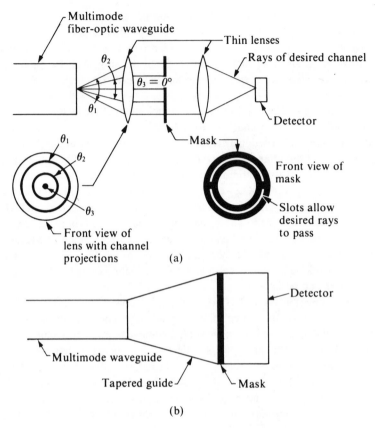

Fig. 12-10. Demultiplexer: (a) Channel separation at detector, and (b) tapered cone and mask separation technique.

ADM DEMULTIPLEXER

A diagram of an ADM demultiplexer is shown in Fig. 12-11. The multiplexer is designed to use liquid-crystal masks. The multiplexer consists of liquid-crystal interface circuitry, latches to hold the mask code, and control for the 1-of-8 logic decoder. The interface performs the necessary decoding logic to produce the mask, which is 1-of-8 in this case. Three bits of the code word select one of the eight digital channels. Thus one code word strobed into the latches selects the digital channel and mask. If this task is performed rapidly enough, the data can be sampled on each angular channel sequentially, thus producing an output of data on all eight channels. Of course, the other possibility is for the microprocessor to select the channel slowly and for the data to be high speed, in a

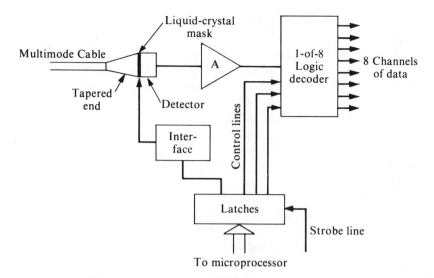

Fig. 12-11. ADM demultiplexer with liquid-crystal mask.

manner similar to a channel tuning technique. These demultiplexing or channel selection techniques are rather simplified. One must take timing into consideration and insure reasonable switching times when designing the actual circuitry. The entire design will not be presented here for lack of space, but this should stimulate the imagination of the reader.

The controller microprocessor required (not shown in the figure) can be a single IC device such as an Intel 8048 or 8051. Some of the single-chip microprocessors (CMOS versions) may be adquate to control the liquid-crystal mask directly. Often these single-chip microprocessors are used in a way similar to UARTs and have a serial bidirectional port as part of their I/O structure. Careful design and selection of parts may yield a very simple circuit.

Information on ADM transmission is at this time rather sketchy because of its short range. There is, however, a large demand for short range LANs for control purposes. Let us examine a few examples.

Aircraft controls could be a candidate for ADM because distances between devices are usually within 100 meters. Fiber-optic LANs may connect local self-contained hydraulic or electric actuator modules to a central computer. The ADM channels can be used to control several operations simultaneously. A second, or redundant, cable can be run through another route aboard the aircraft to the same point. Two fiber-optic switches may be used to remove the primary cable from service if it becomes damaged. If cable data is Manchester-encoded, sensing no transitions on the cable for longer than one bit time would indicate a failure.

Some of the ADM channels can be used to monitor test points throughout the avionics system, and other channels for maintenance purposes. One may logically ask, why not use TDM? If the bandwidth requirements are not very high or signals are not in analog form, then TDM may be adequate. The next question is, why not use FDM? If FDM is implemented, high-frequency modems and other RF components will be more costly than ADM components. In avaiation, reliability and safety are prime factors in design.

In the auto industry, however, where all components are very cost sensitive, ADM LANs could offer a viable approach to engine control, ATE, safety equipment monitors, etc. The very short cable distances in automotive applications make ADM particularly appealing. Since the control function in automobile applications operates slowly compared to that in aircraft, the control mechanisms can afford to be much less sophisticated.

Other areas where ADM can be applied are the machine tool industry (as a central control for lathes, milling machines, shapers, etc.); robotics (as a local control for mechanical arms, visual inspection devices, TV monitors, etc.); and nuclear waste-handling systems (such as fueling cars, waste storage pools, etc.). All of the aforementioned applications required a short-range LAN for control.

An application that may prove useful is to implement ADM in the visible spectrum as multicolored indicator lights. These would be useful for remote displays as failure indicators.

One area where investigation is warranted is the use of edge-emitter LEDs for ADM purposes. These devices exhibit good SNR as comapred to lasers. One should recall that an ADM system 1 km in length is crosstalk- or SNR-limited (in video applications). The spectral width will result in a reduction in bandwidth, but, if system bandwidth can be sacrificed for SNR, the trade-off will be justified.

Another area not yet considered is optical enhancement of a system. Innovations in optical systems may perhaps allow lasers to be replaced with less expensive LEDs.

A final area that may yield improvements is waveguide design. For example, multi-concentric-core waveguides with single channels on each could result in longer transmission distances. The possibilities are tremendous.

The possible advances in fiber-optic technology are so vast because it is a technology that has only recently become suitable for communications applications (within the last 10 years). The need for improved components such as switches, couplers of all types, single-mode lasers, etc., is great. Many avenues of exploration are indeed open for the resourceful engineer.

REVIEW PROBLEMS

1. Given a transmission link with 100/140 cable and an NA of 0.25, what is the largest number of ADM channels it will support at wavelengths of 840 nm, 1300 nm, and 1550 nm?

2. For a value of $V = 1240$, what is the reset time per meter of the cable described in question 12.1?

3. If an ADM channel has the following channel frequencies in its annular ring at the receiver—$f_1 = 5$ MHz (10 μW), $f_2 = 10$ MHz (0.1 μW), $f_3 = 1$ MHz (0.13 μW)— what is the total crosstalk?

4. What are the losses in transmitter coupling for ADM channels with the following channel angles: 2°, 5°, and 10°?

5. Would any of the channels be adequate for a video channel if all transmitters had a 57-dB SNR?

6. Derive an equation for optimal optical beam width from Eq. 12-28, where $n = 12$.

7. Design a total ADM link using a stepped-index waveguide, NA = 0.26, and 100/140-μm cable. Do a complete loss budget similar to that in Table 12-1. The components can be selected from Ref. 4.

REFERENCES

1. Herskowitz, G. J.; Kobrinski, H.; and Levy, V. Angular Division Multiplexing in Optical. *IEEE Communications*, Feb. 1985.

2. Wolf, H. F. *Handbook of Fiber Optics Theory and Applications*. Garland Press, 1979.

3. Amazigo, J. C., and Rubenfeld, L. A. *Advanced Calculus and Its Applications to the Engineering and Physical Sciences*. New York: John Wiley & Sons, 1980.

4. *Fiber Optics Handbook & Buyers Guide*. Boston, MA: Gatekeepers, Inc., 1984.

INDEX